Advanced Nanocatalysis for Organic Synthesis and Electroanalysis

Edited by

Vijai K. Rai

Department of Chemistry
Guru Ghasidas Vishwavidyalaya
Bilaspur, Chhattisgarh-495009, India

Manorama Singh

Department of Chemistry
Guru Ghasidas Vishwavidyalaya
Bilaspur, Chhattisgarh-495009, India

&

Ankita Rai

School of Physical Sciences
Jawaharlal Nehru University
New Delhi-110067, India

Advanced Namocatalysis for Organic Synthesis and Electroanalysis

Editor: Vijai K. Rai, Manorama Singh and Ankita Rai

ISBN (Online): 978-981-5040-16-6

ISBN (Print): 978-981-5040-17-3

ISBN (Paperback): 978-981-5040-18-0

Bentham Science Publishers Pte. Ltd.
80 Robinson Road #02-00
Singapore 068898
Singapore
Email: subscriptions@benthamscience.net

BENTHAM SCIENCE

CONTENTS

PREFACE

To design and develop a green, sustainable and economical reaction is one of the major challenges in chemistry. Besides the traditional need for efficient and selective catalytic reactions, recent chemical synthesis strives for designing new and efficient catalytic systems with high rates of catalyst recovery. The initial papers in the field of nanocatalysis were published as early as 1941 on palladium and platinum nanoparticles as catalyst, which was prepared by reduction of the metal salts. This research of Rampino was evocative of the work of Prof. Paul Sabatier (Chemistry Nobel Prize in 1912), who discovered catalyzed hydrogenation using finely divided nickel particles prepared upon reduction of nickel oxide or hydroxide. Then, in 1987, Haruta *et al.* made another breakthrough in the field of nanocatalysis by reporting the catalytic activity of gold nanoparticles smaller than 5 nm towards oxidation of CO.

Inspired by the above initial discoveries of nanoparticles as catalyst, nanostructured materials have attracted the chemical community and are now recognized as efficient heterogeneous catalysts for various organic transformations and electroanalytical processes. Efficiency, selectivity and recyclability of nanocatalysts depend on their size, shape, composition and assembly, which further enhance the appeal of well-defined nanostructured materials as green and sustainable heterogeneous catalysts in a wide variety of organic transformations as well as electroanalytical processes. Role of nanocatalyst in organic synthesis and electroanalyses helps control the chemical reactions by varying their shape and size, chemical composition, dimensionality, *etc.* to improve the kinetics of the reaction. Several catalytic sites are explored due to variations in shape, size and composition of nanocatalysts because a specific site shows good selectivity towards a specific reaction pathway.

This book offers an exclusive link between the domains of nanocatalyst and their exploitation in organic syntheses as well as electrochemistry using nanotechnology-based catalysts and electrode structures, respectively. The book is aimed to prepare to deliver a quick and highly compiled knowledge which, in turn, is expected to promote further advances in the field of nanaocatalysis. Also, the book will open up new dimensions for designing novel nanocatalysts for unexplored chemical reactions important for academia and industries both. The topic chosen in the proposed book will be beneficial for a broad range of readers such as graduate, postgraduate, Ph.D. students, faculty members, research & development (R & D) personals, working in these areas as well.

Vijai K. Rai
Department of Chemistry
Guru Ghasidas Vishwavidyalaya
Bilaspur, Chhattisgarh-495009
India

Manorama Singh
Department of Chemistry
Guru Ghasidas Vishwavidyalaya
Bilaspur, Chhattisgarh-495009
India

&

Ankita Rai
School of Physical Sciences
Jawaharlal Nehru University
New Delhi-110067 India

List of Contributors

Amrit Puzari	Department of Chemistry, National Institute of Technology Nagaland, Chumukedima, Dimapur, Nagaland, India
Anshu Dandia	Centre of Advanced Studies, Department of Chemistry, University of Rajasthan, Jaipur-302004, India
Aschalew Tadesse	Department of Applied Chemistry, School of Applied Natural Science, Adama Science and Technology University, P O Box 1888, Adama, Ethiopia
Athira Krishnan	Department of Chemistry, Amrita Vishwa Vidyapeetham, Amritapuri, Kerala-690 525, India
Bidyutjyoti Dutta	Department of Chemistry, Dibrugarh University, Dibrugarh-786004, Assam, India
C.R. Ravikumar	Research Center, Department of Science, East West Institute of Technology, VTU, Bengaluru 560091, India
Dhanalakshmi Muniswamy	Department of Physics, Government Science College (Nrupathunga University), Bengaluru 560 001, India
Diganta Sarma	Department of Chemistry, Dibrugarh University, Dibrugarh-786004, Assam, India
Dinesh K. Mahawar	Centre of Advanced Studies, Department of Chemistry, University of Rajasthan, Jaipur-302004, India
Dipen K. Sureja	L. M. College of Pharmacy, Navrangpura, Ahmedabad 380 009, Gujrat, India
Gezahegn Tadesse Ayanie	Department of Chemistry, School of Applied Natural Science, Adama Science and Technology University, Adama P.O. Box.1888, Ethiopia
H.C. Ananda Murthy	Department of Chemistry, School of Applied Natural Science, Adama Science and Technology University, Adama P.O. Box.1888, Ethiopia
J. Anakha	Department of Chemistry, Amrita Vishwa Vidyapeetham, Amritapuri, Kerala-690 525, India
Kalyanjyoti Deori	Department of Chemistry, Dibrugarh University, Dibrugarh-786004, Assam, India
Krishan Kumar	Centre of Advanced Studies, Department of Chemistry, University of Rajasthan, Jaipur-302004, India
Kunjan B. Bodiwala	L. M. College of Pharmacy, Navrangpura, Ahmedabad 380 009, Gujrat, India
Leena Khanna	University School of Basic and Applied Sciences, Guru Gobind Singh Indraprastha University, Dwarka, New Delhi-110078, India
Mansi	University School of Basic and Applied Sciences, Guru Gobind Singh Indraprastha University, Dwarka, New Delhi-110078, India
Merangmenla	Department of Chemistry, National Institute of Technology Nagaland, Chumukedima, Dimapur, Nagaland, India
Muthu Karuppasamy	Department of Chemistry and Chemical Sciences, Central University of Jammu, Rahya-Suchani (Bagla), District-Samba, Jammu-181143, J&K, India
Normi D. Gajjar	L.M. College of Pharmacy, Navrangpura, Ahmedabad 380 009, Gujarat, India

Pankaj Khanna	Department of Chemistry, Acharya Narendra Dev College, University of Delhi, Kalkaji, New Delhi-110019, India
Prantika Bhattacharjee	Department of Chemical Sciences, Tezpur University, Napaam, Tezpur, Assam, 784028, India
Pratibha Saini	Centre of Advanced Studies, Department of Chemistry, University of Rajasthan, Jaipur-302004, India
Priyanka Gogoi	Department of Chemistry, Dibrugarh University, Dibrugarh, Assam, India
R. Ajay	Department of Chemistry, Amrita Vishwa Vidyapeetham, Amritapuri, Kerala-690 525, India
Rajvi H. Amin	L. M. College of Pharmacy, Navrangpura, Ahmedabad 380 009, Gujrat, India
Roktopol Hazarika	Department of Chemistry, Dibrugarh University, Dibrugarh-786004, Assam, India
Rutvi J. Patel	L. M. College of Pharmacy, Navrangpura, Ahmedabad 380 009, Gujrat, India
Savita Meena	Centre of Advanced Studies, Department of Chemistry, University of Rajasthan, Jaipur-302004, India
Shyamal Baruah	Department of Chemistry, National Institute of Technology Nagaland, Chumukedima, Dimapur, Nagaland, India
Soniya Agarwal	Department of Chemistry, Dibrugarh University, Dibrugarh-786004, Assam, India
Surenda Saini	Centre of Advanced Studies, Department of Chemistry, University of Rajasthan, Jaipur-302004, India
Suresh Ghotekar	Department of Chemistry, Smt. Devkiba Mohansinhji Chauhan College of Commerce and Science College, Silvassa- 396 230, Dadra and Nagar Haveli (UT), India
Swati Meena	Centre of Advanced Studies, Department of Chemistry, University of Rajasthan, Jaipur-302004, India
Tanvi Jandial	Department of Chemistry and Chemical Sciences, Central University of Jammu, Rahya-Suchani (Bagla), District-Samba, Jammu-181143, J&K, India
Tegene Desalegn	Department of Chemistry, School of Applied Natural Science, Adama Science and Technology University, Adama, P.O. Box.1888, Ethiopia
Tejas M. Dhameliya	L. M. College of Pharmacy, Navrangpura, Ahmedabad 380 009, Gujrat, India
U.S. Krishnanunni Namboothiri	Department of Chemistry, Amrita Vishwa Vidyapeetham, Amritapuri, Kerala-690 525, India
Utpal Bora	Department of Chemical Sciences, Tezpur University, Napaam, Tezpur, Assam, 784028, India
Vellaisamy Sridharan	Department of Chemistry and Chemical Sciences, Central University of Jammu, Rahya-Suchani (Bagla), District-Samba, Jammu-181143, J&K, India
Vijay Parewa	Centre of Advanced Studies, Department of Chemistry, University of Rajasthan, Jaipur-302004, India

CHAPTER 1

Principles and Concepts of Nanocatalysis

Pratibha Saini[1], Swati Meena[1], Dinesh K. Mahawar[1], Anshu Dandia[1,*] and **Vijay Parewa[1,*]**

[1] *Centre of Advanced Studies, Department of Chemistry, University of Rajasthan, Jaipur-302004, India*

Abstract: Catalysis is one of the fundamental principles of the twelve principles of "Green Chemistry". Over the past few years, nanocatalysis has emerged as a growing field of catalysis in the construction, food, medical, pharmaceutical, energy, and water treatment sectors due to its high activity, selectivity, and productivity. Nanoparticles are different from their bulk counterparts and exhibit unique properties as compared to traditional catalysts, for example, simple and cheap approaches of production, good selectivity, high surface to volume ratio, high catalytic activity, ease of recovery, the possibility of being reused, enhanced mixing with reactants, easy separation, and presence of a large number of active sites. The nanoscale size (1nm = 10^{-9} m), shape, and remarkably large surface area to volume ratio impart inimitable properties to nanoparticles. This chapter presents the principles and concepts of nanocatalysis and discusses the inimitable structure and catalytic properties of monometallic nanoparticles and bimetallic nanomaterials, magnetic nanoparticles, nanocomposites, carbon-based nanomaterials, and nanophotocatalysts in various organic transformations.

Keywords: Bimetallic NPs, Carbon-based nanomaterials, Green chemistry, Magnetic NPs, Monometallic NPs, Nanocatalysis, Nanocomposites, Nanophotocatalysts.

INTRODUCTION

Catalysis

Humans are reliant on endless value-added chemicals and the associated chemical transformations to obtain advanced intermediates, pharmaceuticals, polymers, agrochemicals, natural products, fine chemicals, and bioactive molecules in drug and chemical industries. Among various pharmaceutical industries, the drug industry is the main industry to achieve a quality lifestyle since it generates value-

[*] **Corresponding Authors Anshu Dandia and Vijay Parewa:** Centre of Advanced Studies, Department of Chemistry, University of Rajasthan, Jaipur-302004, India; Tel: +911412702306; E-mails: parewavijay.parewa@gmail.com, parewavijay@uniraj.ac.in

added chemicals for battling infections/diseases. The significance of the drug industry is also highlighted in the "Covid-19" (Coronavirus Disease 2019) pandemic. The pharmaceutical and other sectors are generally dependent on chemical reactions in which a lot of chemical substances, catalyst/photocatalysts, energy (such as heat), and solvents are utilized, delivering an enormous amount of chemical waste and by-products [1 - 4]. Therefore, the innovative period of chemistry (especially medical chemistry) is shifting towards the path of inventive greener/sustainable technologies for the production of value-added chemicals, which essentially concentrate on environmental aspects [5 - 7]. "Catalysis" is a key technology and is located at the midpoint of numerous chemical transformations, which from a green and sustainable chemistry viewpoint need catalysts to accelerate transformations for maximum manufacturing chemical transformations [8, 9].

Green and Sustainable Catalysis

Chemical transformation can be designed in greener and eco-friendly ways by the development and attentive utilization of catalysts. Catalysts fulfill green chemistry aims in various languages of atom economy, elevated yields, low energy requirements, and ease of separation due to increased selectivity at minimal waste. As identified to all, the definite catalytic sites on the surface of catalysts perform a crucial character to regulate various chemical transformations. Traditionally, catalysts can be divided into two main types - heterogeneous and homogeneous [10 - 12]. Transition metals are usually utilised as homogeneous and heterogeneous catalysts in a majority of chemical reactions as they have variable oxidation states and good adsorption properties. On the one hand, heterogeneous catalysts are easy to recover, however, they have a few downsides. For example, in harsh environments, they need to be effective, the mass transportation issues, the susceptibility of metal leaching, and the reaction rate aredelimited because of their inadequate exterior area. In contrast, homogeneous catalysts with good solubility in reaction media are identified for their advanced catalytic activity and selectivity, but the isolation of costly metal-containing catalysts from the final compounds remains a central problem across chemical industries [13].

Disadvantages in homogeneous/heterogeneous catalysts demand innovative catalytic materials for disabling the boundaries related to both types. Due to their large surface area, NPs have originated to link "homogeneous (low loadings and good selectivity) and heterogeneous (recovery and recyclability) catalysis" processes. Both catalysis processes have specific benefits and undesirable properties (Fig. **1**). Nanocatalysts go beyond the boundaries of homogeneous and heterogeneous catalysts through nano-effects, which are completely unwritten yet.

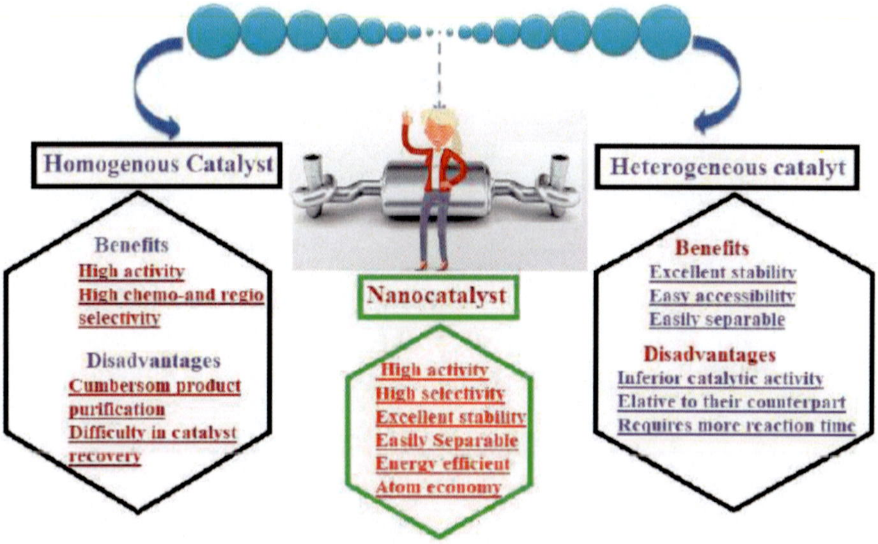

Fig. (1). Relative efficacy of homogeneous, heterogeneous, and nanocatalyst.

Nanocatalysis

In the growing field of catalysis, catalytic reactions and nanocatalysis have been intimately connected for a long time period and modern developments have made it far more stimulating. To advance eco-friendly chemical reactions having the goals of green chemistry, "Nanocatalysis" is an innovative Green Chemistry era [14, 15]. The prefix "nano" is taken from the Greek word 'Nanos' which means 'dwarf" and alludes to things of one billionth in size. Owing to the inimitable properties, "Nanomaterials" are considered to be the building blocks of the next generations of catalysis, especially because they meet the goals of green chemistry [16, 17].

Nanotechnology is the application of science to control matter at the atomic or molecular level. The words 'nanoscale' or 'nanostructured' substances (sub-micron moieties) are any solid that has a nanometer dimension (1nm = 10^{-9} m). Unquestionably, the utilization of nanometer-sized solid particles in catalysis are eye-catching choices to traditional catalysts, for the reason that when the size of the "Nanomaterials" is reduced to a molecular level, the large surface-to-volume ratio is ominously improved and defects are introduced into these materials. NPS are identified to be much more efficient as a large part of active catalytic sites are provided to the reacting molecules, thus, raising its productivity manifold as compared to traditional catalysts.

SILENT FEATURES OF NANOPARTICLES

Nanoparticles have unique chemical properties like good selectivity, catalytic activity, stable activity, ease of recovery, possibility to be reused, enhanced mixing with reactants, ease of separation, and physical properties like photoemission, electrical, or heat conductivities. They also exhibit novel characteristics of quantum size effects [18, 19] (Fig. **2**).

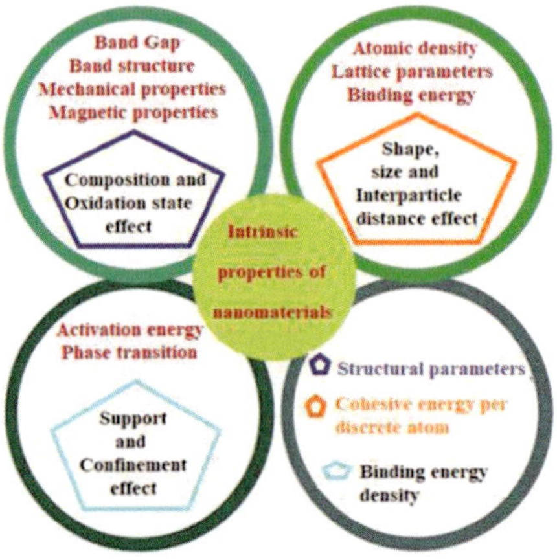

Fig. (2). Intrinsic properties of nanomaterials.

Nanomaterials have characteristics in between those of the bulk substance and the molecular/atomic structures. Additionally, NPs give extra reactant functionalities owing to their fascinating intrinsic properties (*e.g.*, photocatalytic activity, nanomagnetism). NPs can be effortlessly recycled and reused during organic transformations for numerous cycles without loss of their reactivity. Compared to traditional catalysis, nanocatalysis has assured inimitable rewards. For these materials, product distributions/selectivity and the turnover rates rely on their shape, size, oxidation states, and their composition. The role of NPs in the process improvement involves not only stimulating the organic transformations, but also advancing the entire procedure, by sinking energy consumption, the creation of unwanted side compounds, and construction development.

General Look into the Synthesis and Applications of Nanoparticles

In the nanocatalysis arena, the construction of nano-engineered materials is a key invention in material science. Nanomaterials can be produced using two major methods: bottom-up method (NPs merge to construct the bulk substance) and top-down method (the bulk substances are cracked down to get NPs). The 1st method is more appropriate than the latter as the probability of impurity is relatively large. Furthermore, their size and morphology can be monitored by using proper coordination of material to command the nucleation and growth procedures. Size control is achieved by employing a variety of capping agents, also called stabilizing agents (ions, polyoxoanions, ligand, surfactant, polymer, dendrimer, *etc.)* which bound the growth of NPs [20].

Over the last decade, innovative synthetic methodologies for chemical transformations using nano-engineered materials such as metallic and oxide nanoparticles, graphene, and their multifunctional nanocomposites, nanofilms, nanowalls (two-dimensional NPs), magnetically separable NPs, nanoclusters, nanophotocatalyst, nanorods/nanofibers (one dimension NPs), nanowires, supported NPs, nanocages, carbon nanotubes, core-shell nanostructures, nanoporous materials/quantum dots/metal NPs (zero-dimensional NPs), and bulk materials like fullerene (3D NPs) have been a stimulating area in nanoscience. Furthermore, NPs have the magnificent opportunity of functionalization, which enhances its usage in all sectors of science, technology and commerce [21]. There is extensive application of "Nanocatalysis" in different fields which make nanoparticles "tiny heroes" that accomplish certain methods which were otherwise unachievable, most common among them are defined as shown in Fig. (**3**). Nanoparticles are utilized in diverse forms depending upon the type of their applications in catalysis including heterogeneous/ electrochemical/ photochemical catalysis and drug delivery and in harvesting solar energy, *etc*. [22 - 26].

Fig. (3). A glimpse of selected applications of nanocatalysis.

Therefore, NPs are anticipated to have a great impression on making the environment safer, cleaner and greener in the future. This thematic topic devoted to nanocatalysis covers five groups of chapter articles: (a) Monometallic and bimetallic nanoparticles, (b) Magnetic Nanoparticles, (c) Nanocomposites, (d) Carbon-based nanomaterials, (e) Nanophotocatalysts.

Herein, we will give an overview of catalytic applications of the nano-engineered materials concerning their current and forthcoming panorama.

Monometallic and Bimetallic Nanoparticles

Metals are one of the most significant substances in the construction of both fundamental and industrial applications. Metal NPs catalysed chemical transformations are valuable over traditional metal catalysed transformations given its shorter reaction time, high atom economy, cheap, enhanced yields, low catalyst load, and reusability of the NPs. Metallic NPs (nanoclusters) can be noncrystalline, aggregates of crystallites, or single crystallites (nanocrystals) [27].

Monometallic NPs (MMNPs), as the name recommends, involve only single metal. MMNPs can be synthesized by various processes, however, the most significant is the chemical process. The structure of MMNPs can be stabilized utilizing numerous functional groups. Their properties depend upon the nature of the constituted metal atom, for example, metallic, magnetic, and transition metal nanoparticles, *etc.* Some metal and non-metals will be included in monometallic catalysts like Ag, Ni, Co, Cu, Zn, Sn, Pd, Pt, Au, Fe, Au, and any other metals [28 - 30]. Bimetallic nanomaterials (BMNPs) are tiny grains – a few dozen to hundreds of atoms in size- that embrace marvellous catalytic ability as catalysts for a number of organic transformations owing to their small-size, surface, and quantum-size effect. A BMNPs catalyst is a combination of two different types of metallic NPs with higher catalytic properties compared to MMNPs, have a definite mixing pattern in a geometry manner, and accomplish unique characteristics. Their properties also depend upon comparative strengths of metal-metal bond, the composition, synthesis process and environments, surface energies of bulk elements, *etc.* Both bimetallic and monometallic NPs can be synthesized by different approaches (especially the Wet-chemical method). BMNPs have various types of architectures (Fig. **4**). Subsequently, the catalytic properties of BMNPs are strictly connected with their architecture, composition, size, and morphology. Their architecture can be controlled by alteration thermodynamic factors (reduction potential, temperature, *etc.*) and kinetic factors (reactant concentration, diffusion, reaction rate, solubility, *etc.*) [31 - 33].

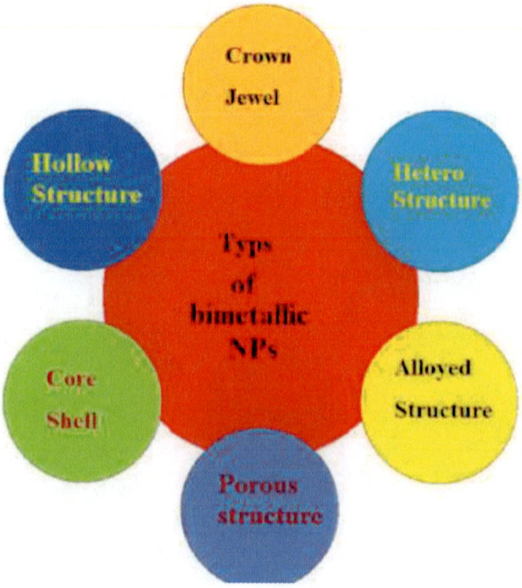

Fig. (4). Various type architectures of BMNPs.

BMNPs have fantastic properties such as different morphologies, mixing designs, component distributions, and, particularly, surface electron structures— for example, Os-Cu, Ru-Cu, Pt-Ir, Pt-Ru, and Ni-Cr. Mostly bimetallic catalysts have good malleability, better activity, electric–magnetic properties, improved selectivity, high mechanical strength and increased stability. BMNPs can potentially attain organic transformations that are unfamiliar with MMNPs because different components of the catalyst have a specific function in the whole reaction pathway. BMNPs show intense enhancement in respect of selectivity, activity, and stability as compared to MMNPs [34, 35]. These outcomes designate that bimetallic NPs display pronounced flexibility and adjustability in nanocatalysis. BMNPs have promising usage in numerous fields, including optical, catalysis, nanomedicine and biosensing, and in disabling environmental pollution. Briefly, BMNPs, as an innovative class of nanomaterials, have excessive capability in catalytic applications [36 - 48] (**Scheme 1**).

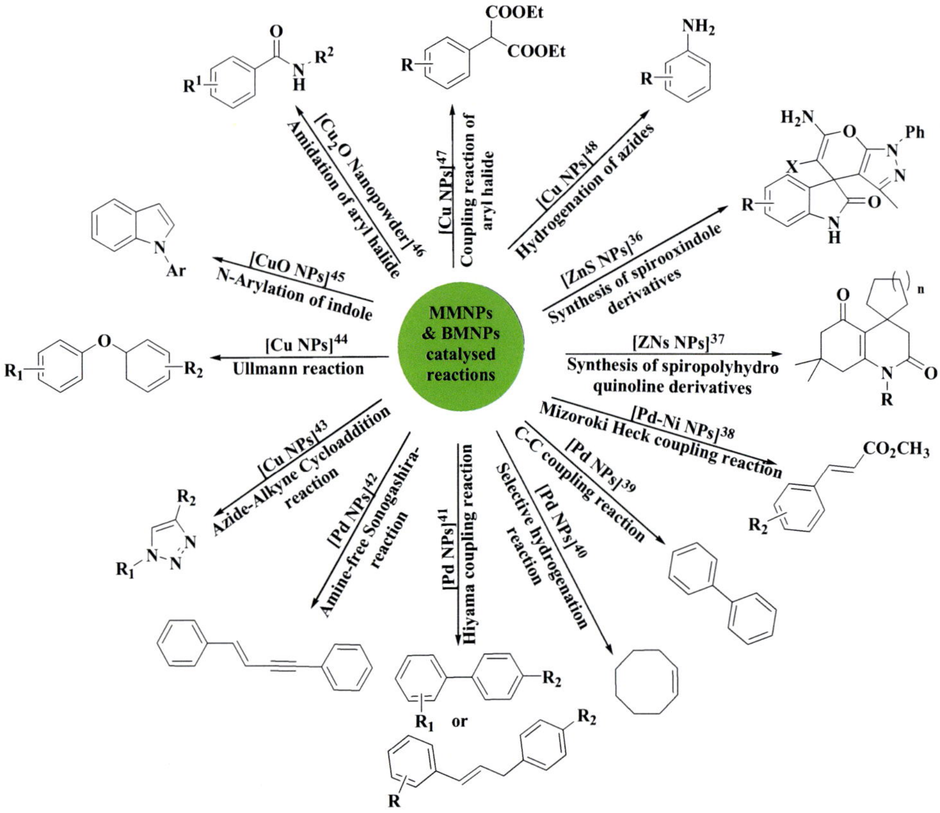

Scheme (1). Different MMNPs and BMNPs catalysed organic transformations.

Magnetic Nanoparticles

Magnetite nanomaterials (MNPs) are tremendously valued materials for bridging the homogeneous and heterogeneous catalysis, thus, containing the worthy characteristics from both. The employ of MNPs as an efficient catalyst and the ability to support various organic transformations has become a topic of penetrating study.MNPs display auspicious functions in heterogeneous catalysis owing to their ease of separation and good reusability [49 - 51].

Magnetic NPs are a class of nanomaterials that can be deployed utilizing magnetic fields. It contains magnetic elements like Fe, Co, and Ni, *etc.,* which can be effortlessly traced by the external magnetic field and can be isolated competently from a mixture *via* a simple magnetic separation. One of the principal characteristics of MNPs is their size-reliant magnetic characteristics. By monitoring their morphology or constituent, the magnetic properties could be well altered. When the size of an MNPs is under a definite acute size, spins of free electrons inside the nanoparticles are related by ferromagnetic coupling into one direction and it behaves as a single-domain magnet. The magnetization saturation values of MNPs principally rely on finite-size and surface properties [52].

Different from the bulk magnet, MNPs display inimitable magnetism, which allows the alteration of their magnetism by organized nano-scale manufacturing. Some magnetism can be included: paramagnetism, diamagnetism, antiferromagnetism, ferrimagnetisms and ferromagnetism. These NPs are mostly used as a catalyst in environmental remediation in the medical field like biomedicine, organic synthesis, magnetic resonance imaging, organic dye removal, and water splitting reactions. They also may act as contaminant carriers. Mostly magnetic NPs have high chemical stability, low Curie temperature, high electrical resistivity/coercivity/magnetic susceptibility, and mechanical hardness. Iron and its oxide are mostly used as MNPs due to their good physical and chemical characteristics. They can be effortlessly prepared by the following methods: Chemical vapor deposition procedures, Co-precipitation process, Hydrothermal/Solvothermal process, Calcination technique, Thermal decomposition procedure, and Microwave-assisted method. Owing to their special properties, MNPs have well-known applications [53 - 56] in various fields such as material science, biotechnology, biomedical, and organic reactions [57 - 67] (**Scheme 2**).

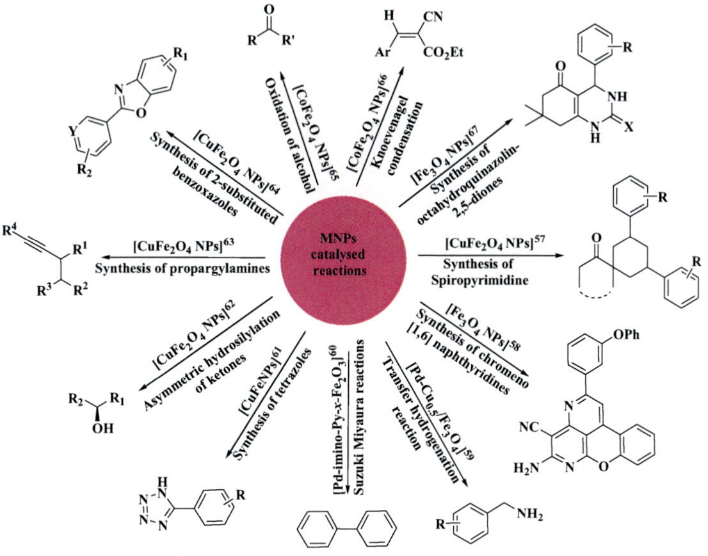

Scheme (2). Various organic transformations using MNPs.

Nanocomposites

To increase the properties of metal nanoparticles, MMNPs/BMNPs are converted into nanocomposites which are designed by supporting them on the inorganic (kaolin or zeolites) and organic (carbonaceous materials) equivalents. The common concept of nanocomposites is created on the concept of making a huge interface between the polymer matrix and the nanomaterials (nanosized particles) with extraordinary flexibility and enhancement in their physical properties. Nanocomposite (multiphase solid materials) is "nano" (dimensions in the range of 10-100 nm) which means small and "composite" means made of two or many different parts like bulk matrix and nanodimensional phase(s).

Nanocomposites' one phase has nanoscale morphology such as nanotubes, NPs, or lamellar nanosheets [68, 69]. These are heterogeneous materials and their structure is usually more complicated than that of microcomposites. Their properties depend on the structure, composition, interfacial interactions, and nature of the components and the size of the interface (Fig. **5**). Usually, nanocomposites can be categorized into three chief groups depending on the material utilized: (a) ceramic-based (b) polymer-based (c) metal-based nanocomposites [70, 71].

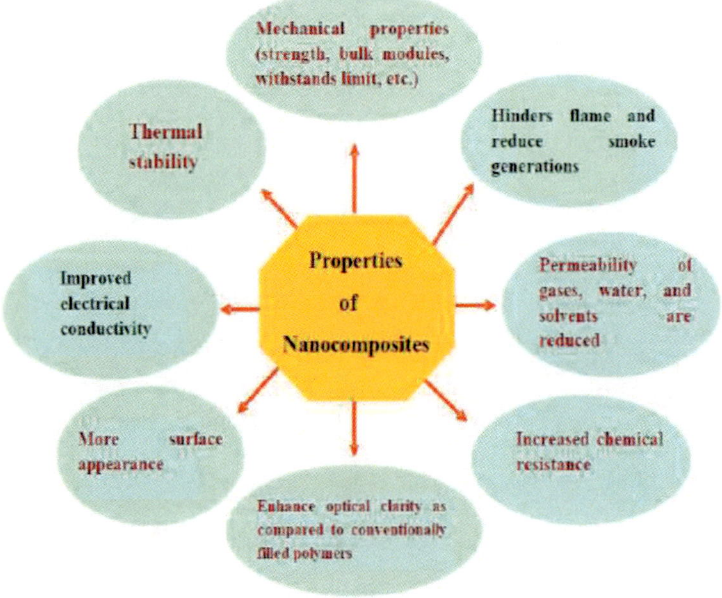

Fig. (5). Properties of nanocomposites.

The nanocomposite is designed for various purposes like bio-medical and polymer/layered silicate. Nanocomposite means nanosized particles (*i.e.* dielectric materials, metals, and semiconductors) fixed in many/different matrix materials like glass, polymers, and ceramics, *etc.* This process consists of two phases, namely, nanocrystalline phase and matrix phase. These phases may be organic-organic, organic-inorganic, and inorganic-inorganic. They have high chemical reactivity, physical sensitivity, surface appearance, decreased gas, water, and hydrocarbon permeability. They are mostly used in organic transformations [72 - 80] (**Scheme 3**).

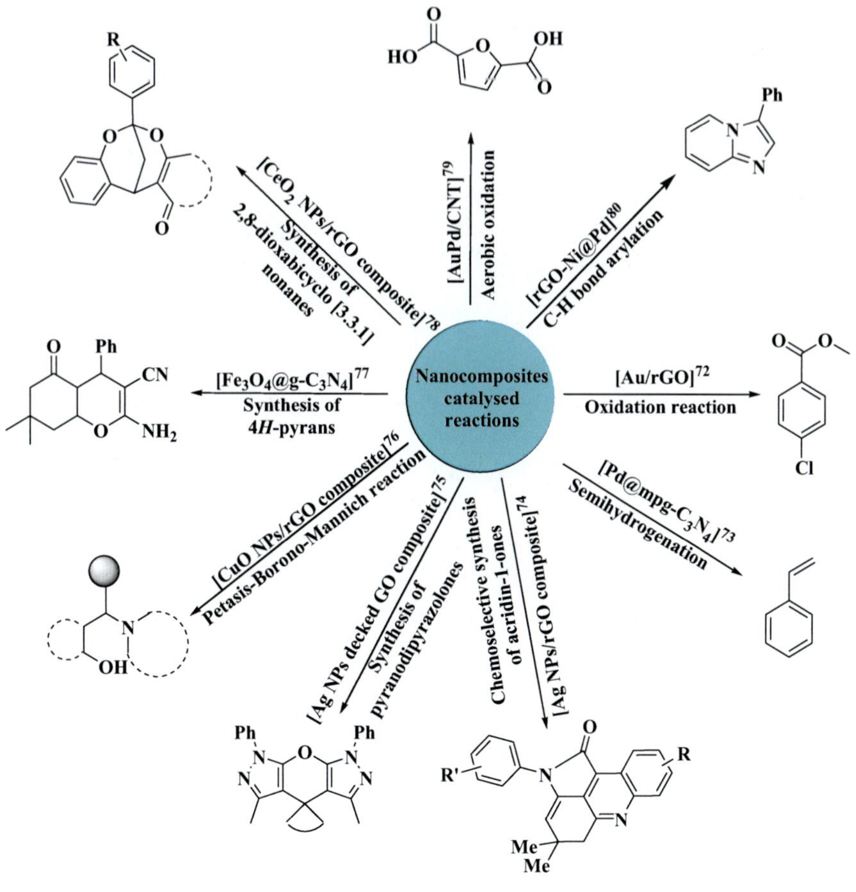

Scheme (3). Nanocomposites as a catalyst for various organic transformations.

Carbon-based Nanomaterials

Given the quick progression of nanocatalysis in the last few years, carbon based nanomaterials are extensively studied owing to their inimitable design opportunities and properties. With an inclination towards the requirement for the expansion of green viable substances, usage of carbon-type substances as a green catalyst in organic synthesis is valuable. Carbon-based nanomaterials presently represent one of the most encouraging material families with tremendous potential for elite applications in various fields depending on their phenomenal electronic, optical, mechanical, and chemical properties. They are cheap and can be effortlessly synthesized on a great level. They have been utilized as supports for metals or other constituents. They have exposed their ability for enlargement of catalytic strategies to nanocatalysis, especially in organic transformations [81, 82].

Carbon nanomaterials are one of the most plenteous and multipurpose materials on the earth, which can exist in diverse forms. Carbon nanomaterials, for example, graphene, carbon nanotubes (CNTs), graphene oxide (GO), activated carbon (AC), graphitic carbon nitrides (GCN), reduced graphene oxide (rGO), carbon black, fullerenes, carbon nanofibers, and carbon-onions, *etc.* have been utilized as a catalyst and support for various chemical reactions. They are very important due to their pure multitude of appealingly attractive structures and exclusive combinations of chemical and physical properties. They have freshly been utilized as catalysts for numerous chemical syntheses because of their inimitable properties for example inexpensive; rich surface chemistry, highly transparent, good thermal conductivity, supreme mechanical resistance, and high charge mobility. Furthermore, they have a large surface area and are biocompatible. Up to this point, the synergist uses of graphene and interconnected carbon nanostructures have concentrated fundamentally on the utilization of these substances as ropes for a lot of metals [83, 84].

Additionally, because of the existence of fully conjugated π-electron systems and high electron mobility, they act as electron reservoirs for various visible-ligh- -induced organic transformations (redox activity). Graphene oxide (GO) has chiefly been utilized as a support for many metals owing to its large distinct surface part, distinctive nanostructure, ready functionalization, reusability, good water-dispersion, high stability to ambient conditions, low cost, ease of synthesis, easy availability, low cytotoxicity, high cell compatibility, the existence of an abundance of oxygen-containing functionalities. C-based nanoparticles can exist in graphite form, carbon nanotubes graphene, fullerene, and diamond. They also have different dimensions like: fullerenes/nanodiamonds/small clusters (0-D), carbon nanotubes (1-D), and graphene and graphane (2-D). They are imperative for various industrial applications, extending from drugs to synthetic materials and various organic transformations [85 - 92] (**Scheme 4**).

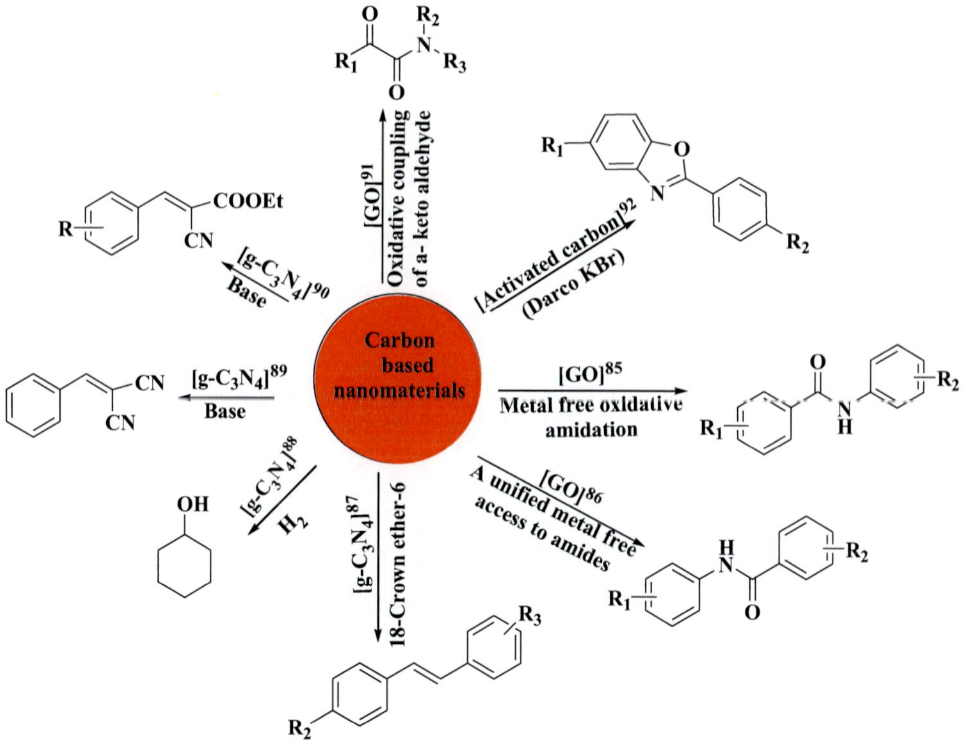

Scheme (4). Carbon-based nanomaterial's for various organic transformations.

Nanophotocatalyst

Natural photosynthetic processes are one of the eventual familiar photocatalytic reactions in "Environment" that can competently grab natural light and translate it quickly into chemical energy to produce sustainable energy resources and chemicals. It has encouraged synthetic scientists to create unnatural photosynthetic methods. Direct solar-to-chemical energy translation has numerous rewards over solar-to-electric energy translation. In recent times, visible-light-triggered organic reactions have found much attention because of their abundance, long-lasting, clean, inexpensive, simple to apply, infinite, green, accessibility, energy sustainability, and eco-friendliness. Likewise, in photocatalysis (a very hot field in heterogeneous nanocatalysis), the energy of photons can be utilized to make several new valuable organic transformations [93].

Compared with the classical thermal chemical transformations, photocatalytic reactions have a prevailing contact for the commitment of green energy and environmental functions. Usually, the employment of photo-sensitizers and photo-catalysts such as organic dyes and transition metals complexes is considered a viable tool in place of dangerous and polluting routes to advance visible-light-mediated photoreactions. Electron–hole (e-h+) pairs are produced from the photoexcitation of a photocatalyst. For photochemical reactions catalyzed by nanomaterials (semiconductors), it is essential to suppress the recombination of e^-/h^+pairs produced following visible-light absorption. This philosophy assisted the scientific communities to originate a variety of usage in green chemistry [94, 95]. Habitually, photoredox catalysis is induced by photocatalyst completely through the single-electron transfer (SET) mechanism (Fig. **6**).

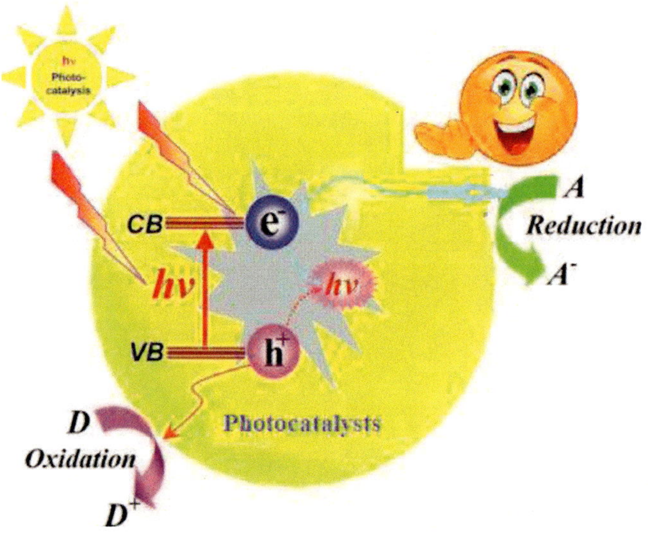

Fig. (6). Schematic representation of photocatalysis.

The effectiveness of photochemical reactions through semiconductor photoredox catalyst depends on four actions- (a) visible-light absorption (b) charge division (c) charge movement (d) charge recombination. The use of photoactivated semiconductor photoredox catalysts to support photochemical reactions has materialized as far-fetched and green approaches for synthetic organic chemistry. Heterocatalysis, particularly photocatalysis with semiconductors has attracted considerable attention. Many semiconducting materials, such as TiO_2, SnO_2, Au/rGO, ZnO, Cu_2O, WO_3, CeO_2, Fe_2O_3, CdS, ZnS, Pd NPs, $BiVO_4$/g-C_3N_4, g-C_3N_4, CuO/TiO_2, Pd/CeO_2 nanocomposites, metal-organic frameworks semiconductors QDs, ABX_3 hybrid perovskite such as $La_2Ti_2O_7$, *etc.* can act as photo responding photocatalytic nano-materials to gain solar light and initiate chemical materials by e^-/h^+ transfer to carry out organic transformations [96, 97].

These nanophotocatalysts have specified a period to grasp and disclose the electron or hole transmits processes (redox reactions) connected with "Natural-photosynthesis" and have astonishing significance due to their utilities in chemical reactions [98 - 106] (**Scheme 5**).

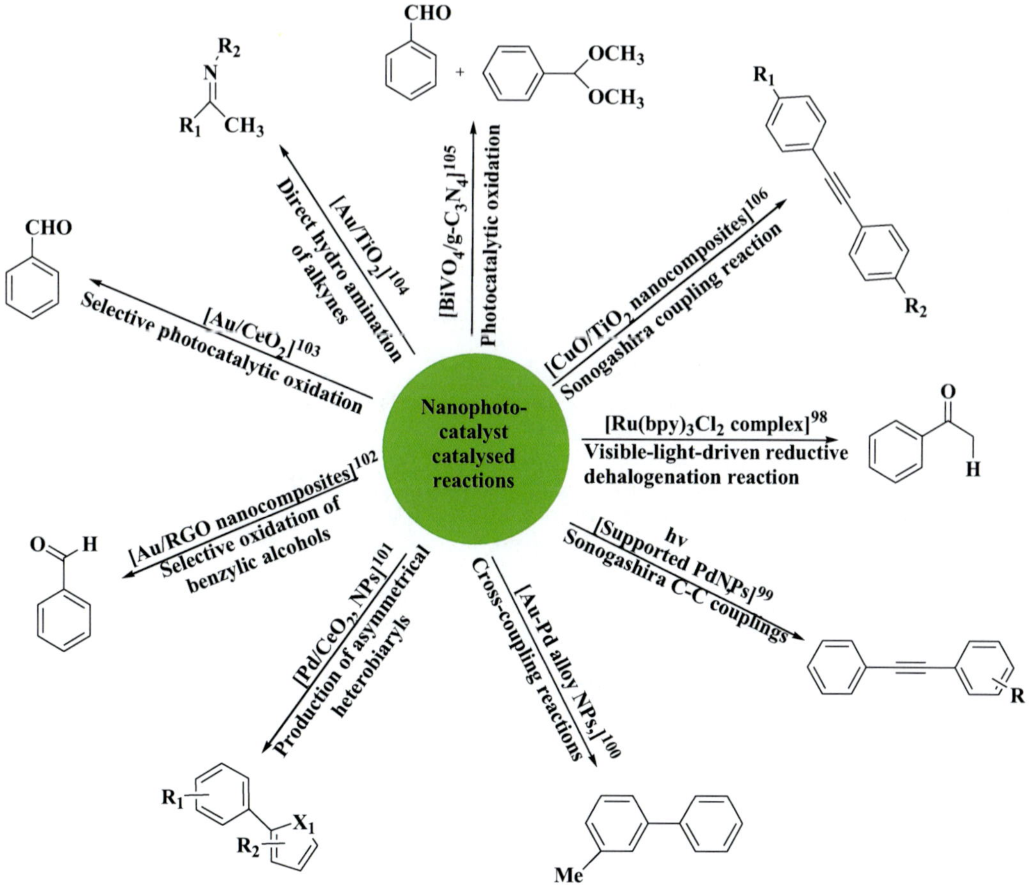

Scheme (5). Nanophotocatalyst as a catalyst for various visible light-induced organic transformations.

CONCLUSION

Chemical transformations can be designed greener and eco-friendly by the development and attentive utilization of catalysts. Due to their large surface area, nanostructured materials have originated to link "homogeneous (low loadings and good selectivity) and heterogeneous (recovery and recyclability) catalysis" processes. Nanocatalysts go beyond the boundaries of homogeneous and heterogeneous catalysts through nano-effects, which are not completely unwritten as yet. An update on catalytic efficiencies of various nanocatalysts such as monometallic nanoparticles and bimetallic nanomaterials, magnetic nanoparticles, nanocomposites, carbon-based nanomaterials, and nanophotocatalysts for various organic transformations have been included in this chapter.

CONSENT FOR PUBLICATION

Not applicable.

CONFLICTS OF INTEREST

The author declares no conflict of interest, financial or otherwise.

ACKNOWLEDGEMENTS

Declared none.

REFERENCES

[1] Dandia, A.; Parihar, S.; Sharma, R.; Rathore, K.S.; Parewa, V. Nanocatalysis in green organic synthesis. In: *Green Sustainable Process for Chemical and Environmental Engineering and Science*; Elsevier, **2020**; pp. 71-103.

[2] Heiz, U.; Landman, U. *Nanocatalysis*; Springer Science & Business Media, **2007**.

[3] Polshettiwar, V.; Varma, R.S. Green chemistry by nano-catalysis. *Green Chem.,* **2010**, *12*(5), 743-754. [http://dx.doi.org/10.1039/b921171c]

[4] Cuenya, B.R.; Behafarid, F. Nanocatalysis: size-and shape-dependent chemisorption and catalytic reactivity. *Surf. Sci. Rep.,* **2015**, *70*(2), 135-187. [http://dx.doi.org/10.1016/j.surfrep.2015.01.001]

[5] Dandia, A.; Parewa, V.; Sharma, A. An approach towards green switch through nanocatalysis for the synthesis of biodynamic heterocycles. In: *Green Chemistry: Synthesis of Bioactive Heterocycles*; Springer: New Delhi, **2014**; pp. 129-161.

[6] Glaser, J. A. Green chemistry with nanocatalysts. **2012**. [http://dx.doi.org/10.1007/s10098-012-0507-0]

[7] Narayanan, R. Synthesis of green nanocatalysts and industrially important green reactions. *Green Chem. Lett. Rev.,* **2012**, *5*(4), 707-725. [http://dx.doi.org/10.1080/17518253.2012.700955]

[8] Bond, G.C. Heterogeneous catalysis. **1987**.

[9] Ertl, G.; Knözinger, H.; Weitkamp, J. Handbook of heterogeneous catalysis. Weinheim: VCH, **1997**.

[10] Astruc, D. Nanoparticles and Catalysis. In: *John Wiley & Sons*; , **2008**.

[11] Astruc, D.; Lu, F.; Aranzaes, J.R. Nanoparticles as recyclable catalysts: the frontier between homogeneous and heterogeneous catalysis. *Angew. Chem. Int. Ed.,* **2005**, *44*(48), 7852-7872. [http://dx.doi.org/10.1002/anie.200500766] [PMID: 16304662]

[12] Corma, A. Attempts to fill the gap between enzymatic, homogeneous, and heterogeneous catalysis. *Catal. Rev., Sci. Eng.,* **2004**, *46*(3-4), 369-417. [http://dx.doi.org/10.1081/CR-200036732]

[13] Cui, X.; Li, W.; Ryabchuk, P.; Junge, K.; Beller, M. Bridging homogeneous and heterogeneous catalysis by heterogeneous single-metal-site catalysts. *Nat. Catal.,* **2018**, *1*(6), 385-397. [http://dx.doi.org/10.1038/s41929-018-0090-9]

[14] Spivey, J. Metal nanoparticles for catalysis: advances and applications *Royal Society of Chemistry*, **2014**.

[15] Hutchings, G. *Nanocatalysis: Synthesis and applications*; John Wiley & Sons, **2013**.

[16] Somwanshi, S.B.; Somvanshi, S.B.; Kharat, P.B. Nanocatalyst: a brief review on synthesis to applications. In: *Journal of Physics: Conference Series*; IOP Publishing, **2020**; p. 012046. [http://dx.doi.org/10.1088/1742-6596/1644/1/012046]

[17] Calvino-Casilda, V.; López-Peinado, A.J.; Martín-Aranda, R.M.; Mayoral, E.P. Nanocatalysis: applications and technologies., **2019**. [http://dx.doi.org/10.1201/9781315202990]

[18] Rotello, V. Nanoparticles: building blocks for nanotechnology. In: *Springer Science & Business Media*; , **2004**. [http://dx.doi.org/10.1007/978-1-4419-9042-6]

[19] Mohanraj, V.J.; Chen, Y. Nanoparticles-a review. *Trop. J. Pharm. Res.*, **2006**, *5*, 561-573.

[20] Hasan, S. A Review on Nanoparticles: Their Synthesis and Types. *Res. J. Recent Sci.*, **2015**, 2277-2502.

[21] Ealia, S.A.M.; Saravanakumar, M.P. A Review on the Classification, Characterisation, Synthesis of Nanoparticles and Their Application. In: *In IOP Conference Series: Materials Science and Engineering*; IOP Publishing, **2017**; 263, p. 032019.

[22] Beydoun, D.; Amal, R.; Low, G.; McEvoy, S. Role of nanoparticles in photocatalysis. *J. Nanopart. Res.*, **1999**, *1*(4), 439-458. [http://dx.doi.org/10.1023/A:1010044830871]

[23] Gupta, S.M.; Tripathi, M. An overview of commonly used semiconductor nanoparticles in photocatalysis. high ener. *Chem*, **2012**, *46*, 1-9.

[24] Dandia, A.; Gupta, S.L.; Saini, P.; Sharma, R.; Meena, S.; Parewa, V. Structure couture and appraisal of catalytic activity of carbon nitride (g-C_3N_4) based materials towards sustainability. *Curr. Res. Green Sustain. Chem.*, **2020**, *3*, 100039. [http://dx.doi.org/10.1016/j.crgsc.2020.100039]

[25] Dandia, A.; Saini, P.; Sharma, R.; Parewa, V. Green Organic Synthesis by Photochemical Protocol. In: *Green Sustainable Process for Chemical and Environmental Engineering and Science*; Elsevier, **2020**; pp. 155-198. [http://dx.doi.org/10.1016/B978-0-12-819539-0.00007-5]

[26] Dandia, A.; Saini, P.; Sharma, R.; Parewa, V. Visible Light Driven Perovskite-Based Photocatalysts: A New candidate for Green Organic Synthesis by Photochemical Protocol. *Curr. Res. Green Sustain. Chem*, **2020**. [http://dx.doi.org/10.1016/j.crgsc.2020.100031]

[27] Rao, C.R.; Kulkarni, G.U.; Thomas, P.J.; Edwards, P.P. Metal nanoparticles and their assemblies. *Chem. Soc. Rev.*, **2000**, *29*(1), 27-35. [http://dx.doi.org/10.1039/a904518j]

[28] Sardar, R.; Funston, A.M.; Mulvaney, P.; Murray, R.W. Gold nanoparticles: past, present, and future. *Langmuir*, **2009**, *25*(24), 13840-13851. [http://dx.doi.org/10.1021/la9019475] [PMID: 19572538]

[29] Mulfinger, L.; Solomon, S.D.; Bahadory, M.; Jeyarajasingam, A.V.; Rutkowsky, S.A.; Boritz, C. Synthesis and study of silver nanoparticles. *J. Chem. Educ.*, **2007**, *84*(2), 322. [http://dx.doi.org/10.1021/ed084p322]

[30] Medina-Cruz, D.; Saleh, B.; Vernet-Crua, A.; Nieto-Argüello, A.; Lomelí-Marroquín, D.; Vélez-Escamilla, L.Y.; Webster, T. Bimetallic nanoparticles for biomedical applications: A review. In: *Rac. Surf*; , **2020**; pp. 397-434.

[http://dx.doi.org/10.1007/978-3-030-34471-9_16]

[31] Cai, S.; Wang, D.; Niu, Z.; Li, Y. Progress in organic reactions catalyzed by bimetallic nanomaterials. *Chin. J. Catal.,* **2013**, *34*(11), 1964-1974.
[http://dx.doi.org/10.1016/S1872-2067(12)60701-3]

[32] Liu, X.; Wang, D.; Li, Y. Synthesis and catalytic properties of bimetallic nanomaterials with various architectures. *Nano Today,* **2012**, *7*(5), 448-466.
[http://dx.doi.org/10.1016/j.nantod.2012.08.003]

[33] Xu, J.; Wilson, A.R.; Rathmell, A.R.; Howe, J.; Chi, M.; Wiley, B.J. Synthesis and catalytic properties of Au-Pd nanoflowers. *ACS Nano,* **2011**, *5*(8), 6119-6127.
[http://dx.doi.org/10.1021/nn201161m] [PMID: 21761821]

[34] Imura, Y.; Tsujimoto, K.; Morita, C.; Kawai, T. Preparation and catalytic activity of Pd and bimetallic Pd-Ni nanowires. *Langmuir,* **2014**, *30*(17), 5026-5030.
[http://dx.doi.org/10.1021/la500811n] [PMID: 24731103]

[35] Blosi, M.; Ortelli, S.; Costa, A.L.; Dondi, M.; Lolli, A.; Andreoli, S.; Benito, P.; Albonetti, S. Bimetallic nanoparticles as efficient catalysts: facile and green microwave synthesis. *Materials (Basel),* **2016**, *9*(7), 550.
[http://dx.doi.org/10.3390/ma9070550] [PMID: 28773672]

[36] Dandia, A.; Parewa, V.; Jain, A.K.; Rathore, K.S. Step-economic, efficient, ZnS nanoparticle-catalyzed synthesis of spirooxindole derivatives in aqueous medium *via* knoevenagel condensation followed by michael addition. *Green Chem.,* **2011**, *13*(8), 2135-2145.
[http://dx.doi.org/10.1039/c1gc15244k]

[37] Dandia, A.; Singh, R.; Gupta, S.L.; Rathore, K.S. ZnS nanoparticle catalysed four component syntheses of novel spiropolyhydroquinoline derivatives in aqueous medium under ultrasonic irradiation. *Proc. Natl. Acad. Sci., India, Sect. A Phys. Sci.,* **2015**, *85*(1), 19-27.
[http://dx.doi.org/10.1007/s40010-014-0172-2]

[38] Kaur, N.; Kaur, G.; Bhalla, A.; Dhau, J.S.; Chaudhary, G.R. Metallosurfactant based Pd–Ni alloy nanoparticles as a proficient catalyst in the mizoroki heck coupling reaction. *Green Chem.,* **2018**, *20*(7), 1506-1514.
[http://dx.doi.org/10.1039/C7GC03877A]

[39] Dewan, A.; Sarmah, M.; Thakur, A.J.; Bharali, P.; Bora, U. Greener biogenic approach for the synthesis of palladium nanoparticles using papaya peel: an eco-friendly catalyst for c-c coupling reaction. *ACS Omega,* **2018**, *3*(5), 5327-5335.
[http://dx.doi.org/10.1021/acsomega.8b00039] [PMID: 31458742]

[40] Huang, T.S.; Wang, Y.H.; Jiang, J.Y.; Jin, Z.L. PEG-stabilized palladium nanoparticles: an efficient and recyclable catalyst for the selective hydrogenation of 1, 5-cyclooctadiene in thermoregulated PEG biphase system. *Chin. Chem. Lett.,* **2008**, *19*(1), 102-104.
[http://dx.doi.org/10.1016/j.cclet.2007.10.042]

[41] Srimani, D.; Bej, A.; Sarkar, A. Palladium nanoparticle catalyzed Hiyama coupling reaction of benzyl halides. *J. Org. Chem.,* **2010**, *75*(12), 4296-4299.
[http://dx.doi.org/10.1021/jo1003373] [PMID: 20491503]

[42] Firouzabadi, H.; Iranpoor, N.; Ghaderi, A. Gelatin as a bioorganic reductant, ligand and support for palladium nanoparticles. Application as a catalyst for ligand- and amine-free Sonogashira-Hagihara reaction. *Org. Biomol. Chem.,* **2011**, *9*(3), 865-871.
[http://dx.doi.org/10.1039/C0OB00253D] [PMID: 21120230]

[43] Pérez-Balderas, F.; Ortega-Muñoz, M.; Morales-Sanfrutos, J.; Hernández-Mateo, F.; Calvo-Flores, F.G.; Calvo-Asín, J.A.; Isac-García, J.; Santoyo-González, F. Multivalent neoglycoconjugates by regiospecific cycloaddition of alkynes and azides using organic-soluble copper catalysts. *Org. Lett.,* **2003**, *5*(11), 1951-1954.
[http://dx.doi.org/10.1021/ol034534r] [PMID: 12762694]

[44] Khalil, A.; Jouiad, M.; Khraisheh, M.; Hashaikeh, R. FacileSynthesis of copper oxide nanoparticles *via* electrospinning. *J. Nanomater.,* **2014**, 438407.

[45] Saha, A.; Saha, D.; Ranu, B.C. Copper nano-catalyst: sustainable phenyl-selenylation of aryl iodides and vinyl bromides in water under ligand free conditions. *Org. Biomol. Chem.,* **2009**, *7*(8), 1652-1657. [http://dx.doi.org/10.1039/b819137a] [PMID: 19343253]

[46] Sarkar, S.; Pal, R.; Roy, M.; Chatterjee, N.; Sarkar, S.; Sen, A.K. Nanodomain cubic copper (I) oxide as reusable catalyst for the synthesis of amides by amidation of aryl halides with isocyanides. *Tetrahedron Lett.,* **2015**, *56*(4), 623-626. [http://dx.doi.org/10.1016/j.tetlet.2014.12.060]

[47] Pai, G.; Chattopadhyay, A.P. New efficient ligand-free, copper nanoparticle catalyzed coupling reactions of aryl halides with diethyl malonate to produce α-arylation of malonates. *Synthesis,* **2013**, *45*(11), 1475-1482. [http://dx.doi.org/10.1055/s-0033-1338437]

[48] Ahammed, S.; Saha, A.; Ranu, B.C. Hydrogenation of azides over copper nanoparticle surface using ammonium formate in water. *J. Org. Chem.,* **2011**, *76*(17), 7235-7239. [http://dx.doi.org/10.1021/jo200915h] [PMID: 21793592]

[49] Gubin, S.P. Magnetic nanoparticles. In: *John Wiley & Sons*; , **2009**.

[50] Indira, T.K.; Lakshmi, P.K. Magnetic nanoparticles–a review. *Int. J. Pharm. Pharm. Sci. Nano.,* **2010**, *3*, 1035-1042.

[51] Gubin, S.P.; Koksharov, Y.A.; Khomutov, G.B.; Yurkov, G.Y. Magnetic nanoparticles: preparation, structure and properties. *Russ. Chem. Rev.,* **2005**, *74*(6), 489-520. [http://dx.doi.org/10.1070/RC2005v074n06ABEH000897]

[52] Liu, S.; Yu, B.; Wang, S.; Shen, Y.; Cong, H. Preparation, surface functionalization and application of Fe_3O_4 magnetic nanoparticles. *Adv. Colloid Interface Sci.,* **2020**, *281*, 102165. [http://dx.doi.org/10.1016/j.cis.2020.102165] [PMID: 32361408]

[53] Katz, E. Magnetic nanoparticles. **2020**. [http://dx.doi.org/10.3390/magnetochemistry6010006]

[54] Baig, R.B.; Varma, R.S. Magnetically retrievable catalysts for organic synthesis. *Chem. Commun. (Camb.),* **2013**, *49*(8), 752-770. [http://dx.doi.org/10.1039/C2CC35663E] [PMID: 23212208]

[55] Majidi, S.; Sehrig, F.Z.; Farkhani, S.M.; Goloujeh, M.S.; Akbarzadeh, A. Current methods for synthesis of magnetic nanoparticles. *Artif. Cells Nanomed. Biotechnol.,* **2016**, *44*(2), 722-734. [http://dx.doi.org/10.3109/21691401.2014.982802] [PMID: 25435409]

[56] Kidwai, M.; Jain, A.; Bhardwaj, S. Magnetic nanoparticles catalyzed synthesis of diverse N-heterocycles. *Mol. Divers.,* **2012**, *16*(1), 121-128. [http://dx.doi.org/10.1007/s11030-011-9336-z] [PMID: 22057791]

[57] Dandia, A.; Jain, A.K.; Sharma, S. $CuFe_2O_4$ Nanoparticles as a Highly Efficient and Magnetically Recoverable Catalyst for the Synthesis of Medicinally Privileged Spiropyrimidine Scaffolds. *RSC Advances,* **2013**, *3*(9), 2924-2934. [http://dx.doi.org/10.1039/c2ra22477a]

[58] Dandia, A.; Parewa, V.; Gupta, S.L.; Sharma, A.; Rathore, K.S.; Jain, A. Microwave-assisted Fe_3O_4 Nanoparticles Catalyzed Synthesis of Chromeno [1, 6] naphthyridines in Aqueous Media. *Catal. Commun.,* **2015**, *61*, 88-91. [http://dx.doi.org/10.1016/j.catcom.2014.12.015]

[59] Liu, L.; Liu, Y.; Ai, Y.; Li, J.; Zhou, J.; Fan, Z.; Bao, H.; Jiang, R.; Hu, Z.; Wang, J.; Jing, K.; Wang, Y.; Liang, Q.; Sun, H. Pd-CuFe catalyst for transfer hydrogenation of nitriles: controllable selectivity to primary amines and secondary amines. *iScience,* **2018**, *8*, 61-73.

[http://dx.doi.org/10.1016/j.isci.2018.09.010] [PMID: 30286395]

[60] Sobhani, S.; Falatooni, Z.M.; Asadi, S.; Honarmand, M. Palladium-schiff base complex immobilized covalently on magnetic nanoparticles as an efficient and recyclable catalyst for heck and suzuki cross-coupling reactions. *Catal. Lett.*, **2016**, *146*(1), 255-268. [http://dx.doi.org/10.1007/s10562-015-1636-y]

[61] Sreedhar, B.; Kumar, A.S.; Yada, D. $CuFe_2O_4$ nanoparticles: a magnetically recoverable and reusable catalyst for the synthesis of 5-substituted 1H-tetrazoles. *Tetrahedron Lett.*, **2011**, *52*(28), 3565-3569. [http://dx.doi.org/10.1016/j.tetlet.2011.04.094]

[62] Kantam, M.L.; Yadav, J.; Laha, S.; Srinivas, P.; Sreedhar, B.; Figueras, F. Asymmetric hydrosilylation of ketones catalyzed by magnetically recoverable and reusable copper ferrite nanoparticles. *J. Org. Chem.*, **2009**, *74*(12), 4608-4611. [http://dx.doi.org/10.1021/jo9002823] [PMID: 19518151]

[63] Kantam, M.L.; Yadav, J.; Laha, S.; Jha, S. Synthesis of propargylaminesby three-component coupling of aldehydes, amines and alkynes catalyzed by magnetically separable copper ferrite nanoparticles. *Synlett,* **2009**, *2*(11), 1791-1794. [http://dx.doi.org/10.1055/s-0029-1217362]

[64] Sarode, S.A.; Bhojane, J.M.; Nagarkar, J.M. An efficient magnetic copper ferrite nanoparticle: for one pot synthesis of 2-substituted benzoxazole *via* redox reactions. *Tetrahedron Lett.*, **2015**, *56*(1), 206-210. [http://dx.doi.org/10.1016/j.tetlet.2014.11.065]

[65] Sadri, F.; Ramazani, A.; Massoudi, A.; Khoobi, M.; Azizkhani, V.; Tarasi, R.; Min, B.K. Magnetic $CoFe_2O_4$ nanoparticles as an efficient catalyst for the oxidation of alcohols to carbonyl compounds in the presence of oxone as an oxidant. *Bull. Korean Chem. Soc.*, **2014**, *35*(7), 2029-2032. [http://dx.doi.org/10.5012/bkcs.2014.35.7.2029]

[66] Senapati, K.K.; Borgohain, C.; Phukan, P. Synthesis of highly stable $CoFe_2O_4$ nanoparticles and their use as magnetically separable catalyst for knoevenagel reaction in aqueous medium. *J. Mol. Catal. Chem.*, **2011**, *339*(1-2), 24-31. [http://dx.doi.org/10.1016/j.molcata.2011.02.007]

[67] Hedayati, V.R.; Ghasemzadeh, M.A. Efficient, One-Pot Synthesis of polyfunctionalised octahydroquinazolin-2, 5-Diones catalysed by Fe_3O_4 nanoparticles. *J. Chem. Res.*, **2015**, *39*(1), 56-61. [http://dx.doi.org/10.3184/174751915X14199511243271]

[68] Komarneni, S. Nanocomposites. *J. Mater. Chem.*, **1992**, *2*(12), 1219-1230. [http://dx.doi.org/10.1039/jm9920201219]

[69] Schaefer, D.W.; Justice, R.S. How nano are nanocomposites? *Macromolecules,* **2007**, *40*(24), 8501-8517. [http://dx.doi.org/10.1021/ma070356w]

[70] Behrens, S.; Appel, I. Magnetic nanocomposites. *Curr. Opin. Biotechnol.*, **2016**, *39*, 89-96. [http://dx.doi.org/10.1016/j.copbio.2016.02.005] [PMID: 26938504]

[71] Prasad, C.; Tang, H.; Bahadur, I. Graphitic carbon nitride based ternary nanocomposites: From synthesis to their applications in photocatalysis: A recent review. *J. Mol. Liq.*, **2019**, *281*, 634-654. [http://dx.doi.org/10.1016/j.molliq.2019.02.068]

[72] Pocklanova, R.; Rathi, A.K.; Gawande, M.B.; Datta, K.K.R.; Ranc, V.; Cepe, K.; Zboril, R. Gold nanoparticle-decorated graphene oxide: synthesis and application in oxidation reactions under benign conditions. *J. Mol. Catal. Chem.*, **2016**, *424*, 121-127. [http://dx.doi.org/10.1016/j.molcata.2016.07.047]

[73] Deng, D.; Yang, Y.; Gong, Y.; Li, Y.; Xu, X.; Wang, Y. Palladium nanoparticles supported on mpg-C_3N_4 as active catalyst forsemihydrogenation of phenylacetylene under mild conditions. *Green Chem.*, **2013**, *15*(9), 2525-2531.

[http://dx.doi.org/10.1039/c3gc40779a]

[74] Dandia, A.; Sharma, A.; Parewa, V.; Kumawat, B.; Rathore, K.S.; Sharma, A. Amidic C–N bond cleavage of isatin: chemoselective synthesis of pyrrolo [2, 3, 4-kl] acridin-1-ones using Ag NPs decorated rGO composite as an efficient and recoverable catalyst under microwave irradiation. *RSC Advances,* **2015**, *5*(111), 91888-91902. [http://dx.doi.org/10.1039/C5RA11747J]

[75] Dandia, A.; Gupta, S.L.; Indora, A.; Saini, P.; Parewa, V.; Rathore, K.S. Ag NPs decked GO composite as a competent and reusable catalyst for 'ON WATER'chemoselective synthesis of pyrano [2, 3-c: 6, 5-c′] dipyrazol]-2-ones. *Tetrahedron Lett.,* **2017**, *58*(12), 1170-1175. [http://dx.doi.org/10.1016/j.tetlet.2017.02.014]

[76] Dandia, A.; Bansal, S.; Sharma, R.; Rathore, K.S.; Parewa, V. Microwave-Assisted Nanocatalysis: A CuoNps/Rgo Composite As An Efficient And Recyclable Catalyst For The Petasis-borono–mannich reaction. *RSC Advances,* **2018**, *8*(53), 30280-30288. [http://dx.doi.org/10.1039/C8RA05203D]

[77] Azizi, N.; Ahooie, T.S.; Hashemi, M.M.; Yavari, I. Magnetic graphitic carbon nitride-catalyzed highly efficient construction of functionalized 4H-pyrans. *Synlett,* **2018**, *29*(5), 645-649. [http://dx.doi.org/10.1055/s-0036-1589145]

[78] Dandia, A.; Sharma, A.; Indora, A.; Rathore, K.S.; Jain, A.; Parewa, V. CeO_2 NPs/rGO composite catalyzed chemoselective synthesis of 2, 8-dioxabicyclo [3.3. 1] nonanes in aqueous medium *via* aldol condensation/michael addition/bicyclization reaction sequence. *Mol Catal.,* **2018**, *459*, 97-105. [http://dx.doi.org/10.1016/j.mcat.2018.08.024]

[79] Wan, X.; Zhou, C.; Chen, J.; Deng, W.; Zhang, Q.; Yang, Y.; Wang, Y. Base-free aerobic oxidation of 5-hydroxymethyl-furfural to 2, 5-furandicarboxylic acid in water catalyzed by functionalized carbon nanotube-supported Au–Pd alloy nanoparticles. *ACS Catal.,* **2014**, *4*(7), 2175-2185. [http://dx.doi.org/10.1021/cs5003096]

[80] Kilic, H.; Turgut, M.; Yilmaz, M.S.; Dalkilic, O.; Metin, O. Monodisperse Ni@ Pd Core@ shell nanoparticles assembled on reduced graphene oxide as a highly efficient and reusable heterogeneous catalyst for the C–H bond arylation of imidazo [1, 2-a] pyridine with Aryl Halides. *ACS Sustain. Chem.& Eng.,* **2018**, *6*(9), 11433-11440. [http://dx.doi.org/10.1021/acssuschemeng.8b01431]

[81] Siqueira, J.R., Jr; Oliveira, O.N., Jr Carbon-based nanomaterials.*Nanostructures*; William Andrew Publishing, **2017**, pp. 233-249. [http://dx.doi.org/10.1016/B978-0-323-49782-4.00009-7]

[82] Compton, O.C.; Nguyen, S.T. Graphene oxide, highly reduced graphene oxide, and graphene: versatile building blocks for carbon-based materials. *Small,* **2010**, *6*(6), 711-723. [http://dx.doi.org/10.1002/smll.200901934] [PMID: 20225186]

[83] Bahuguna, A.; Kumar, A.; Krishnan, V. Carbon support based heterogeneous nanocatalysts: synthesis and applications in organic reactions. *Asian J. Org. Chem.,* **2019**, *8*(8), 1263-1305. [http://dx.doi.org/10.1002/ajoc.201900259]

[84] Schaetz, A.; Zeltner, M.; Stark, W.J. Carbon modifications and surfaces for catalytic organic transformations. *ACS Catal.,* **2012**, *2*(6), 1267-1284. [http://dx.doi.org/10.1021/cs300014k]

[85] Dandia, A.; Parihar, S.; Saini, P.; Rathore, K.S.; Parewa, V. Metal-free sustainable synthesis of amides *via* oxidative amidation using graphene oxide as carbocatalyst in aqueous edium. *Catal. Lett.,* **2019**, *149*(11), 3169-3175. [http://dx.doi.org/10.1007/s10562-019-02878-5]

[86] Dandia, A.; Mahawar, D.K.; Sharma, R.; Badgoti, R.S.; Rathore, K.S.; Parewa, V. Graphene oxide-catalyzed CSp3–H activation of methylarenes in aqueous medium: A unified metal-free access to amides and benzimidazoles. *Appl. Organomet. Chem.,* **2019**, *33*(11), e5232.

[http://dx.doi.org/10.1002/aoc.5232]

[87] Sharma, P.; Sasson, Y. Highly active gC_3N_4 as a solid base catalyst for knoevenagel condensation reaction under phase transfer conditions. *RSC Advances,* **2017**, *7*(41), 25589-25596. [http://dx.doi.org/10.1039/C7RA03051G]

[88] Akhmedov, V.M.; Ahmadov, I.D.; Nurullayev, H.G.; Ahmadov, V.M. New metal-free catalysts for the selective hydrogenation of multible bonds İn aromatic hydrocarbons based on graphitic carbon nitrides. *J. Azerb. Chem.,* **2016**, *2*, 867-874.

[89] Wang, Y.; Yao, J.; Li, H.; Su, D.; Antonietti, M. Highly selective hydrogenation of phenol and derivatives over a Pd@carbon nitride catalyst in aqueous media. *J. Am. Chem. Soc.,* **2011**, *133*(8), 2362-2365. [http://dx.doi.org/10.1021/ja109856y] [PMID: 21294506]

[90] Shirini, F.; Kamali, F. $Fe_3O_4/g\text{-}C_3N_4$ nanocomposites as a reusable catalyst for the synthesis of 5-arylidenepyrimidine-2,4,6-(1H,3H,5H)-trione and pyrano-pyrimidinone derivatives in aqueous media. *J. Nanosci. Nanotechnol.,* **2020**, *20*(9), 5433-5444. [http://dx.doi.org/10.1166/jnn.2020.17863] [PMID: 32331115]

[91] Majumdar, B.; Sarma, D.; Bhattacharya, T.; Sarma, T.K. Graphene oxide as metal-free catalyst in oxidative dehydrogenative C–N coupling leading to α-Ketoamides: importance of dual catalytic activity. *ACS Sustain. Chem.& Eng.,* **2017**, *5*(10), 9286-9294. [http://dx.doi.org/10.1021/acssuschemeng.7b02267]

[92] Kawashita, Y.; Nakamichi, N.; Kawabata, H.; Hayashi, M. Direct and practical synthesis of 2-arylbenzoxazoles promoted by activated carbon. *Org. Lett.,* **2003**, *5*(20), 3713-3715. [http://dx.doi.org/10.1021/ol035393w] [PMID: 14507212]

[93] Marzo, L.; Pagire, S.K.; Reiser, O.; König, B. Visible-light photocatalysis: does it make a difference in organic synthesis? *Angew. Chem. Int. Ed. Engl.,* **2018**, *57*(32), 10034-10072. [http://dx.doi.org/10.1002/anie.201709766] [PMID: 29457971]

[94] König, B. Photocatalysis in organic synthesis–past, present, and future. *Eur. J. Org. Chem.,* **2017**, *2*(15), 1979-1981. [http://dx.doi.org/10.1002/ejoc.201700420]

[95] Stephenson, C.R.; Yoon, T.P.; MacMillan, D.W. Visible light photocatalysis in organic chemistry. In: *John Wiley & Sons*; , **2018**. [http://dx.doi.org/10.1002/9783527674145]

[96] Gisbertz, S.; Pieber, B. Heterogeneous Photocatalysis in Organic Synthesis. *ChemPhotoChem,* **2020**, *4*(7), 456-475. [http://dx.doi.org/10.1002/cptc.202000014]

[97] Dandia, A.; Saini, P.; Sharma, R.; Parewa, V. Green organic synthesis by photochemical protocol.*Green Sustainable Process for Chemical and Environmental Engineering and Science*; Elsevier, **2020**, pp. 155-198. [http://dx.doi.org/10.1016/B978-0-12-819539-0.00007-5]

[98] Li, X.; Hao, Z.; Zhang, F.; Li, H. Reduced graphene oxide-immobilized tris (bipyridine) ruthenium (II) complex for efficient visible-light-driven reductive dehalogenation reaction. *ACS Appl. Mater. Interfaces,* **2016**, *8*(19), 12141-12148. [http://dx.doi.org/10.1021/acsami.6b01100] [PMID: 27104739]

[99] Elhage, A.; Lanterna, A.E.; Scaiano, J.C. Light-induced sonogashira C–C coupling under mild conditions using supported palladium nanoparticles. *ACS Sustain. Chem.& Eng.,* **2018**, *6*(2), 1717-1722. [http://dx.doi.org/10.1021/acssuschemeng.7b02992]

[100] Xiao, Q.; Sarina, S.; Bo, A.; Jia, J.; Liu, H.; Arnold, D.P.; Zhu, H. Visible light-driven cross-coupling reactions at lower temperatures using a photocatalyst of palladium and gold alloy nanoparticles. *ACS*

Catal., **2014**, *4*(6), 1725-1734.
[http://dx.doi.org/10.1021/cs5000284]

[101] Ge, Y.; Diao, P.; Xu, C.; Zhang, N.; Guo, C. Visible light induced cross-coupling synthesis of asymmetrical heterobiaryls using Pd/CeO_2 nanocomposite photocatalyst. *Chin. Chem. Lett.,* **2018**, *29*(6), 903-906.
[http://dx.doi.org/10.1016/j.cclet.2018.01.002]

[102] Zhang, Y.; Guo, H.; Weng, W.; Fu, M.L. The surface plasmon resonance, thermal, support and size effect induced photocatalytic activity enhancement of Au/reduced graphene oxide for selective oxidation of benzylic alcohols. *Phys. Chem. Chem. Phys.,* **2017**, *19*(46), 31389-31398.
[http://dx.doi.org/10.1039/C7CP05378A] [PMID: 29155914]

[103] Tanaka, A.; Hashimoto, K.; Kominami, H. Selective photocatalytic oxidation of aromatic alcohols to aldehydes in an aqueous suspension of gold nanoparticles supported on cerium(IV) oxide under irradiation of green light. *Chem. Commun. (Camb.),* **2011**, *47*(37), 10446-10448.
[http://dx.doi.org/10.1039/c1cc13801d] [PMID: 21847491]

[104] Zhao, J.; Zheng, Z.; Bottle, S.; Chou, A.; Sarina, S.; Zhu, H. Highly efficient and selective photocatalytic hydroamination of alkynes by supported gold nanoparticles using visible light at ambient temperature. *Chem. Commun. (Camb.),* **2013**, *49*(26), 2676-2678.
[http://dx.doi.org/10.1039/c3cc38985e] [PMID: 23435475]

[105] Samanta, S.; Khilari, S.; Pradhan, D.; Srivastava, R. An efficient, visible light driven, selective oxidation of aromatic alcohols and amines with O_2 using $BiVO_4$/g-C_3N_4nanocomposite: a systematic and comprehensive study toward the development of a photocatalytic process. *ACS Sustain. Chem.& Eng.,* **2017**, *5*(3), 2562-2577.
[http://dx.doi.org/10.1021/acssuschemeng.6b02902]

[106] Li, Y.; Feng, X.; Li, Z. Visible-light-initiated sonogashira coupling reactions reactions over CuO/TiO_2 nanocomposites. *Catal. Sci. Technol.,* **2019**, *9*(2), 377-383.
[http://dx.doi.org/10.1039/C8CY02333F]

CHAPTER 2

Nanocatalytic Carbon-Carbon Coupling Reactions

Prantika Bhattacharjee[1] and **Utpal Bora**[1,*]

[1] *Department of Chemical Sciences, Tezpur University, Napaam, Tezpur, Assam, 784028, India*

Abstract: This chapter provides a brief overview of the general aspects and recent advances in the chemistry of carbon-carbon coupling reactions. Nanoclusters, their synthesis, stability, and catalytic performance in various C-C bond formation reactions are discussed. The interest in nanoparticles originates from the nanoscale size, shape, and large surface area to volume ratio, imparting exceptional surface properties compared to bulk catalysts. In recent years, many detailed investigations on nanocatalytic carbon-carbon coupling reactions have been documented. We tried to summarize some of the significant achievements in this flourishing field. Most of the nanocatalysts outlined in this chapter allow convenient and green synthetic pathways for the construction of carbon-carbon bonds.

Keywords: Bi-metallic, Bio-genic nanoparticles, Cross-coupling, Homocoupling, Multi-metallic, Nanocatalysis, Supported nanoparticles.

INTRODUCTION

The formation of carbon-carbon (C-C) bonds continues to be the vital synthetic paradigm in organic chemistry since its beginning in the 19th century [1]. “Coupling reaction,” which conventionally referred to the transition metal-catalyzed reaction of an unsaturated organic halide electrophile with an organometallic nucleophile [2], is a comprehensive term today encompassing many advanced synthetic transformations. The very idea of C(sp^2)-C(sp^2) coupling reactions emerged from the eminent observations in the 1940s by studying the unique ability of simple transition-metal salts to activate various organic compounds and allowing the formation of new bonds through homocoupling of Grignard reagents [3]; hence, marking the foundation of a new era in shaping chemical synthesis. Since then, C-C coupling reactions have been one of the vastly researched domains of study for organic chemists.

The initial phase of development witnessed the utility of copper (Cu) as the sole metal for aryl-aryl bond formation, giving rise to the famous Cu mediated Glaser

* **Corresponding author Utpal Bora:** Department of Chemical Sciences, Tezpur University, Napaam, Tezpur, Assam, 784028, India; Tel: (+91) 03712 275067; Fax: +91-3712-267005; E-mail: ubora@tezu.ernet.in & utbora@yahoo.co.in

[4], Ullman homo-coupling [5], and Cadiot-Chodkiewicz [6] cross-coupling reactions. However, the introduction of palladium (Pd) catalysis revolutionized the field of organic synthesis. The chemistry of Pd in catalytic Wacker-type processes [7] inspired the researchers to exploit its activity in the reaction of carbon nucleophiles to construct C-C bonds. Accordingly, the classical cross-coupling reactions using olefin (Heck) [8], terminal alkyne (Sonogashira) [9], organozinc (Negishi) [10], organoboron (Suzuki-Miyaura) [11], organotin (Stille) [12], organosilicon (Hiyama) [13] or organomagnesium (Kumada-Corriu) [14] nucleophiles were devised (Scheme **1**). Apart from its phenomenal achievement in the construction of C-C bond, Pd catalysis gained popularity due to the mild conditions associated with the reactions along with its commendable tolerance towards different functional groups, high reactivity, and the diversity of organometallic reagents utilized in these reactions. These astonishing advances in the field of transition-metal catalyzed cross-coupling reaction [15, 16] have presented an ample number of synthetic alternatives for the construction of complex carbon frameworks of diversified interests and are conveniently extendable to large scale synthesis for the application in natural products, pharmaceuticals, material, and industrial processes and nanotechnology [17, 18].

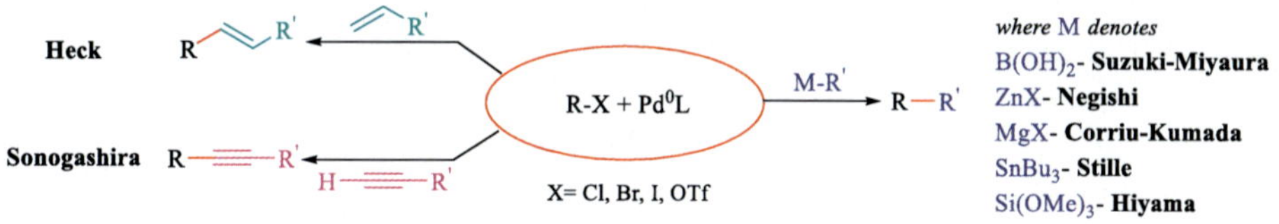

Scheme (1). Pd catalyzed common cross-coupling reactions.

The past decades have seen numerous developments in homogeneous and heterogeneous metal-catalyzed cross-coupling strategies in organic as well as aqueous media [19]. However, the issue of recovery and reusability limits the practical utility of soluble homogeneous catalyst species. Those issues with homogeneous catalysis were addressed by their heterogenization on solid supports with high surface area [20]. While a majority of the solid-supported catalysts are known to follow heterogeneous pathways [21], yet under specific conditions, they are found to liberate highly active nanoparticle species into the reaction media [22], thereby acting as a pre-catalyst and enhancing the overall selectivity and performance in C-C coupling reactions. Hence, driving a necessary shift from conventional catalysis to nano-catalysis in the quest for an improved and advanced catalytic pathway [23]. The chemistry of metal particles at the nanoscale level (1-100 nm) has materialized as an interesting and promising field with increasing applications in industrial and chemical synthesis [24]. Owing to their properties between homogeneous and heterogeneous, these species have attracted great attention for the time being, and recently, much of their endeavors have been realised on carbon-carbon coupling reactions [25, 26].

Morphology Controlled Nanostructures for C-C Coupling Reactions

The concept behind the outstanding catalytic performance of nanocatalysts can be comprehended by accounting for the fundamental material properties of nanostructures on catalysis. The appropriate tunning of size, shape, and morphology of nanocatalyst allows the rational design of nanomaterials by a specific catalytic application [27]. Such catalytic dependence on the shape of nanomaterials has been significantly observed in C-C coupling reactions.

El-Sayed and co-workers prepared Pt nanoparticles (NPs) of different shapes namely, spherical, cubic, and tetrahedral as observed in Transmission electron microscopy (TEM) images, using the hydrogen reduction method [28]. When testing their catalytic activity in Suzuki cross-coupling reactions between iodobenzene and phenylboronic acid, they found that while moving from nearly spherical to cubic to tetrahedral Pt NPs, there is a surprising increase in their catalytic performance [29]. The tetrahedral NPs proved to be the most catalytically active followed by the spherical NPs while the cubic NPs showed the least catalytic activity with the highest activation energy. It was observed that tetrahedral Pt NPs can successfully catalyze the Suzuki reaction forming biphenyl, but itself transforms the shape from tetrahedral to spherical due to the high reaction temperature. When the recovered NPs are employed to catalyze the second cycle of the Suzuki reaction, a sharp reduction in the catalytic activity is observed. The reason behind such a difference in catalytic behavior is due to the difference in the number of catalytically active surface atoms on the nanomaterials. Tetrahedral NPs which are composed of facets have a large fraction of surface atoms on edges and corners compared to the cubic NPs, hence they are catalytically more active.

In a similar stroke, Zhang and co-workers [30] demonstrated a one-pot shape-selective synthesis of Pd-Rh nanocrystals of different shapes and morphologies namely hollow nanocubes (NC), nanoicosahedrons (NI), and nanotruncated octahedrons (NTO). The catalytic performance of the nanohybrids was evaluated in Suzuki cross-coupling reaction between aryl halides and phenylboronic acids. It was observed that Pd-Rh hollow NCs displayed the highest reactivity in comparison to other Pd-based solid nanocrystals, including Pd−Rh NIs, Pd−Rh NTOs, Pd−Rh NCs, Pd NCs, and commercial Pd/C, with both iodo and bromobenzene as the reactant (Fig. **1a**). The catalytic activities were also checked with Pd-Rh hollow NCs of varying Pd/Rh ratios, such as $Pd_{0.22}Rh_{0.78}$, $Pd_{0.32}Rh_{0.68}$, $Pd_{0.49}Rh_{0.51}$, and $Pd_{0.68}Rh_{0.32}$, however, no direct correlation was found (Fig. **1b**).

The results, therefore, indicate that the morphology rather than the composition held a dominant effect on the catalytic performance of the prepared nanocatalysts in C-C coupling reactions. It is assumed that the presence of rich kink and step sites on the inner rough faces of Pd–Rh hollow nanocubes is responsible for its high catalytic activity.

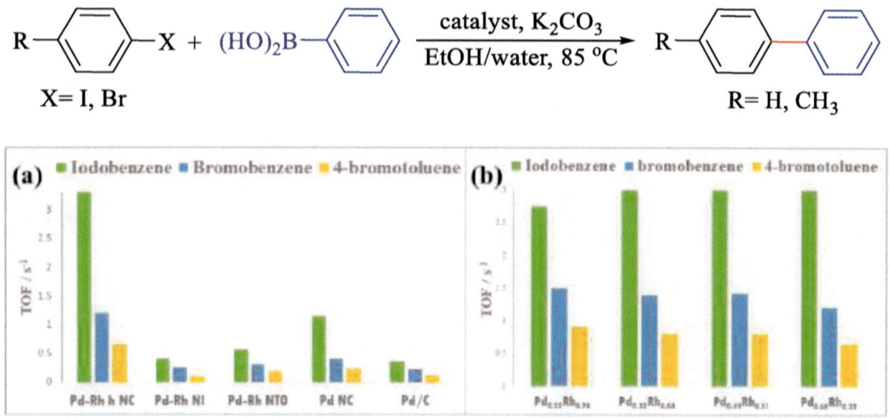

Fig. (1). TOFs per surface Pd atom for Suzuki cross-coupling reaction with **(a)** Pd-based nanocrystals of different morphology; **(b)** Pd–Rh hollow NCs in various Pd/Rh ratios.

Choice of Nanomaterial Supports for C-C Coupling Reactions

The small size and probable solubility of the nanomaterials often make its separation difficult from the reaction media. To combat the issue, NPs are usually anchored into solid support. A wide range of support materials has been used for the adsorption of NPs, acting as a stabilizer to prevent the agglomeration of supported NPs [31]. The supports do not exhibit any catalytic activity of its own, however, cooperates with the NPs to assist in promoting a particular transformation. Hence, the choice of appropriate supports is a topic of great importance. The most common types of nanomaterial support that have been used for incorporating Pd NPs to be suitable in various C-C coupling reactions are discussed herewith.

Silica is one of the most abundant and widely used families of known materials. It becomes a good choice of solid support, rendering to its wide accessibility, excellent stability, better surface reactivity, and the inert environment facilitating the immobilization of the metal NPs [32]. The high surface area and volume of SBA-15 [33] and MCM-41 [34] make them the most common types of mesoporous silica as supports in catalysis. Pd NPs supported on these ordered mesoporous supports are found to be highly active and recyclable in Suzuki,

Sonogashira, and Heck coupling reactions [35]. A representative example of the Heck coupling reaction between aryl halides and alkenes catalyzed by Pd doped silica is shown in Scheme **2a**. One of the cheapest and most eco-friendly sources of silica is rice husk containing about 90-97% of its composition. Reusable Pd immobilized on rising husk-derived nanosilica has been successfully used as a catalyst in Suzuki-Miyaura cross-coupling reactions between aryl halides and arylboronic acids in aqueous media (Scheme **2(b)**) [36].

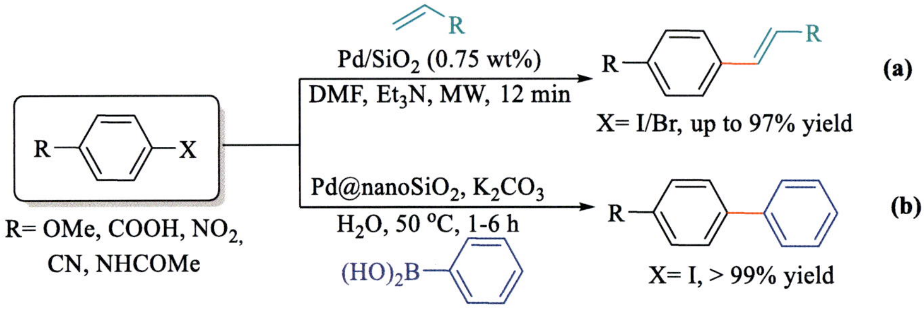

Scheme (2). Pd/SiO_2 catalyzed coupling of aryl halides with **(a)** olefins (Heck coupling), and **(b)** phenylboronic acids (Suzuki coupling).

Likewise, many metal oxides and hydroxides have also been used as supports for different C-C coupling reactions. For example, the catalytic activity of ZnO-supported Pd NP catalyst was studied for Hiyama coupling between phenyl trimethoxysilane and aryl halides in the presence of K_2CO_3 as the base in ethylene glycol (Scheme **3**) [37]. The corresponding biaryls were obtained in high yields (up to 96%). The catalytic system also proved highly efficient for organo-boronates in Suzuki coupling reactions. Similarly, MgO, ZrO_2, CuO, and layered doubled hydroxides have been used to support Pd nanocatalysts in various cross-coupling reactions [25].

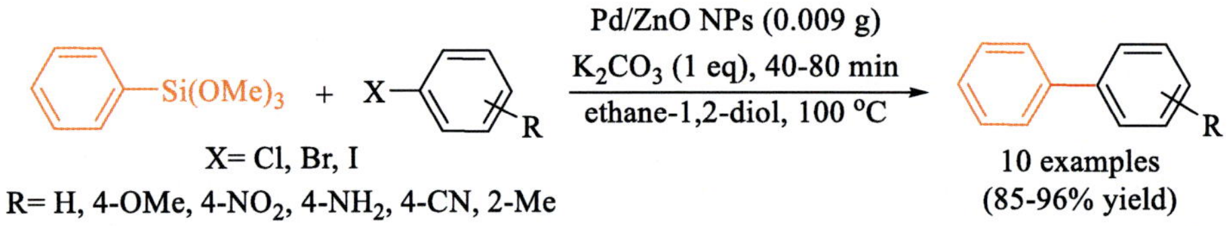

Scheme (3). Pd/ZnO NPs catalyzed coupling of aryl halides with phenyl trimethoxysilane (Hiyama coupling).

Pd NPs have been successfully immobilized onto functionalized single and multiwalled carbon nanotubes (CNTs) [38]. Several methodologies have been developed for the decoration of CNTs with Pd NPs including thermal, chemical, electrical deposition techniques. Recently, Shaabani and co-workers [39] have designed a multicomponent reaction (MCR) for the covalent functionalization of multi-walled carbon nanotubes (MWCNT) (Scheme **4**). For this purpose, aminated-MWCNT **4** which was earlier prepared from commercially available MWCNT **1** *via* oxidization followed by amination with *p*-phenylenediamine **3**, was reacted with 4-methoxy benzaldehyde **5**, Meldrum's acid **6**, cyclohexyl isocyanide **7** in a one-pot vessel to produce pentaamide scaffolds covalently attached on the sidewalls of MWCNTs **8** at room temperature. Finally, H_2PdCl_4 solution was added along with the reducing agent for the preparation and immobilization of Pd NPs on the pentaamide ligands coated support **8** generating the resultant Pd NPs@pentaamide-*f*-MWCNT nanocomposite. The catalytic activities of the nanocatalyst were examined in diverse C-C coupling reactions (Scheme **5**) for the formation of Csp^2-Csp^2, Csp-Csp^2, and Csp-Csp bonds through Ullman homocoupling of aryl halides, Sonogashira cross-coupling between aryl halides and terminal acetylene and Glacer homocoupling of alkynes respectively in bio-based reaction media albeit at a higher reaction temperature of 100 °C in each case.

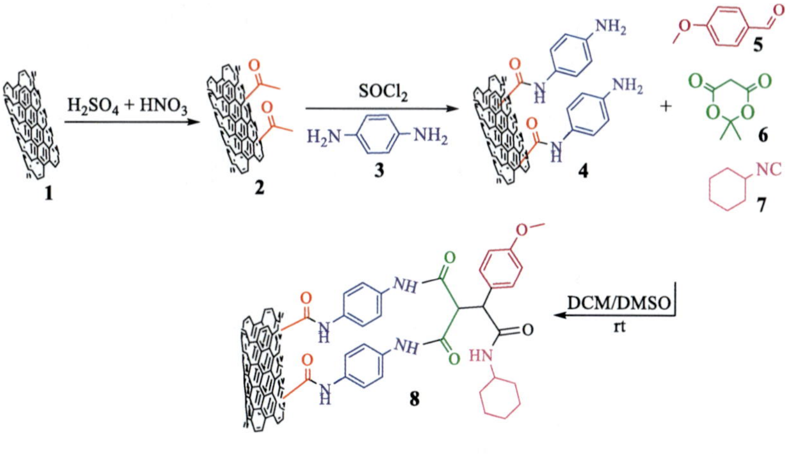

Scheme (4). Synthesis of pentaamide-functionalized MWCNT nanocomposite with isocyanide based pseudo-five component reaction for Pd NPs immobilization.

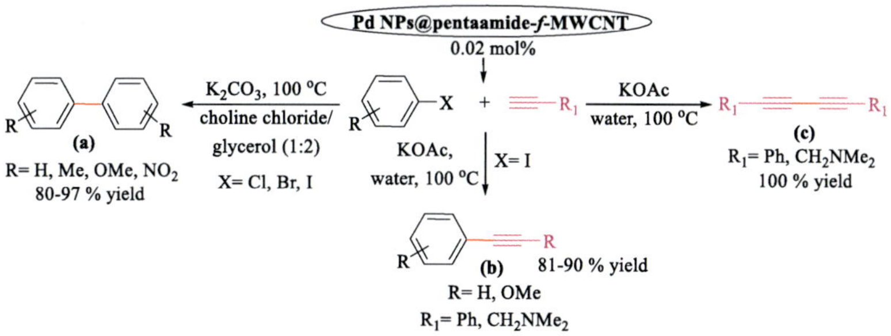

Scheme (5). Pd NPs@pentaamide-*f*-MWCNT catalyzed **(a)** Csp^2-Csp^2 coupling of aryl halides (Ullman coupling); **(b)** Csp^2-Csp coupling of aryl iodides and terminal acetylene (Sonogashira coupling); and (c) Csp-Csp coupling of alkynes (Glacer coupling).

To provide further insight into the catalytic versatility of the nanocomposite, a platform was designed for intramolecular cross-coupling reaction (Scheme **6**). A carboxamide structure **5** was generated through an Ugi four-component reaction [39] by the condensation of 2-iodoaniline **1**, 2-bromobenzaldehyde **2**, benzoic acid **3,** and cyclohexyl isocyanide **4**. The resultant Ugi product **5** contains two potential nucleophilic sites, the α-C of the secondary amide and the NH of the secondary amide, besides the iodo and bromo functionality in the benzene ring. When used as a substrate in our catalytic system, it underwent an intramolecular Ullmann cross-coupling reaction in a chemoselective manner.

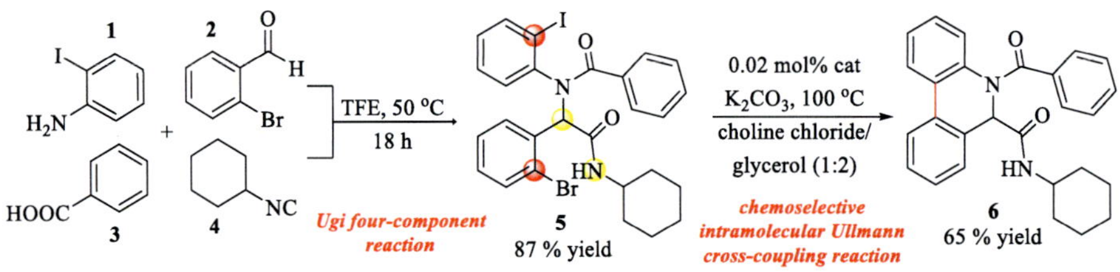

Scheme (6). Pd NPs@pentaamide-*f*-MWCNT catalyzed intramolecular Ullmann cross-coupling reaction.

More recently, functionalized polymer-based supports have also been utilized for immobilizing metal catalysts in C-C coupling reactions. Pd NPs dispersed within a polymeric matrix are receiving intense attention due to their novel and versatile material properties. Sureshbabu and co-workers described the preparation of polyvinyl chloride (PVC)-supported Pd NPs through the reduction of $PdCl_2$ by employing sodium formate and NaOH in the ethanol-water system [40]. The prepared PVC-Pd(0) displayed outstanding catalytic performance in Sonogashira coupling reaction between iodobenzene and phenylacetylene at room temperature under Cu and phosphine-free conditions (Scheme **7**).

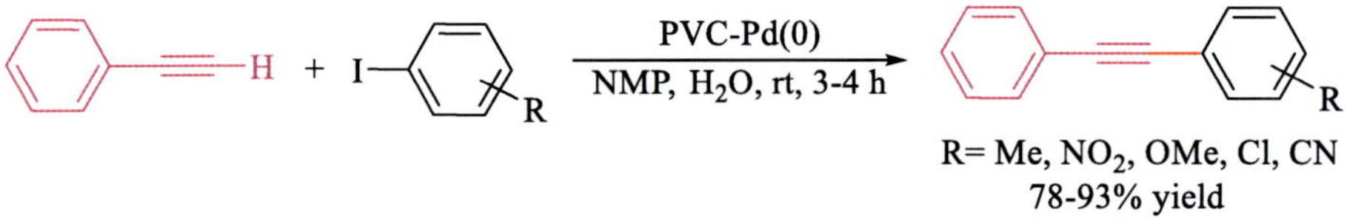

Scheme (7). PVC-Pd(0) catalyzed coupling of aryl iodides with phenylacetylene (Sonogashira coupling).

In addition to increased catalyst activity, the enhanced recyclability of the catalyst is also an important factor demanding attention and consideration. On that ground, magnetically retrievable nanocatalyst supports have gained interest as catalyst carriers due to easy magnetic separation of the catalyst from the reaction mixture [41]. Hence, minimizing the number of purifications, filtration, and centrifugation steps as required for isolating and re-using other non-magnetic catalysts. The application of such magnetic nanomaterials in C-C coupling reactions has been steadily rising. Recently, Saha and co-workers [42] have used starch-coated magnetic ferrite NPs as the heterogeneous support for Pd(II) catalyst. The nanomaterials were simply prepared by dissolving the Fe^{3+} and Fe^{2+} salts in an aqueous starch solution. The subsequent black NPs were used for immobilizing the Pd(II) by stirring the methanolic suspension of the NPs with $PdCl_2$ at room temperature. The resultant biopolymer coated magnetic catalyst was further explored to perform C-C cross-coupling reactions of organosilanes with terminal alkenes (oxidative Heck coupling) and allylic acetates (Tsuji-Trost coupling) and displayed outstanding catalytic activity under open aerial conditions (Scheme **8**). Thus, disclosing the high moisture and air stability of the magnetic nanomaterials. In addition to the above-mentioned conventional solid supports, Pd nanocatalyst systems based on fluorous and ionic-liquid-based media [25] have also been exploited in various C-C coupling reactions.

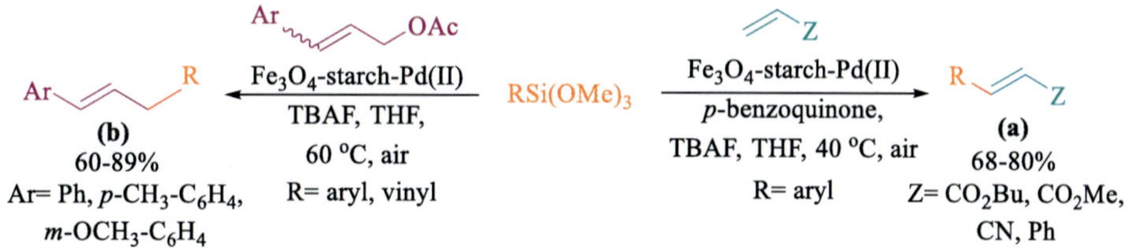

Scheme (8). Magnetic Fe_3O_4-starch-Pd(II) catalyzed C-C cross-coupling reactions of organosilanes with **(a)** terminal alkenes (Heck coupling); and **(b)** allylic acetates (Tsuji-Trost coupling).

Beyond Pd Catalysis: New Trends in C-C Coupling Reactions

The domain of C-C coupling reactions has been dominated by Pd catalysis, either as metal complexes or as nanoparticles. The reason for this general catalytic activity of Pd arises from its facile swing from Pd(0) to Pd(II) in oxidative addition and the reverse in reductive elimination [43], which is considered crucial for the catalytic performance in the coupling reactions. In comparison to Pd, the application of other transition metals in this field has still remained uncommon, though some beautiful advances have been possible. This section highlights the application of metal NPs other than Pd in catalyzing various C-C coupling reactions.

After Pd, the next extensively studied metal has been gold (Au) NPs in various C-C couplings, including Ullman homocoupling of halides, oxidative homocoupling of organoboronates, Suzuki and Sonogashira cross-coupling reactions [44]. Parvulescu and co-workers [45] have synthesized Au NPs of size 20-40 nm grafted on graphene films (Au/G). They found that when tested on the Suzuki-

Miyaura coupling reaction of aryl halides with phenylboronic acids, the Au particles exhibited reactivity three times higher than the analogous Pd-graphene catalyst (Pd/G). An interesting outcome in the reactivity order of halobenzenes in the coupling reaction has been observed. The product yields in Suzuki-Miyaura coupling catalyzed by Au NPs are found to follow the order: chlorobenzene > bromobenzene > iodobenzene, which is in stark contradiction to Pd catalysis (Table **1**). This reversal in order for Au *vs* Pd catalysis can be derived from the strong adsorption ability of iodide on Au followed by bromide, which leads to Au catalyst deactivation, hence lowering the reactivity in the coupling reaction. Likewise, supported silver (Ag) NP catalysts could also be utilized for cross-coupling of aryl halides and phenylboronic acid [46].

Table 1. Comparative study of Au *vs* Pd on graphene in Suzuki coupling reactions.

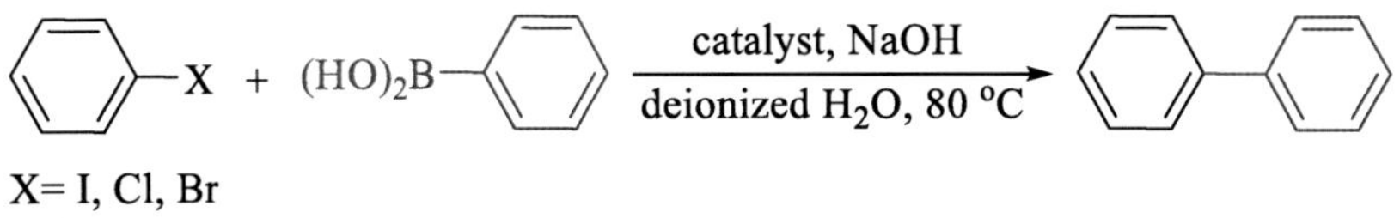

X	Catalyst	Conversion (%)
I	Au/G	2.4
	Pd/G	70.5
Br	Au/G	90.9
	Pd/G	37.0
Cl	Au/G	100
	Pd/G	8.0

However, the high cost of noble metals like Pd, Ag, Au limits the practical scope of the coupling reactions. Thus, recent attention in the scientific community has been concerned with switching to alternative low-cost, economical, and environment-friendly metal catalysts. The role of Cu in C-C coupling reactions has been known since its beginning. It has been extensively explored as a co-catalyst in Pd catalyzed Sonogashira cross-coupling reactions [47]. However, lately, Cu has shown notable performance in Pd free Sonogashira cross-coupling strategies. Obora and co-workers [48] have prepared Cu NPs by simple DMF-reduction method from precursor Cu(II) salt, where DMF played multiple roles in acting as a solvent, reductant, and stabilizer for the Cu NPs (<10 nm) generated. The catalyst displayed outstanding catalytic activity in promoting the reaction of terminal alkynes with aryl halides at low catalyst loadings (0.01 mol%) (Scheme **9A**). The results inspired researchers to exploit the activity of other non-toxic, readily available, and low-cost metal complexes immobilized on solid supports in the field of C-C bond formation. In the present context, coupling strategies have been designed by various research groups employing environmental benign catalytic systems based on cobalt (Co) [49] (Scheme **9B**), iron (Fe) [50] (Scheme **9C**), and nickel (Ni) NPs [51] (Scheme **9A**). All the heterogeneous supported metal NPs exhibited remarkable catalytic activity in Heck and Sonogashira cross-coupling reactions of a variety of aryl halides. However, the superior catalytic activeness of Pd NPs in C-C coupling becomes obvious from the requirement of higher reaction temperatures (Scheme **9**) in each of the cases for complete reaction conversions.

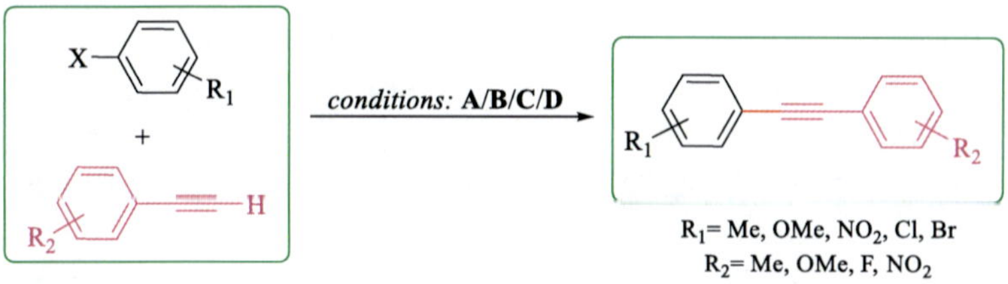

R_1= Me, OMe, NO_2, Cl, Br
R_2= Me, OMe, F, NO_2

Conditions:

A: Cu NPs/PPh_3, K_2CO_3, DMF, 135 °C, X= I, Br

B: MS-CS@Co, KOH, DMSO, 140°C, X= I, Br, Cl
MS-CS: chitosan functionalized with methyl salicylate

C: Fe(III)@SiO_2-DABCO, H_2O/DMF, Cs_2CO_3, 110 °C, X= I

D: Ni NPs-rGO, NMP, CuI, K_2CO_3, 120 °C, X= I, Br, Cl
rGO: reduced graphene oxide

Scheme (9). Coupling of aryl halides with phenylacetylene (Sonogashira coupling) using different metal NPs.

Advances in Bi-Metallic and Multi-Metallic Nanocatalysts for C-C Coupling Reactions

Bi-metallic nanoparticles have raised significant concern worldwide in recent years, providing some interesting insights into the field of nanotechnology. The combination of two different metals permits tuning of physical and chemical properties leading to an enhancement of the features of the nanostructure in comparison with its monometallic configurations [52]. The field of heterogeneous catalysis involving bi-metallic, tri-metallic, and multi-metallic metal nanoparticles as catalysts in different types of C-C coupling reactions such as Suzuki, Heck, and Sonogashira reactions have been taken much interest in the past few years [53]. Though Pd is still considered the most active metal for C-C coupling reactions, remarkable improvements have been observed from monometallic Pd NP catalyst to the bi- and multi-metallic NPs for C-C coupling reactions.

Different combinations of noble and non-noble metal-based catalysts have been explored for C-C coupling reactions [53]. However, the most extensively studied combination has been Pd-Au-based bi-metallic systems. The intermetallic atomic-level mixing of Au and Pd is found to be excellent and the high electronegative nature of Au allows the flow of electronic charge transfer from Pd to Au [54]. Tang and co-workers [55] prepared core-shell and alloy Au-Pd bimetallic nanocrystals confined in nanoscale cavities of silica nanorattles. Three different compositions of the Au-Pd alloy core was synthesized hydrothermally by simple tuning of the ratio of Au-Pd precursor concentration from 3:1 to 1:1 to 1:3. The catalysts were successfully employed in Suzuki coupling between aryl halides and phenylboronic acids (Scheme **10(a)**). The catalytic efficiency of the bimetallic catalysts was found in the order: alloy Au_3Pd_1 > Au_1Pd_1 > Au_1Pd_3 > core-shell Au-Pd > Pd > Au, revealing the highest activity of the alloy Au-Pd@SiO_2 with the lowest Pd content for both iodobenzene and bromobenzene in Suzuki coupling reaction. The mesoporous walls of the silica nanorattles are expected to protect the bi-metallic core from aggregation and prevent the leaching of Pd, hence providing the required stability to the bi-metallic nanohybrid. Moreover, mixing of Pd with other noble metals such as Ag, Rh, Ru has also proved to be highly active for cross-coupling reactions [53].

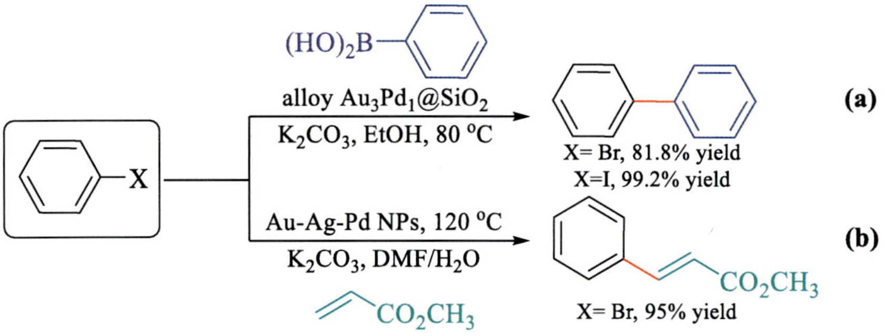

Scheme (10). (a) Au-Pd bimetallic catalyst in the coupling of an aryl halide with phenylboronic acids (Suzuki coupling), **(b)** Au-Ag-Pd tri-metallic catalyst in the coupling of methyl acrylate with bromobenzene (Heck coupling).

To further enhance the catalytic activity of Au-Pd alloy nanoparticle catalyst, Santhanalakshmi and co-workers [56] synthesised Au-Ag-Pd tri-metallic nanoparticle catalyst in a ratio of 1:1:1 and explored their catalytic activity for Heck reaction between methyl acrylate and bromobenzene in DMF/H_2O at 120 °C (Scheme **10(b)**). The trimetallic core was found to exhibit better catalytic efficiency compared to their monometallic and bi-metallic counterparts, which can be attributed to the electronic effects of the Au-Ag core on the Pd shell atoms.

Limited reserves and the high cost of noble metals have also encouraged researchers to incorporate other inexpensive non-noble metals such as Ni, Co, Cu, Fe with or without Pd and employ them in various C-C coupling reactions. Recently, Allahresani and the group prepared a new magnetically recyclable Co-

Cu bi-metallic catalyst by anchoring imidazolium moiety and PEG chains on the Fe_3O_4 NPs [57]. The Fe_3O_4@PEG/Cu-Co catalyst displayed outstanding catalytic performance in base/Pd free Suzuki and Sonogashira cross-coupling reactions in water (Scheme **11**). The presence of PEG chains rendered the nanohybrid to be water-dispersible thereby increasing the efficiency of coupling reactions in water. Hence, the development of cost-effective and noble-metal free based multi-metallic combinations should get consideration for future developments.

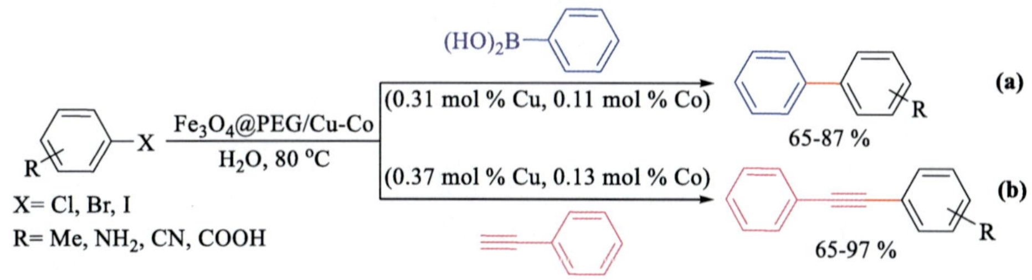

Scheme (11). Co-Cu bi-metallic catalyst in the coupling of aryl halides with **(a)** phenylboronic acid (Suzuki coupling), and **(b)** phenylacetylene (Sonogashira coupling).

Mechanism of C-C Coupling Reactions using Nanocatalysts

Even though the nanomaterial catalysts are in high demand, details about the exact mechanism in action at the nanoscale level are unclear. Several reports claim that the nanocatalysts remain as metal particles during the entire catalytic cycle and the reaction occurs at the periphery of nanoparticles [58]. While others accept the fact that metals 'leach out' from the nanoclusters and initiate a homogeneous catalytic cycle [21]. In general, the mechanism of Pd-catalyzed cross-coupling is proposed to involve a Pd(0)/ Pd(II) catalytic cycle (Scheme **12**), where the elementary steps involved are oxidative addition of the electrophile to the catalytic site Pd(0), transmetalation of the nucleophile, and reductive elimination to form the product [43]. Finally completing the catalytic cycle by regenerating the Pd(0) catalyst, reabsorbed into the support. There is now extensive experimental and computational evidence to support this pathway. However, not much information is available for other metal NPs and bi- and multi-metallic combinations.

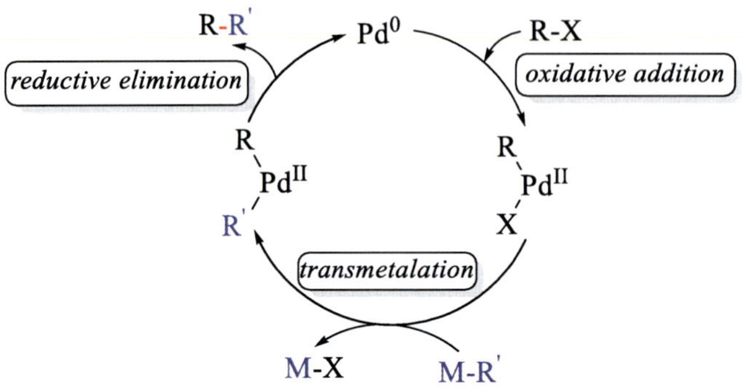

Scheme (12). General mechanism for Pd(0) catalyzed C-C coupling reaction.

Green Bio-Based Nanoparticles for C-C Coupling Reactions

Discussion on nanocatalytic C-C coupling reactions would be incomplete without the introduction of biological methods for nanomaterial preparation. Recently, the green and eco-friendly synthesis of metallic nanoparticles has attracted tremendous attention due to the rising environmental concerns instigated by the conventional chemical methods. Several biological feedstocks such as microorganisms, marine organisms, biomolecule extracts and plant materials have been used as natural precursors for the synthesis of NPs [59]. Among the biological sources, plant-based materials emerge as the preferred choice due to the ease of handling, availability, and cost-effectiveness compared to the other organisms making them suitable for large-scale bio-synthesis of metal NPs. The presence of a large number of active phytochemicals in the plant extracts such as alkaloids, phenolics, proteins, enzymes, vitamins, *etc* makes them perform as good bio-reductants and stabilizing agents for the transformation of bulk transition metals to nano dimensional particles avoiding any chemical reagent [60]. Plant parts such as leaf [61, 62], root [63], branch [63], seed [64], fruit [65] and peel [66 - 68] have been used for extracting the desired components. The liquid extraction of the plants has been carried out by using various solvent systems such as water, ethanol, petroleum ether, *etc.* Several plant species that have been studied for the biogenic preparation of metal NPs displaying high catalytic efficiency in various C-C coupling reactions are summarized in Table **2** [61 - 70].

Table 2. Biogenic nanoparticles in various C-C coupling reactions.

Sl. no	Species	Part used	Type of NPs	Application in the C-C coupling reaction
1 [61]	*Boswellia sarrata*	Leaf	Pd	Suzuki and Heck coupling

(Table 2) cont.....

Sl. no	Species	Part used	Type of NPs	Application in the C-C coupling reaction
2 [62]	*Ocimum sanctum*	Leaf	Pd	Suzuki and Sonogashira coupling
3 [63]	*Glycyrrhiza glabra*	Root and branch	Pd	Suzuki coupling
4 [64]	*Piper nigrum*	Seed	Pd	Hiyama coupling
5 [65]	*Piper longum*	Fruit	Pd	Sonogashira coupling
6 [66]	*Carica papaya*	Peel	Pd	Suzuki and Sonogashira coupling
7 [67]	*Punica granatum*	Peel	Pd	Suzuki and Sonogashira coupling
8 [68]	*Juglans regia*	Walnut shell	Cu	Ullman homocoupling
9 [69]	*Pulicaria glutinosa*	Whole plant	Pd	Suzuki coupling
10 [70]	*Shewanella oneidensis*	Bacterium	Pd/Au	Suzuki coupling

CONCLUDING REMARKS

This chapter is dedicated to the progressive and rapidly growing field of nanocatalysis in diverse carbon-carbon bond formation reactions. Over the past decades, significant research and development have been possible which demonstrates the extraordinary catalytic activity of the nanomaterials in catalysis. The field of supported metal nanoparticles continues to grow at a rapid rate giving rise to many interesting types of supports that have been used to immobilize metal nanoparticles. These supported nanocatalysts have shown several advantages over conventional systems, including good activity, selectivity, and enhanced stability in C-C coupling reactions. The use of bi-metallic, tri-metallic, and multi-metallic nanoparticles is a new area of interest and provides massive scope in the coming times. Also, the combination of nanocatalysis and green chemistry principles has emerged as a boon for the industry and environment, introducing greener and highly efficient coupling strategies.

At the current juncture, Pd has been mostly explored as catalysts with only a handful of examples with other metals. Also, the toxicity and environmental impact of most of these materials are not well studied. Thus, more emphasis should be given to the development of non-noble metal-based catalysts as cheap and economical alternatives to Pd catalysts. However, most of the research on the mechanism of these nanocatalyzed reactions has been focused on whether the reaction proceeds in solution or on the surface of NPs. The exact reaction routes remain unknown in most cases. Hence, more detailed studies are desired to understand the actual mechanisms in play for each reaction. Considering these aspects in the future advances will result in catalytic systems with the highest activity while lowering the production costs of the nanocatalysts.

CONSENT FOR PUBLICATION

Not applicable.

CONFLICT OF INTEREST

The author declares no conflict of interest, financial or otherwise.

ACKNOWLEDGEMENTS

The authors gratefully acknowledge Tezpur University, Assam and SERB, New Delhi for the financial support (No. EMR/2016/005944) and CSIR, New Delhi for providing the CSIR-Junior Research Fellowship to PB.

REFERENCES

[1] Gilman, H.; Lichtenwalter, M. Relative Reactivities of Organometallic Compounds. XXV. Coupling Reaction With Halides of Group VIII Metals. *J. Am. Chem. Soc.,* **1939**, *61*(4), 957-959. [http://dx.doi.org/10.1021/ja01873a056]

[2] Corbet, J.P.; Mignani, G. Selected patented cross-coupling reaction technologies. *Chem. Rev.,* **2006**, *106*(7), 2651-2710. [http://dx.doi.org/10.1021/cr0505268] [PMID: 16836296]

[3] Kharasch, M.S.; Fields, E.K. Factors determining the course and mechanisms of grignard reactions. iv. the effect of metallic halides on the reaction of aryl grignard reagents and organic halides. *J. Am. Chem. Soc.,* **1941**, *63*(9), 2316-2320. [http://dx.doi.org/10.1021/ja01854a006]

[4] Siemsen, P.; Livingston, R.C.; Diederich, F. Acetylenic coupling: a powerful tool in molecular construction. *Angew. Chem. Int. Ed. Engl.,* **2000**, *39*(15), 2632-2657. [http://dx.doi.org/10.1002/1521-3773(20000804)39:15<2632::AID-ANIE2632>3.0.CO;2-F] [PMID: 10934391]

[5] Hennings, D.D.; Iwama, T.; Rawal, V.H. Palladium-catalyzed (Ullmann-Type) homocoupling of aryl halides: A convenient and general synthesis of symmetrical biaryls *via* inter-and intramolecular coupling reactions. *Org. Lett.,* **1999**, *1*(8), 1205-1208. [http://dx.doi.org/10.1021/ol990872+]

[6] Marino, J.P.; Nguyen, H.N. Bulky trialkylsilyl acetylenes in the cadiot-chodkiewicz cross-coupling reaction. *J. Org. Chem.,* **2002**, *67*(19), 6841-6844. [http://dx.doi.org/10.1021/jo025745x] [PMID: 12227823]

[7] Smidt, J.; Hafner, W.; Jira, R.; Sieber, R.; Sedlmeier, J.; Sabel, A. The oxidation of olefins with palladium chloride catalysts. *Angew. Chem. Int. Ed. Engl.,* **1962**, *1*(2), 80-88. [http://dx.doi.org/10.1002/anie.196200801]

[8] Arvela, R.K.; Leadbeater, N.E. Microwave-promoted Heck coupling using ultralow metal catalyst concentrations. *J. Org. Chem.,* **2005**, *70*(5), 1786-1790. [http://dx.doi.org/10.1021/jo048052k] [PMID: 15730302]

[9] Elangovan, A.; Wang, Y.H.; Ho, T.I. Sonogashira coupling reaction with diminished homocoupling. *Org. Lett.,* **2003**, *5*(11), 1841-1844. [http://dx.doi.org/10.1021/ol034320+] [PMID: 12762666]

[10] Negishi, E.I.; Hu, Q.; Huang, Z.; Qian, M.; Wang, G.; Brown, H. Palladium-catalyzed alkenylation by the Negishi coupling. *Aldrichim Acta,* **2005**, *38*, 71-87.

[11] Molander, G.A.; Ellis, N. Organotrifluoroborates: protected boronic acids that expand the versatility of the Suzuki coupling reaction. *Acc. Chem. Res.,* **2007**, *40*(4), 275-286. [http://dx.doi.org/10.1021/ar050199q] [PMID: 17256882]

[12] Tierney, S.; Heeney, M.; McCulloch, I. Microwave-assisted synthesis of polythiophenes *via* the Stille coupling. *Synth. Met.,* **2005**, *148*(2), 195-198. [http://dx.doi.org/10.1016/j.synthmet.2004.09.015]

[13] Hiyama, T. How I came across the silicon-based cross-coupling reaction. *J. Organomet. Chem.,* **2002**, *653*(1-2), 58-61. [http://dx.doi.org/10.1016/S0022-328X(02)01157-9]

[14] Martin, R.; Buchwald, S.L. Pd-catalyzed Kumada-Corriu cross-coupling reactions at low temperatures allow the use of Knochel-type Grignard reagents. *J. Am. Chem. Soc.,* **2007**, *129*(13), 3844-3845. [http://dx.doi.org/10.1021/ja070830d] [PMID: 17352483]

[15] Campeau, L.C.; Hazari, N. Cross-coupling and related reactions: Connecting past success to the development of new reactions for the future. *Organometallics,* **2019**, *38*(1), 3-35. [http://dx.doi.org/10.1021/acs.organomet.8b00720] [PMID: 31741548]

[16] Pérez Sestelo, J.; Sarandeses, L.A. Advances in Cross-Coupling Reactions. *Molecules,* **2020**, *25*(19), 4500. [http://dx.doi.org/10.3390/molecules25194500] [PMID: 33019540]

[17] Biajoli, A.F.; Schwalm, C.S.; Limberger, J.; Claudino, T.S.; Monteiro, A.L. Recent progress in the use of Pd-catalyzed C-C cross-coupling reactions in the synthesis of pharmaceutical compounds. *J. Braz. Chem. Soc.,* **2014**, *25*, 2186-2214. [http://dx.doi.org/10.5935/0103-5053.20140255]

[18] Budarin, V.L.; Shuttleworth, P.S.; Clark, J.H.; Luque, R. Industrial applications of C-C coupling reactions. *Curr. Org. Synth.,* **2010**, *7*, 614-627. [http://dx.doi.org/10.2174/157017910794328529]

[19] Pagliaro, M.; Pandarus, V.; Ciriminna, R.; Béland, F.; Demma Carà, P. Heterogeneous *versus* homogeneous palladium catalysts for cross-coupling reactions. *ChemCatChem,* **2012**, *4*(4), 432-445. [http://dx.doi.org/10.1002/cctc.201100422]

[20] Modak, A.; Mondal, J.; Sasidharan, M.; Bhaumik, A. Triazine functionalized ordered mesoporous polymer: a novel solid support for Pd-mediated C-C cross-coupling reactions in water. *Green Chem.,* **2011**, *13*(5), 1317-1331. [http://dx.doi.org/10.1039/c1gc15045f]

[21] Pavia, C.; Giacalone, F.; Bivona, L.A.; Salvo, A.M.P.; Petrucci, C.; Strappaveccia, G.; Vaccaro, L.; Aprile, C.; Gruttadauria, M. Evidences of release and catch mechanism in the Heck reaction catalyzed by palladium immobilized on highly cross-linked-supported imidazolium salts. *J. Mol. Catal. Chem.,* **2014**, *387*, 57-62. [http://dx.doi.org/10.1016/j.molcata.2014.02.025]

[22] Mondal, M.; Dewan, A.; Begum, T.; Gogoi, P.K.; Thakur, A.J.; Bora, U. Suzuki-miyaura cross-coupling in aqueous medium using recyclable palladium/amide-silica catalyst. *Catal. Lett.,* **2016**, *146*(9), 1718-1728. [http://dx.doi.org/10.1007/s10562-016-1796-4]

[23] Prinsen, P.; Luque, R. *Nanoparticle Design and Characterization for Catalytic Applications in Sustainable Chemistry*; The Royal Society of Chemistry: Cambridge, UK, **2019**, pp. 1-36.

[24] Astruc, D. Introduction: Nanoparticles in Catalysis. *Chem. Rev.,* **2020**, *120*(2), 461-463. [http://dx.doi.org/10.1021/acs.chemrev.8b00696] [PMID: 31964144]

[25] Fihri, A.; Bouhrara, M.; Nekoueishahraki, B.; Basset, J.M.; Polshettiwar, V. Nanocatalysts for Suzuki cross-coupling reactions. *Chem. Soc. Rev.,* **2011**, *40*(10), 5181-5203. [http://dx.doi.org/10.1039/c1cs15079k] [PMID: 21804997]

[26] Ohtaka, A. Transition-metal Nanoparticles Catalyzed Carbon-Carbon Coupling Reactions in Water. *Curr. Org. Chem.,* **2019**, *23*(6), 689-703.
[http://dx.doi.org/10.2174/1385272823666190419211714]

[27] Cuenya, B.R. Synthesis and catalytic properties of metal nanoparticles: Size, shape, support, composition, and oxidation state effects. *Thin Solid Films,* **2010**, *518*(12), 3127-3150.
[http://dx.doi.org/10.1016/j.tsf.2010.01.018]

[28] Ahmadi, T.S.; Wang, Z.L.; Green, T.C.; Henglein, A.; El-Sayed, M.A. Shape-controlled synthesis of colloidal platinum nanoparticles. *Science,* **1996**, *272*(5270), 1924-1926.
[http://dx.doi.org/10.1126/science.272.5270.1924] [PMID: 8662492]

[29] Narayanan, R.; El-Sayed, M.A. Effect of colloidal nanocatalysis on the metallic nanoparticle shape: the Suzuki reaction. *Langmuir,* **2005**, *21*(5), 2027-2033.
[http://dx.doi.org/10.1021/la047600m] [PMID: 15723506]

[30] Wang, S.B.; Zhu, W.; Ke, J.; Lin, M.; Zhang, Y.W. Pd–Rh nanocrystals with tunable morphologies and compositions as efficient catalysts toward Suzuki cross-coupling reactions. *ACS Catal.,* **2014**, *4*(7), 2298-2306.
[http://dx.doi.org/10.1021/cs5001625]

[31] Ndolomingo, M.J.; Bingwa, N.; Meijboom, R. Review of supported metal nanoparticles: synthesis methodologies, advantages and application as catalysts. *J. Mater. Sci.,* **2020**, *55*(15), 6195-6241.
[http://dx.doi.org/10.1007/s10853-020-04415-x]

[32] Da'na, E. Adsorption of heavy metals on functionalized-mesoporous silica: A review. *Microporous Mesoporous Mater.,* **2017**, *247*, 145-157.
[http://dx.doi.org/10.1016/j.micromeso.2017.03.050]

[33] Han, P.; Wang, X.; Qiu, X.; Ji, X.; Gao, L. One-step synthesis of palladium/SBA-15 nanocomposites and its catalytic application. *J. Mol. Catal. Chem.,* **2007**, *272*(1-2), 136-141.
[http://dx.doi.org/10.1016/j.molcata.2007.03.006]

[34] Das, D.D.; Sayari, A. Applications of pore-expanded mesoporous silica 6. Novel synthesis of monodispersed supported palladium nanoparticles and their catalytic activity for Suzuki reaction. *J. Catal.,* **2007**, *246*(1), 60-65.
[http://dx.doi.org/10.1016/j.jcat.2006.11.020]

[35] Kotadia, D.A.; Patel, U.H.; Gandhi, S.; Soni, S.S. Pd-doped SiO_2 nanoparticles: an efficient recyclable catalyst for Suzuki, Heck and Sonogashira reactions. *RSC Advances,* **2014**, *4*(62), 32826-32833.
[http://dx.doi.org/10.1039/C4RA01813C]

[36] Gogoi, N.; Bora, U.; Borah, G.; Gogoi, P.K. Nanosilica-anchored Pd(II)-Schiff base complex as efficient heterogeneous catalyst for activation of aryl halides in Suzuki-Miyaura cross-coupling reaction in water. *Appl. Organomet. Chem.,* **2017**, *31*(9), 3686.
[http://dx.doi.org/10.1002/aoc.3686]

[37] Hosseini-Sarvari, M.; Razmi, Z. Palladium supported on zinc oxide nanoparticles as efficient heterogeneous catalyst for suzuki-miyaura and hiyama reactions under normal laboratory conditions. *Helv. Chim. Acta,* **2015**, *98*(6), 805-818.
[http://dx.doi.org/10.1002/hlca.201400331]

[38] Labulo, A.H.; Martincigh, B.S.; Omondi, B.; Nyamori, V.O. Advances in carbon nanotubes as efficacious supports for palladium-catalysed carbon-carbon cross-coupling reactions. *J. Mater. Sci.,* **2017**, *52*(16), 9225-9248.
[http://dx.doi.org/10.1007/s10853-017-1128-0]

[39] Afshari, R.; Hooshmand, S.E.; Atharnezhad, M.; Shaabani, A. An insight into the novel covalent functionalization of multi-wall carbon nanotubes with pseudopeptide backbones for palladium nanoparticles immobilization: A versatile catalyst towards diverse cross-coupling reactions in bio-based solvents. *Polyhedron,* **2020**, *175*, 114238.

[http://dx.doi.org/10.1016/j.poly.2019.114238]

[40] Samarasimhareddy, M.; Prabhu, G.; Sureshbabu, V.V. A simple and greener approach for the synthesis of PVC supported Pd(0): application to Heck and Sonogashira reactions in water. *Tetrahedron Lett.,* **2014**, *55*(14), 2256-2260. [http://dx.doi.org/10.1016/j.tetlet.2014.02.083]

[41] Goonesinghe, C.; Shaik, M.; Ratnaweera, R.; Nalin De Silva, K.M.; De Silva, R.M. A magnetically retrievable air and moisture stable gold and palladium nanocatalyst for efficient C-C coupling reactions. *R. Soc. Open Sci.,* **2020**, *7*(9), 200916. [http://dx.doi.org/10.1098/rsos.200916] [PMID: 33047052]

[42] Patra, D.; Panja, S.; Saha, A. C-C Cross-Coupling Reactions of Organosilanes with Terminal Alkenes and Allylic Acetates Using Pd(II) Catalyst Supported on Starch Coated Magnetic Nanoparticles. *Eur. J. Org. Chem.,* **2020**, *2020*(7), 878-883. [http://dx.doi.org/10.1002/ejoc.201901812]

[43] Trzeciak, A.M.; Augustyniak, A.W. The role of palladium nanoparticles in catalytic C-C cross-coupling reactions. *Coord. Chem. Rev.,* **2019**, *384*, 1-20. [http://dx.doi.org/10.1016/j.ccr.2019.01.008]

[44] Li, G.; Jin, R. Catalysis by gold nanoparticles: carbon-carbon coupling reactions. *Nanotechnol. Rev.,* **2013**, *2*(5), 529-545. [http://dx.doi.org/10.1515/ntrev-2013-0020]

[45] Candu, N.; Dhakshinamoorthy, A.; Apostol, N.; Teodorescu, C.; Corma, A.; Garcia, H.; Parvulescu, V.I. Oriented Au nanoplatelets on graphene promote Suzuki-Miyaura coupling with higher efficiency and different reactivity pattern than supported palladium. *J. Catal.,* **2017**, *352*, 59-66. [http://dx.doi.org/10.1016/j.jcat.2017.04.034]

[46] Chen, Y.; Feng, L. Silver nanoparticles doped TiO_2 catalyzed Suzuki-coupling of bromoaryl with phenylboronic acid under visible light. *J. Photochem. Photobiol. B,* **2020**, *205*, 111807. [http://dx.doi.org/10.1016/j.jphotobiol.2020.111807] [PMID: 32120182]

[47] Nasrollahzadeh, M.; Atarod, M.; Alizadeh, M.; Hatamifard, A.; Sajadi, S.M. Recent advances in the application of heterogeneous nanocatalysts for Sonogashira coupling reactions. *Curr. Org. Chem.,* **2017**, *21*(8), 708-749. [http://dx.doi.org/10.2174/1385272820666161021102943]

[48] Oka, H.; Kitai, K.; Suzuki, T.; Obora, Y.N. N-Dimethylformamide-stabilized copper nanoparticles as a catalyst precursor for Sonogashira–Hagihara cross coupling. *RSC Advances,* **2017**, *7*(37), 22869-22874. [http://dx.doi.org/10.1039/C6RA27910D]

[49] Hajipour, A.R.; Rezaei, F.; Khorsandi, Z. Pd/Cu-free Heck and Sonogashira cross-coupling reaction by Co nanoparticles immobilized on magnetic chitosan as reusable catalyst. *Green Chem.,* **2017**, *19*(5), 1353-1361. [http://dx.doi.org/10.1039/C6GC03377F]

[50] Hajipour, A.R.; Abolfathi, P.; Tavangar-Rizi, Z. Iron-catalyzed cross-coupling reaction: Heterogeneous palladium and copper-free Heck and Sonogashira cross-coupling reactions catalyzed by a reusable Fe(III) complex. *Appl. Organomet. Chem.,* **2018**, *32*(6), 4353. [http://dx.doi.org/10.1002/aoc.4353]

[51] Hussain, N.; Gogoi, P.; Khare, P.; Das, M.R. Nickel nanoparticles supported on reduced graphene oxide sheets: a phosphine free, magnetically recoverable and cost effective catalyst for Sonogashira cross-coupling reactions. *RSC Advances,* **2015**, *5*(125), 103105-103115. [http://dx.doi.org/10.1039/C5RA22601E]

[52] Fan, J.; Du, H.; Zhao, Y.; Wang, Q.; Liu, Y.; Li, D.; Feng, J. Recent progress on rational design of bimetallic pd based catalysts and their advanced catalysis. *ACS Catal.,* **2020**, *10*(22), 13560-13583. [http://dx.doi.org/10.1021/acscatal.0c03280]

[53] Rai, R.K.; Tyagi, D.; Gupta, K.; Singh, S.K. Activated nanostructured bimetallic catalysts for C–C coupling reactions: recent progress. *Catal. Sci. Technol.,* **2016**, *6*(10), 3341-3361. [http://dx.doi.org/10.1039/C5CY02225H]

[54] Han, S.; Mullins, C.B. Catalytic Reactions on Pd-Au Bimetallic Model Catalysts. *Acc. Chem. Res.,* **2021**, *54*(2), 379-387. [http://dx.doi.org/10.1021/acs.accounts.0c00642] [PMID: 33371669]

[55] Tan, L.; Wu, X.; Chen, D.; Liu, H.; Meng, X.; Tang, F. Confining alloy or core-shell Au-Pd bimetallic nanocrystals in silica nanorattles for enhanced catalytic performance. *J. Mater. Chem. A Mater. Energy Sustain.,* **2013**, *1*(35), 10382-10388. [http://dx.doi.org/10.1039/c3ta11749a]

[56] Venkatesan, P.; Santhanalakshmi, J. Synthesis and characterization of surfactant stabilized trimetallic Au-Ag-Pd nanoparticles for heck coupling reaction. *Phys. Chem.,* **2012**, *2*(1), 12-15. [http://dx.doi.org/10.5923/j.pc.20120201.03]

[57] Nasseri, M.A.; Rezazadeh, Z.; Kazemnejadi, M.; Allahresani, A. A Co-Cu bimetallic magnetic nanocatalyst with synergistic and bifunctional performance for the base-free Suzuki, Sonogashira, and C-N cross-coupling reactions in water. *Dalton Trans.,* **2020**, *49*(30), 10645-10660. [http://dx.doi.org/10.1039/D0DT01846E] [PMID: 32699870]

[58] Sun, B.; Ning, L.; Zeng, H.C. Confirmation of Suzuki-Miyaura cross-coupling reaction mechanism through synthetic architecture of nanocatalysts. *J. Am. Chem. Soc.,* **2020**, *142*(32), 13823-13832. [http://dx.doi.org/10.1021/jacs.0c04804] [PMID: 32667195]

[59] Singh, P.; Kim, Y.J.; Zhang, D.; Yang, D.C. Biological synthesis of nanoparticles from plants and microorganisms. *Trends Biotechnol.,* **2016**, *34*(7), 588-599. [http://dx.doi.org/10.1016/j.tibtech.2016.02.006] [PMID: 26944794]

[60] Kharissova, O.V.; Dias, H.V.; Kharisov, B.I.; Pérez, B.O.; Pérez, V.M.J. The greener synthesis of nanoparticles. *Trends Biotechnol.,* **2013**, *31*(4), 240-248. [http://dx.doi.org/10.1016/j.tibtech.2013.01.003] [PMID: 23434153]

[61] Arde, S.M.; Rashinkar, G.S.; Jadhav, S.N.; Patil, A.D.; Salunkhe, R.S. Biogenic synthesis of palladium nanoparticles using Boswellia sarrata and their applications in cross-coupling reactions. *Appl. Organomet. Chem.,* **2020**, *34*(12), 6012. [http://dx.doi.org/10.1002/aoc.6012]

[62] Sarmah, M.; Neog, A.B.; Boruah, P.K.; Das, M.R.; Bharali, P.; Bora, U. Effect of substrates on catalytic activity of biogenic palladium nanoparticles in C-C cross-coupling reactions. *ACS Omega,* **2019**, *4*(2), 3329-3340. [http://dx.doi.org/10.1021/acsomega.8b02697] [PMID: 31459548]

[63] Veisi, H.; Adib, M.; Karimi-Nami, R.; Yasaei, Z.; Tajik, M.; Mosavat, T.S.; Hemmati, S. Suzuki-Miyaura coupling catalyzed by palladium nanoparticles biosynthesized using Glycyrrhiza glabra as reducing and stabilyzing agent. *Appl. Organomet. Chem.,* **2018**, *32*(3), 4138. [http://dx.doi.org/10.1002/aoc.4138]

[64] Kandathil, V.; Dateer, R.B.; Sasidhar, B.S.; Patil, S.A.; Patil, S.A. Green synthesis of palladium nanoparticles: applications in aryl halide cyanation and hiyama cross-coupling reaction under ligand free conditions. *Catal. Lett.,* **2018**, *148*(6), 1562-1578. [http://dx.doi.org/10.1007/s10562-018-2369-5]

[65] Nasrollahzadeh, M.; Sajadi, S.M.; Maham, M.; Ehsani, A. Facile and surfactant-free synthesis of Pd nanoparticles by the extract of the fruits of Piper longum and their catalytic performance for the Sonogashira coupling reaction in water under ligand-and copper-free conditions. *RSC Advances,* **2015**, *5*(4), 2562-2567. [http://dx.doi.org/10.1039/C4RA12875C]

[66] Dewan, A.; Sarmah, M.; Thakur, A.J.; Bharali, P.; Bora, U. Greener biogenic approach for the

synthesis of palladium nanoparticles using papaya peel: An eco-friendly catalyst for C-C coupling reaction. *ACS Omega,* **2018**, *3*(5), 5327-5335. [http://dx.doi.org/10.1021/acsomega.8b00039] [PMID: 31458742]

[67] Dewan, A.; Sarmah, M.; Bharali, P.; Thakur, A.J.; Boruah, P.K.; Das, M.R.; Bora, U. Pd nanoparticles-loaded honeycomb-structured bio-nanocellulose as a heterogeneous catalyst for heteroaryl cross-coupling reaction. *ACS Sustain. Chem.& Eng.,* **2021**, *9*(2), 954-966. [http://dx.doi.org/10.1021/acssuschemeng.0c08315]

[68] Zamani, A.; Poursattar Marjani, A.; Nikoo, A.; Heidarpour, M.; Dehghan, A. Synthesis and characterization of copper nanoparticles on walnut shell for catalytic reduction and C-C coupling reaction. *Inorg. Nano Met. Chem.,* **2018**, *48*(3), 176-181. [http://dx.doi.org/10.1080/24701556.2018.1503676]

[69] Khan, M.; Khan, M.; Kuniyil, M.; Adil, S.F.; Al-Warthan, A.; Alkhathlan, H.Z.; Tremel, W.; Tahir, M.N.; Siddiqui, M.R.H. Biogenic synthesis of palladium nanoparticles using *pulicaria glutinosa* extract and their catalytic activity towards the suzuki coupling reaction. *Dalton Trans.,* **2014**, *43*(24), 9026-9031. [http://dx.doi.org/10.1039/C3DT53554A] [PMID: 24619034]

[70] Heugebaert, T.S.; De Corte, S.; Sabbe, T.; Hennebel, T.; Verstraete, W.; Boon, N.; Stevens, C.V. Biodeposited Pd/Au bimetallic nanoparticles as novel Suzuki catalysts. *Tetrahedron Lett.,* **2012**, *53*(11), 1410-1412. [http://dx.doi.org/10.1016/j.tetlet.2012.01.030]

CHAPTER 3

Nanocatalysis in Aryl Carbon-Hetero Atom Coupling Reactions: Special Emphasis on Copper Free Protocols

Roktopol Hazarika[1], **Bidyutjyoti Dutta**[1] and **Diganta Sarma**[1,*]

[1] *Department of Chemistry, Dibrugarh University, Dibrugarh-786004, Assam, India*

Abstract: Nanomaterials have emerged as a keystone of nanoscience and nanotechnology that cover a broad area in the field of research and development worldwide. A set of substances having at least one dimension less than around 100 nanometres are termed nanoscale materials. At such a range, these nanomaterials possess some unique optical, electrical, and magnetic properties; they have substantial impacts in electronics, medicine as well as various organic synthesis. A wide variety of nanoparticles, such as carbon-based/polymer-based/lipid-based/ceramic/metal-based, *etc*., are found depending on composition, size, chemical and physical properties, and morphology of nanoparticles. Metal nanoparticles have received special attraction for organic synthesis nowadays. Due to their stronger ability to transfer electrons, large surface area to volume ratio, easy recovery, and recyclability, metal nanoparticles have become an attractive alternative for both homogenous as well as heterogeneous catalysis in synthetic organic chemistry. The aryl-heteroatom bond formation in organic synthesis is very important as the molecules with the generality of this bond are of pharmaceutical, chemical, and biological interest. Synthesizing such molecules with metal nanoparticles has the advantage of high atom economy, easy recyclability, and reusability of the catalyst, which is far better than the conventional procedures that demand toxic ligands and tedious procedures. The use of copper-free metal nanoparticles for the synthesis of such moieties is a greener approach as copper has a toxic mechanism of action.

Keywords: Catalysis, Copper Free Protocols, Metal-Based Nanoparticles, N-arylation, O-arylation, S-arylation.

INTRODUCTION

The nature around us involves numerous spontaneous chemical processes such as photosynthesis, degradation, oxidation-reduction, *etc*. The presence of sunlight, air, moisture, and some other natural factors make these processes feasible.

* **Corresponding author Diganta Sarma**: Department of Chemistry, Dibrugarh University, Dibrugarh-786004, Assam, India; E-mail: dsarma22@gmail.com., dsarma22@dibru.ac.in.

Perhaps, these processes mainly regulate the lives on this planet. However, these natural processes are not enough to standardize the life of every human being along with the other living organisms. Therefore, we are greatly dependent on synthetic procedures for mass production of most of the necessary commodities like medicines, groceries, fibres, *etc*. To enhance the productivities of the synthetic procedures, the catalyst must be used; in other words, catalysts play a vital role in the process of standardisation. Catalysts provide an alternative pathway for the reaction to occur easily by decreasing the activation energy. They have done it by exposing their surfaces, by physical interactions with the starting materials, by changing their physiologies, or by protecting the reactants in particular shapes and geometries. Though there exist different catalytic processes like homogenous, heterogeneous, photocatalytic, *etc*., we will only discuss nanocatalysis of aryl carbon hetero bond forming reaction. Moreover, as per literature, the construction of carbon-heteroatom bond is mostly accompanied by transition metal catalysts, among which copper assists as the leading domain. In this context, quite comprehensive literature is available on copper both in bulk as well as nano states [1]. As there is always a toxicity issue attached with copper, in this chapter, we have tried to cover the literature other than copper to highlight the cheap, less or nontoxic copper-free protocol to support environment, as well as green chemistry in the construction of carbon hetero atom bond.

NANOCATALYSIS: AN INTRODUCTORY OVERVIEW

The first reports published on nanocatalysis were in early 1941 based on poly(vinyl) alcohol-protected palladium and platinum nanoparticles (NPs) catalyzed hydrogenation reaction. The nanoparticles were prepared by the reduction of the metal salts by H_2 [2]. This work was very significantly inspired by the work of Paul Sabatier, which was a hydrogenation reaction catalysed by finely divided nickel particles prepared through the reduction of NiO_2 or $Ni(OH)_2$ by H_2.

Nanocatalysis revolutionized the chemical procedures for the synthesis of active molecular scaffolds in a far better way compared to the traditional procedures. The nanomaterial-based catalysts are heterogeneous type catalysts that are finely divided into smaller pieces with diameters between 1 and 100 nanometres (nm) [3]. As the particles' size decreases, more surface areas are exposed, which increases the catalytic activity. Moreover, nanoparticles can easily be separated and recycled [4].

The variety in shapes and geometry of the nanoparticles makes them a suitable partner for interaction with the starting reagents, which play a very significant factor in the enhancement of their catalytic activity. The nanocatalyst mainly combines the advantages of both homogeneous and heterogeneous catalysts and, therefore, can provide better selectivity, prominent efficiency, good to high stability, and easy recovery or recyclability. Moreover, specific reactivity can also be anticipated for nanoparticles that cannot be achieved with regular, bulk materials. From this point of view, since the 1990s, nanocatalysis has undoubtedly emerged as a vast domain that offers a wide range of properties that meet the demanding conditions for catalyst improvement [5].

Arylation Reaction

In the context of aryl carbon heteroatom bond-forming reaction, the arylation reactions play a vital role in demanding a better choice for its construction. Among all the arylation reactions, N-arylation, O-arylation, and S-arylation are common because the products obtained *via* these reactions possess a variety of important properties. For instance, N-Arylindoles are very much effective as antiestrogen, analgesic, antiallergic as well as effective against HIV-1 [6]. Similarly, diaryl ethers, which are O-arylation products, are present as an essential part of most of the natural and synthetic compounds having medicinal properties such as antibacterial, antibiotic, antifungal, antitumor, *etc.* [7]. Similar to that, S-arylated products also find tremendous applications in pharmaceutical industries as some of important drug molecules such as butoconazole (antifungal), Nelfinavir (anti HIV), Axitinib (anticancer) contain the aryl carbon sulphur bond in their core structures [8]. Some of the representative moieties are given in Fig. (**1**). Therefore, we have chosen this particular reaction for our discussion.

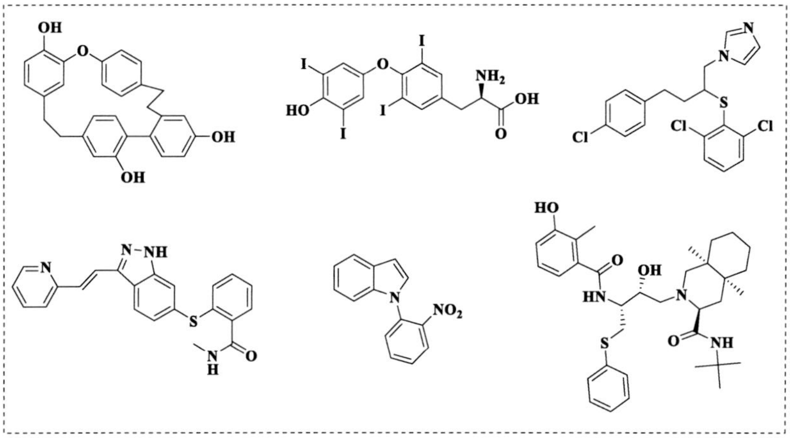

Fig. (1). Some biologically active aryl carbon hetero atom bond containing molecules.

N-Arylation

A new approach for the synthesis of quinoline *via* intermolecular C-N bond formation is introduced by Wu *et al.* [9]. They presented ligand-free nano Fe_3O_4 to be an active, stable as well as reusable catalyst for the intermolecular C-N cross-coupling reaction and developed different 1,4-dihydroquinoline derivatives starting from o-halobenzaldehyde and a wide range of ß-enaminones (Scheme **1**). For the inactive aryl chlorides and arylamine derivatives, the iron nanocatalyst is highly efficient. The catalyst is suitable for subsequent industrial processes due to its recyclability.

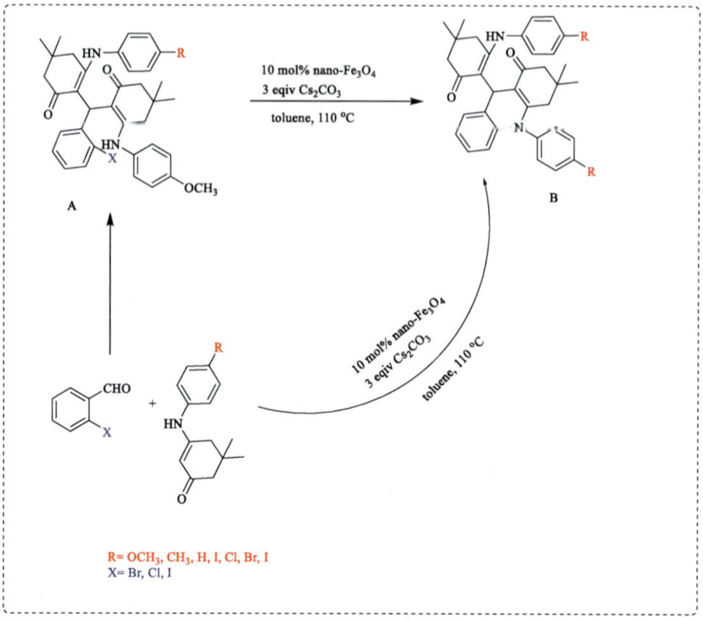

Scheme (1). Synthesis of quinoline derivatives B using iron nanoparticle.

C-N bond formation reaction using Fe@Pd nanowires was also reported [10]. The Fe@Pd nanowire was synthesized in two steps. Using arc discharge of iron electrodes in deionised water followed by dispersion on glass substrates and drying in air, they synthesized Fe nanowires of several micron lengths. By a simple drop-drying process palladium nanoparticles (Pd NPs) were overlaid on the surface of Fe-nanowire. They employed the synthesized Fe@Pd nanowire for amination of aryl boronic acid with various amines in an aqueous medium under ligand-free conditions at room temperature (Scheme **2**). The catalyst is easily separable using a magnet and can be reused up to several times without significant loss in catalytic activity.

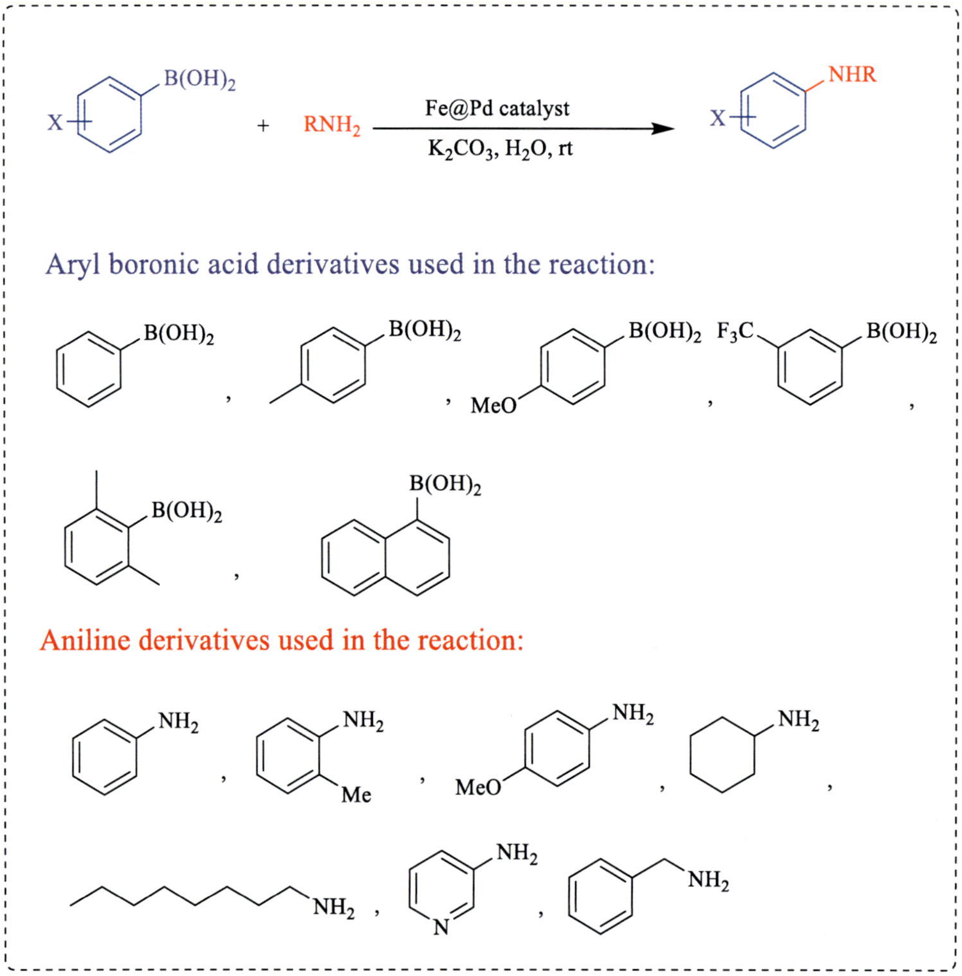

Scheme (2). Amination reaction using Fe@Pd nanowires.

Roopan *et al.* reported an efficient regioselective N-alkylation of 4(3H)-pyrimidone with various electrophiles using Fe(0) nanoparticle [11]. The newly developed nanocatalyst was found to be inexpensive, non-volatile, recyclable, non-explosive, easy to handle and eco-friendly. N-alkylation of amides by alkyl chlorides was initiated by the synthesized nanocatalyst. 3-[(2-chloroquinoli--3-yl]methyl)pyrimidin-4(3H)ones is formed from the reaction of equimolar 4(3H)-pyrimidone and 2-chloro-3-(chloromethyl)quinolines in presence of KOH and Fe nanoparticle (5 mol %) in DMSO solution under reflux condition (Scheme **3**).

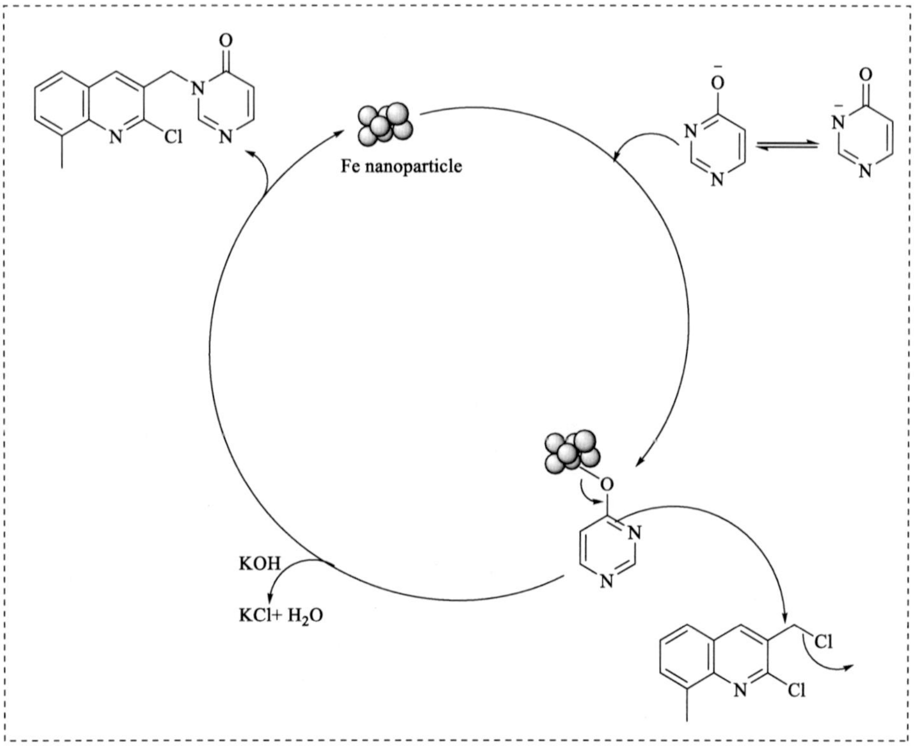

Scheme (3). Role of Fe nanoparticles in synthesis of 3-((2-chloroquinolin-3-yl)methyl)pyrimidin-4(3H)-one.

Carbon nitride supported palladium nanoparticle Pd-CN was reported as an active catalytic system for amination of aryl and pyridyl chlorides [12]. The reaction proceeded in presence of dialkyl amine under mild conditions. The nanocatalyst was synthesised *via* a two-step methodology using the precursor's urea and potassium tetrachloropalladate (II) respectively (Scheme **4**). 5 mg of Pd-CN with 5 wt.% of Pd loading is found to be the optimized condition for the reaction. Compared to $Pd(OAc)_2$ and $Pd(PPh_3)_4$; the reaction displays better performance with Pd-CN nanoparticles. The Pd particles were found to be dispersed on CN support within the size range of 5-20 nm. The catalyst was recyclable up to the 5th cycle without significant loss in catalytic activity. After the sixth cycle an amount of enlargement (14-35nm) was observed. Ostwald ripening effect is the best way to illustrate the reason for enlargement.

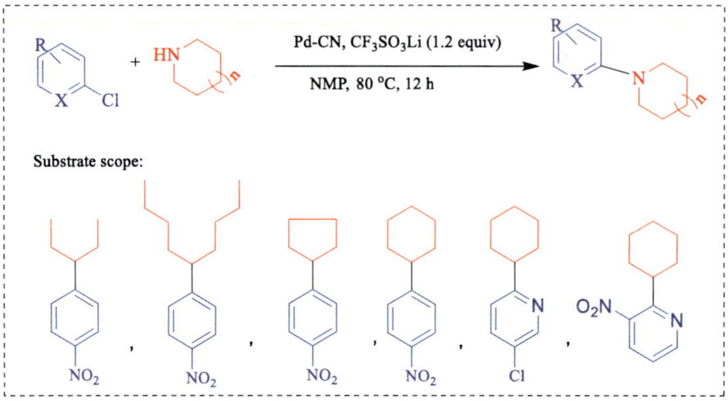

Scheme (4). Aryl-amine coupled product using the Pd-CN catalyst.

A new heterogeneous nickel ferrite nanocatalyst was reported in a C-N cross-coupling reaction with great efficiency [13]. This nanocatalyst is found to generalise the C-N bond formation reaction *i.e.,* different types of amines such as aliphatic or aromatic, cyclic or acyclic, primary or secondary *etc.* are found to couple with a wide variety of aryl halides containing both electron-donating or withdrawing substituents using this nanocatalyst (Scheme **5**). A few heterocyclic amines and amides confirmed the generality of the procedure too. 5 mol% of the nanocatalyst is found to catalyse the reaction in a short reaction time. The catalyst is superparamagnetic in nature and hence is easily separable and can be reused up to the 5th cycle.

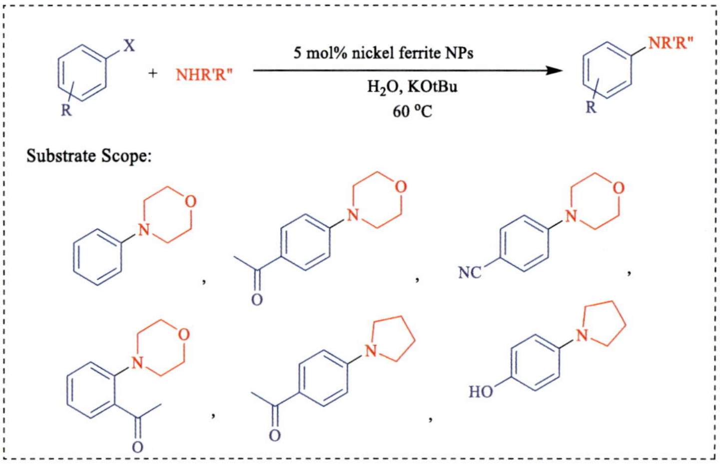

Scheme (5). Amination reaction of various aryl halides using nickel ferrite NPs.

Pd NPs synthesized from green alga Chlorella Vulgaris is reported in Buchwald-Hartwig reactions of aryl halides with mono protected piperazine [14]. A benefit of the synthesis of Pd nanoparticles from the green algae is that it acts as a green ligand for reduction of Pd (II) to Pd(0) *i.e.*, the reduction in the size from bulk to nanoparticles. This Pd nanoparticle-green ligand system is found to be efficient for the synthesis of several N-aryl piperazines in good to excellent yields (Scheme **6**). The advantage of this system over other Pd nanoparticle catalysed Buchwald-Hartwig reaction is that the green algae used for the reduction of Pd metal to Pd(0) nanoparticles makes it an environmentally benign and inexpensive methodology. The nanocatalyst can be reused up to the 5th cycle without significant loss in activity.

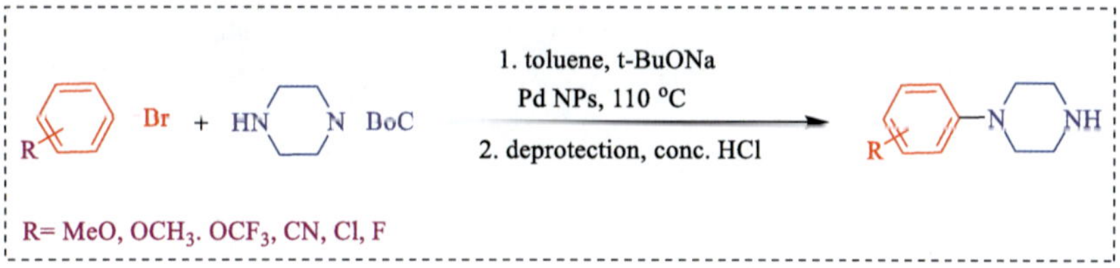

Scheme (6). Buchwald-Hartwig reaction of aryl halides with N-Boc piperazine using Pd NPs synthesized from *C. vulgaris*.

A palladium-based nanoparticle is reported for the N-arylation of indoles, pyrroles and benzyl amines (Scheme **7**) [15]. The reaction is done under ligand-free conditions and found effective for several functional groups with electron-donating as well as electron-withdrawing ability. By photolyzing a solution of $Pd(OAc)_2$ and (E)-3-(2-methoxynaphthalen-1-yl)-1-(4-methoxyphenyl)prop-2-en-1-one in methanol (MeOH) under anaerobic conditions using a medium pressure mercury UV lamp, the palladium nanoparticles were synthesized. The catalyst can be recycled up to 7th times but after the 4th cycle, the yield of the reaction was found to be dropped rapidly and quite significantly. This may be because of the aggregation of the Pd nanoparticles *via* each catalytic cycle.

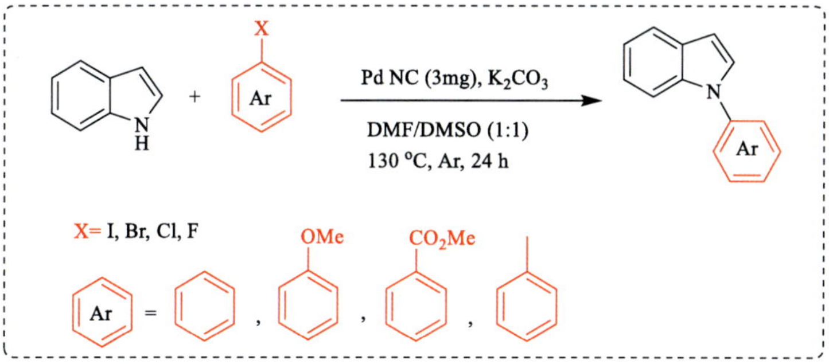

Scheme (7a). PdNC catalyzed N-arylation of indoles.

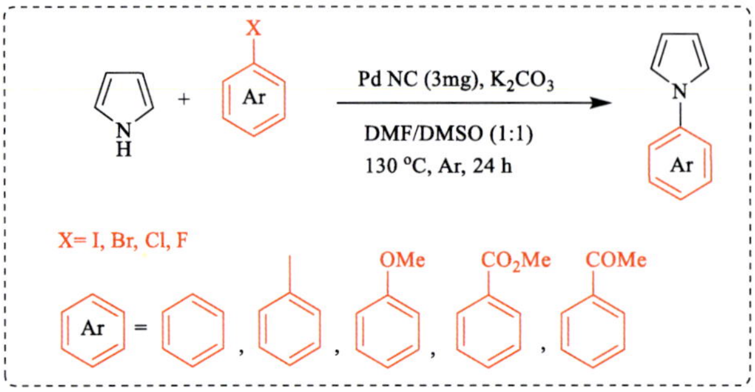

Scheme (7b). PdNC catalyzed N-arylation of pyrroles.

Scheme (7c). PdNC catalyzed N-arylation of benzylamine.

S-Arylation Reaction

A magnetically separable copper-free system mesoporous polyaniline and porous magnetic Fe_3O_4 (mPANI/pFe_3O_4) nanocomposite emerged as an efficient catalyst for S–arylation reaction (Scheme **8**). In presence of different bases and solvents, the catalyst mPANI/pFe_3O_4 produced the S-arylated product even with the more challenging substrate such as 4-methyl-chlorobenzene. The high reaction yield was due to the nano size dimension of the catalyst with high porous surface of Fe_3O_4. The polar solvents *viz.* DMF, DMSO, water *etc.* offered better reaction yield as compared to the non-polar solvents such as toluene, THF *etc.* Among different bases studied, KOH was found as the best choice for the reaction in water. Thus, 5 mol% of mPANI/pFe_3O_4 as catalyst and KOH as the base under reflux condition was considered as the optimized condition for the S-arylation of 4-methyl-chlorobenzene and thiophenol in water [16]. The preparation of the catalyst was depicted in Fig. (**2**). Another ferrite nanoparticle based on Ni was also employed as the catalytic system in the S-arylation reaction. In the case of ferrite particles, the catalytic properties can easily be adjusted due to the transfer of electrons between Fe^{2+} and Fe^{3+} on octahedral sites [17 - 19]. That is one of the reasons why ferrite nanoparticles are always a suitable option for nanocatalysis. The inverse spinel structure of the nickel ferrite possesses ferrimagnetism properties, have low cost and are air-stable by nature. In this catalytic system, the nickel ferrite particles were coated with silica for its surface stabilization to prevent aggregation. Silica was chosen based on its adequate properties like chemical inertness, high stability in an aqueous medium, easy regulation of the coating process, controlled porosity, processability and optical transparency. As claimed by Iqubal *et. al.,* this catalytic system $NiFe_2O_4$/Silica diminished the use of phosphine ligand, additives, higher temperature and at the same time reduced the reaction time compared to the earlier methods. In this method, the model reaction was the S-arylation reaction between iodobenzene and thiophenol performed under aerial conditions. KO^tBu was turned out to be the superior base to KOH, K_2CO_3 and K_3PO_4 which might be due to the strong basicity and high solubility of the base in water. There was not present any pronounced electronic effect as both electron withdrawing and donating groups in the starting reagents provided similar results (Scheme **9**) [20].

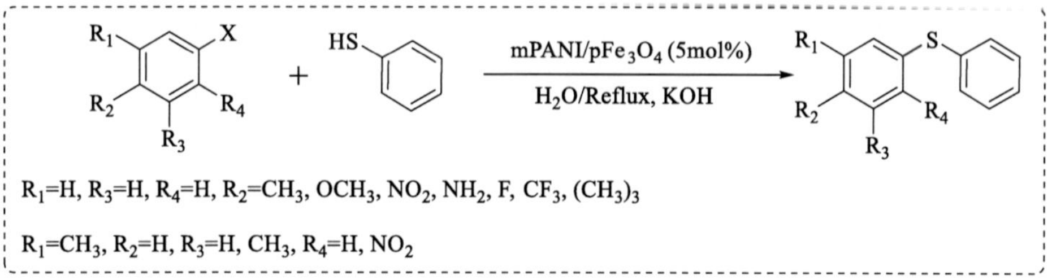

Scheme (8). Synthetic routes for S-arylated product with different starting materials.

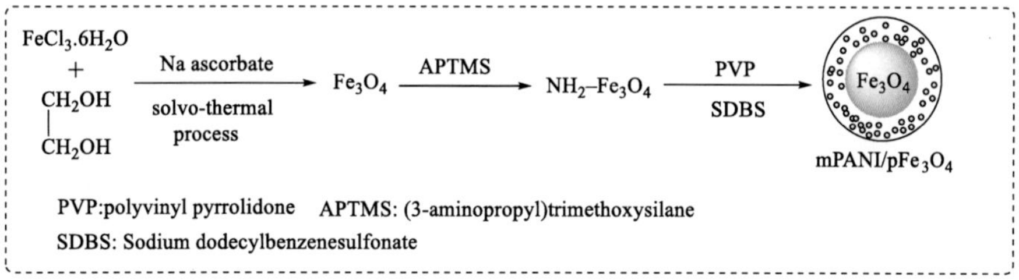

Fig. (2). Preparative method of the desired catalyst.

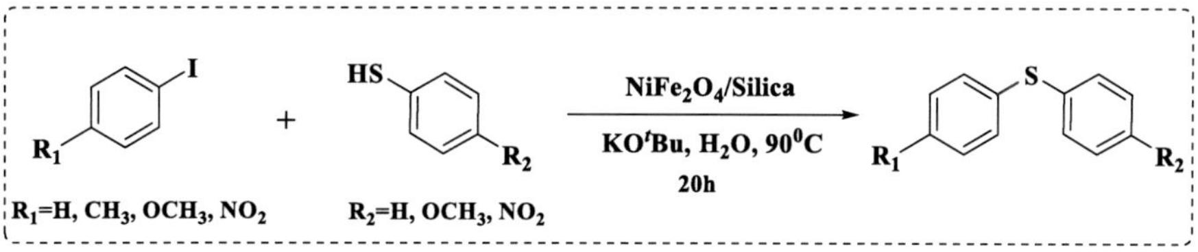

Scheme (9). Ni ferrite catalysed S-arylation reaction.

O-Arylation

Ferrosoferric oxide supported on mesoporous polyaniline (Fe_3O_4@mesoporousPANI) was also applied as an effective catalyst in O-arylation reaction by Arundhathi and co-workers [21]. For this investigation, the coupling of aryl chloride with different substituted phenols were tested and quite comprehensive results were obtained. The electron-deficient phenols slightly slowed down the reaction whereas, with electron-rich phenols, the O-arylated products were obtained in excellent yields. Along with it, a challenging substrate such as nitrophenol also produced the desired product in good yield. The work was extended by exploring the alcohol functionalities *viz.* benzyl alcohol, cyclohexanol, 4-hydroxyacetophenone as coupling partners with various substituted aryl chlorides. There was a little electronic effect observed in this reaction. In both the reaction strategies, K_2CO_3 and dimethylformamide (DMF) was used as the optimized base and solvent. The methodology turned out to be advantageous compared to the traditional methods as it required adequate reaction temperature, shorter reaction time, ligand-free environment and offered recyclability of the catalyst up to 5th times. The catalyst was prepared as depicted in Scheme (**10a - 10c**).

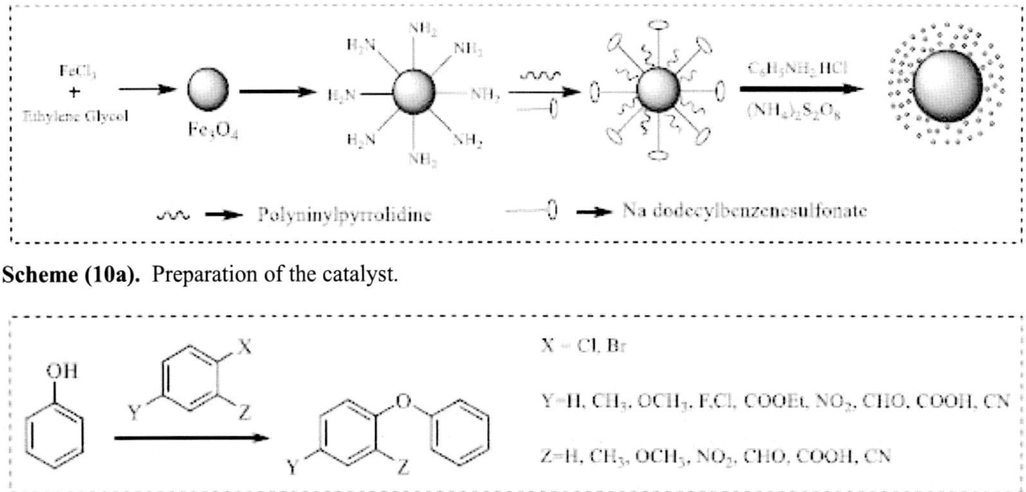

Scheme (10a). Preparation of the catalyst.

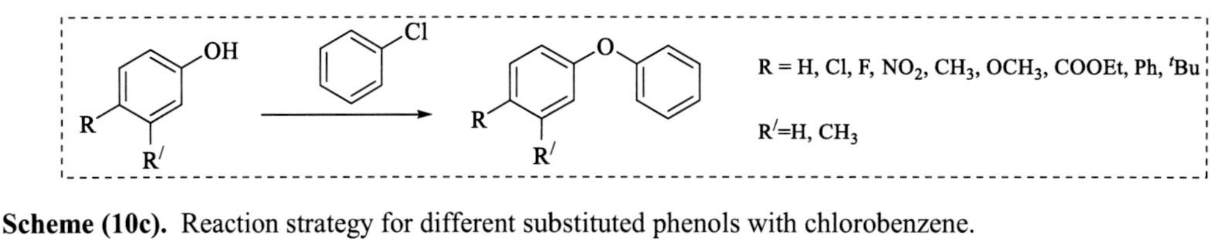

Scheme (10b). Reaction of phenol with different starting halogen derivatives.

Scheme (10c). Reaction strategy for different substituted phenols with chlorobenzene.

Moreover, the catalytic protocol was tolerable against alcohol, naphthol and biaryl based hydroxy compounds with a quantitative yield of the products. The products were summarized here in Fig. (**3**).

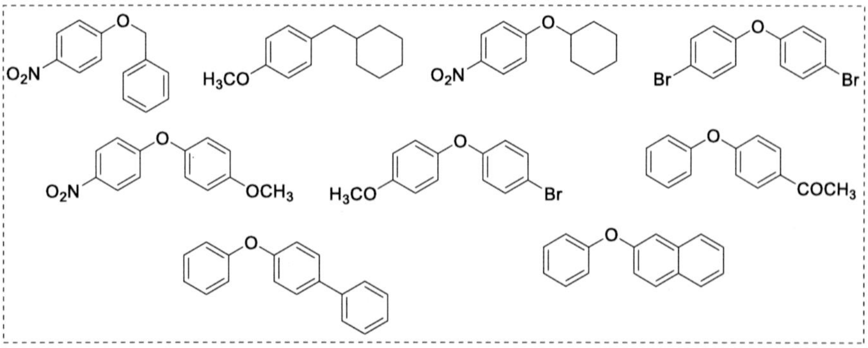

Fig. (3). Some more desired products following reaction scheme 10.

A cerium based nanocatalyst was reported by S. M. Agawane *et. al.* in 2001 for O-arylation reaction. They prepared the nanoceria (CeO_2) by following a literature report [22] as shown schematically in Fig. (**4**) and applied it to Ulmann synthesis.

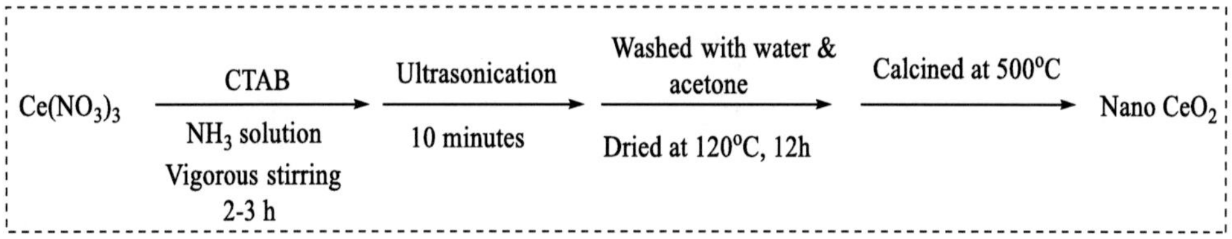

Fig. (4). Schematic representation of the preparation of nanoceria.

According to the authors, the optimized reaction condition for this transformation as depicted in Scheme (**11**) was found to be 110 °C temperature with 2.5 mol % catalyst loading where KOH served as the basic additive. The reactivity order for this transformation for different chlorobenzene derivatives indicated that 4-NO_2 substituted reactant gave superior results followed by 4-CN and 4-CHO substituted reactant. The formation of an excellent amount of product in the case of 2-nitrochlorobenzene ruled out the existence of any kind of steric effect. But the unsubstituted chloro, iodo and bromobenzene were ineffective for the reaction. Along with O-arylation, the catalyst was also employed for N-arylation as well as S-arylation and they found that various substituted phenols, amines, heterocyclic amines and thiophenols successfully coupled with 4-nitrochlorobenzene to obtain the desired products. Moreover, the reusability up to three cycles added more value to their developed protocol [23].

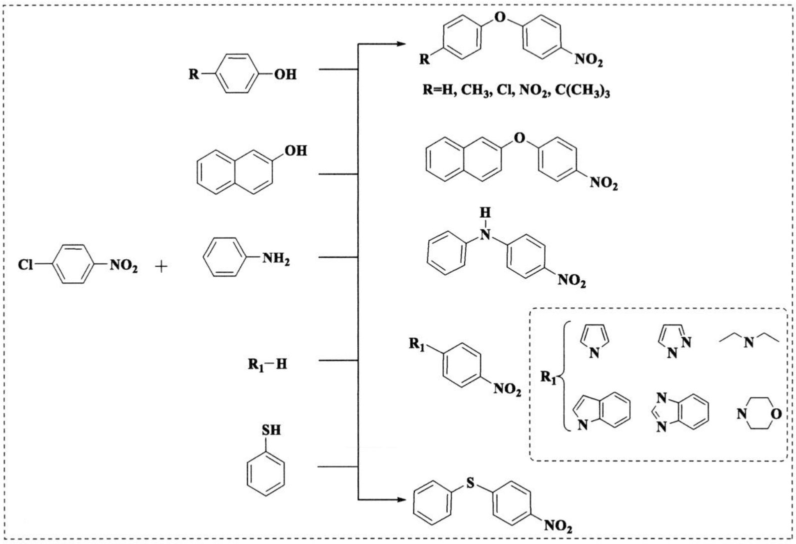

Scheme (11). Different product structures formed *via* the above-mentioned methodology.

BRIEF COMPARISON AND CRITICAL REMARKS

The construction of aryl carbon hetero atom bond in organic synthesis gains prime importance in the last few decades as it leads to the development of various medicinal as well as natural molecular scaffolds. The use of nanoparticles in this process is always appreciated as it combines the merits of both homogeneous and heterogeneous catalysts along with greater reactivity, selectivity and reusability. Again, as per environmental concern, the limited use of toxic reagents is highly recommendable. Therefore, in our chapter, we have discussed only the copper-free protocols as copper is always related to the cytotoxicity effect. All the discussed methods of this chapter possess their own merits and demerits. Among all the employed catalysts discussed here, ferrosoferric oxide supported on mesoporous polyaniline is found to be more effective for both S-arylation [16] and O-arylation [21] reaction. The product yields and the reaction times are also quite impressive. Moreover, reusability and cheaper precursors of the starting reagents make this catalytic system very cost-effective. Low toxicity associated with this catalyst indicates the environment-friendliness of the aforesaid catalyst.

CONCLUDING REMARKS

As conclusive remarks, we have highlighted here some methodologies regarding aryl carbon hetero bond formation. The purpose of this chapter is too widespread the utilization of copper-free nanocatalytic system in the text of the aryl carbon heteroatom bond formation. Since toxicity is always associated with copper metal, so it is better to minimize the use of it or to avoid it. Hence, we have focused here only on the copper-free protocols to create environmental awareness.

CONSENT FOR PUBLICATION

Not applicable.

CONFLICT OF INTEREST

The author declares no conflict of interest, financial or otherwise.

ACKNOWLEDGEMENTS

Declared none.

REFERENCES

[1] Mitrofanov, A.Y.; Murashkina, A.V.; Garcia, I.M.; Alonso, F.; Beletskaya, I.P. Formation of C–C, C–S and C–N bonds catalysed by supported copper nanoparticles. *Catal. Sci. Technol.,* **2017**, *7*(19), 4401-4412.
[http://dx.doi.org/10.1039/C7CY01343D]

[2] (a)Rampino, L.D.; Nord, F.F. Preparation of palladium and platinum synthetic high polymer catalysts and the relationship between particle size and rate of hydrogenation. *J. Am. Chem. Soc.,* **1941**, *63*(10), 2745-2749.
[http://dx.doi.org/10.1021/ja01855a070] (b)Sabatier, P. In La Catalyse en Chimie Organique. In: *Beranger,* 1st C; French: Paris, Liége, **1913**.

[3] Vert, M.; Doi, Y.; Hellwich, K.H.; Hess, M.; Hodge, P.; Kubisa, P.; Rinaudo, M.; Schué, F.O. Terminology for biorelated polymers and applications. *Pure Appl. Chem.,* **2012**, *84*(2), 377-410.
[http://dx.doi.org/10.1351/PAC-REC-10-12-04]

[4] (a) Bahrami, F.; Panahi, F.; Daneshgar, F.; Yousefi, R.; Shahsavani, M.B.; Khalafi-nezhad, A. Synthesis of new α-aminophosphonate derivatives incorporating benzimidazole, theophylline and adenine nucleobases using l-cysteine functionalized magnetic nanoparticles (LCMNP) as magnetic reusable catalyst: evaluation of their anticancer properties. *RSC Advances,* **2016**, *6*(7), 5915-5924.
[http://dx.doi.org/10.1039/C5RA21419J] (b) Fukui, T.; Murata, K.; Ohara, S.; Abe, H.; Naito, M.; Nogi, K. Morphology control of Ni–YSZ cermet anode for lower temperature operation of SOFCs. *J. Power Sources,* **2004**, *125*(1), 17-21.
[http://dx.doi.org/10.1016/S0378-7753(03)00817-6] (c) Barbaro, P.; Liguori, F. Heterogenized homogeneous catalysts for fine chemicals production: materials and processes. **2010**, *33*.
[http://dx.doi.org/10.1016/S0378-7753(03)00817-6]

[5] (a)Thomas, J.M. The advantages of exploring the interface between heterogeneous and homogeneous catalysis. *ChemCatChem,* **2010**, *2*(2), 127-132.
[http://dx.doi.org/10.1002/cctc.200900275] (b)Astruc, D. *Nanoparticles and Catalysis*; Astruc, D., Ed.; Wiley-VCH Verlag GmbH: Weinheim, Germany, **2008**, pp. 1-49.

[6] Xu, H. Advances on N-Arylation of Indoles by Cross-Coupling Reactions. *Mini Rev. Org. Chem.,* **2009**, *6*(4), 367-377.
[http://dx.doi.org/10.2174/157019309789371613]

[7] Nejati, K.; Ahmadi, S.; Nikpassand, M.; Nezhada, P.D.K.; Vessally, E. Diaryl ethers synthesis: nano-catalysts in carbonoxygen cross-coupling reactions. *RSC Advances,* **2018**, *8*(34), 19125-19143.
[http://dx.doi.org/10.1039/C8RA02818D]

[8] Vessally, E.; Didehban, K.; Mohammadi, R.; Hosseinian, A.; Babazadeh, M. Recent advantages in the metal (bulk and nano)- catalyzed S-arylation reactions of thiols with aryl halides in water: a perfect synergy for ecocompatible preparation of aromatic thioethers. *J. Sulfur Chem.*, **2018**, *39*(3), 332-349. [http://dx.doi.org/10.1080/17415993.2018.1436711]

[9] Wu, X.J.; Jiang, R.; Wu, B.; Xu, X.P.; Xi, S.J. Nano-Iron Oxide as a Recyclable Catalyst for Intramolecular C-N Cross-Coupling Reactions under Ligand-Free Conditions: One-Pot Synthesis of 1,4-Dihydroquinoline Derivatives. *Adv. Synth. Catal.*, **2009**, *351*(18), 3150-3156. [http://dx.doi.org/10.1002/adsc.200900481]

[10] Nasrollahzadeh, M.; Azarian, M.; Ehsani, A.; Zahraei, A. Facile synthesis of Fe@Pd nanowires and their catalytic activity in ligand-free C-N bond formation in water. *Tetrahedron Lett.*, **2010**, *51*, 2309-2311.

[11] Roopan, S.M.; Khan, F.R.N.; Mandal, B. Fe nano particles mediated C–N bond-forming reaction: Regioselective synthesis of 3-[(2-chloroquinolin-3-yl)methyl]pyrimidin-4(3H)ones. *Tetrahedron Lett.*, **2010**, *51*(17), 2309-2311. [http://dx.doi.org/10.1016/j.tetlet.2010.02.128]

[12] Nandi, D.; Islam, R.U.; Devi, N.; Siwal, S.; Mallick, K. Palladium nanoparticle catalyzed aryl-amine coupling reaction: High performance of aryl and pyridyl chlorides as the coupling partner. *New J. Chem.*, **2018**, *42*(2), 812-816. [http://dx.doi.org/10.1039/C7NJ03447D]

[13] Moghaddam, F.M.; Tavakoli, G.; Moafi, A.; Saberi, V.; Rezvani, H.R. C-N bond formation using highly effective and reusable nickel ferrite nanoparticles in water. *ChemCatChem*, **2014**, *6*(12), 3474-3481. [http://dx.doi.org/10.1002/cctc.201402556]

[14] Mishra, V.; Arya, A.; Chundawat, T.S. High catalytic activity of PD nanoparticles synthesized from green alga *Chlorella vulgaris* in buchwald-hartwig synthesis of n-aryl piperazines. *Curr. Organocatal.*, **2020**, *7*(1), 23-33. [http://dx.doi.org/10.2174/2213337206666190515091945]

[15] Paul, A.; Chatterjee, D.; Banerjee, S.; Yadav, S. Ligand and Cu free N-arylation of indoles, pyrroles and benzylamines with aryl halides catalyzed by a Pd nanocatalyst. *New J. Chem.*, **2020**, *44*(34), 14447-14452. [http://dx.doi.org/10.1039/D0NJ02129F]

[16] Damodara, D.; Arundhathia, R.; Likhar, P.R. High surface and magnetically recoverable mPANI/pFe_3O_4 nanocomposites for C–S bond formation in water. *Catal. Sci. Technol.*, **2013**, *3*(3), 797-802. [http://dx.doi.org/10.1039/C2CY20624B]

[17] Ramankutty, C.G.; Sugunan, S. Surface properties and catalytic activity of ferrospinels of nickel, cobalt and copper, prepared by soft chemical methods. *Appl. Catal. A Gen.*, **2001**, *218*(1-2), 39-51. [http://dx.doi.org/10.1016/S0926-860X(01)00610-X]

[18] Jacobs, J.P.; Maltha, A.; Reitjes, J.G.H.; Drimal, J.; Ponec, V.; Brongersma, H.H. The surface of catalytically active spinels. *J. Catal.*, **1994**, *47*(1), 294-300. [http://dx.doi.org/10.1006/jcat.1994.1140]

[19] Garade, A.C. Bharadwaj, M.; Bhagwat, S.V.; Athawale, A. A.; Rode, C. V. An efficient γ-Fe_2O_3 catalyst for liquid phase air oxidation of p-hydroxybenzyl alcohol under mild conditions. *Catal. Commun.*, **2009**, *10*, 485-489. [http://dx.doi.org/10.1016/j.catcom.2008.10.044]

[20] Iqubal, Md. A.; Islam, S. S.; Ghosh, K.; Molla, R. A.; Kamaluddin; Islam, S. M. Silica functionalized magnetic nickel ferrite nanoparticles as an efficient recyclable catalyst for S-arylation in aqueous medium. *J Inorg Organomet Polym*, **2017**, *27*, 1730-1739.

[21] Arundhathi, R.; Damodara, D.; Likhar, P. R.; Kantam, M. L.; Saravanan, P.; Magdaleno, T.; Kwon, S.H. Corrigendum: Fe_3O_4@mesoporouspolyaniline: a highly efficient and magnetically separable catalyst for cross-coupling of aryl chlorides and phenols. *Adv. Synth. Catal.,* **2011**, *353*, 1591-1600.

[22] (a) Agawane, S. M.; Nagarkar, J. M. Nano ceria catalyzed synthesis of α-aminophosphonates under ultrasonication. Tetrahedron Lett., 2011, 52, 3499–3504. (b) Terribile, D.; Trovarelli, A.; Llorca, J.; Leitenburg, C.; Dolcetti, G. The synthesis and characterization of mesoporous high-surface area ceria prepared using a hybrid organic/inorganic route. *J. Catal.,* **1998**, *178*, 299-308.

[23] Agawane, S.M.; Nagarkar, J.M.. Nano ceria catalyzed Ullmann type coupling reactions. *Tetrahedron Lett,* **2011**, *52*, 5220-5223.

CHAPTER 4

Organic Carbon-Heteroatom Cross-Coupling Reactions Mediated by Supported Transition Metal-Based Catalysts

Shyamal Baruah[1], **Merangmenla**[1] and **Amrit Puzari**[1,*]

[1] *Department of Chemistry, National Institute of Technology Nagaland, Chumukedima, Dimapur, Nagaland, India*

Abstract: Organic cross-coupling reactions have been broadly utilized in the synthesis of various organic products and valuable compounds. It is considered the backbone for the synthesis of various medicinally significant compounds. Although numerous organic compounds are made out of C-C bonds, however, the presence of heteroatoms, like nitrogen, oxygen and sulfur, makes them utilitarian. Conductive polymers and drugs often contain amine C-N bonds, and practically, the majority of natural products contain ether, ketone, or ester C-O bonds. Heterocyclic compounds containing C-N, C-O, or C-S bonds in the ring structure are found in all applications of chemistry. As a result of the developing significance of these cross-coupling reactions, the advancement of new extemporized strategies for the synthesis of such items has become the most elevated need for some researchers. Researchers are zeroing in on the advancement of green synthetic methods for the synthesis of such compounds. Catalysis is steadily overwhelming in the field of synthesis because of the expanding capacity to plan new materials with improved synergist exhibitions. Catalysis is proclaimed as a more up-to-date process that shows stability, durability, higher activity, and cost *via*bility analysis. High-yielding cross-coupling reactions using supported catalysts are also reported, resulting in the development of a large number of very effective catalysts for organic cross cross-coupling processes.. Heterogeneous catalysis has the advantage of easy separation and compared to homogeneous catalysis, product isolation requires less amount of the solvent. However, sometimes there may be a compromise in the selectivity process. This chapter contains a detailed discussion on different carbon-heteroatom cross-coupling reactions mediated by supported transition metal-based catalysts. The scope of these cross-coupling reactions in the pharmaceutical and industrial sectors has been investigated. Exceptional methodologies and significant advances accomplished in the field of carbon-heteroatom cross-coupling reactions have been talked about.

Keywords: Cross-coupling reaction, Green methodology, Heterogeneous catalyst, Pharmaceutical, Transition metals.

* **Corresponding author Amrit Puzari:** Department of Chemistry, National Institute of Technology Nagaland, Chumukedima, Dimapur, Nagaland, India; Tel: +918638514575; E-mail: amrit09us@yahoo.com.

INTRODUCTION

Cross-coupling reaction involves a combination of an organometallic reagent (that generally contain a main group metal atom) with an organic electrophile in the presence of metal catalysts (group 8-10 metals) to accomplish a C–H, C–N, C–O, C–C, C–S, C–P, or C–M bond formation [1]. This was coined in early 1970s by Kumada, Kochi, Corriu, *etc.* and since then, many organometallic reagents have been developed which even revolutionized the field of organic cross coupling reactions. Examples of successful organometallic catalysts include many organoboron, organotin, organozinc, or organosilicon compounds. Various types of electrophiles and metal complexes have been successfully employed in these types of reactions, resulting in a plethora of synthetic methods for molecular assemblies [2]. Because of their milder reaction conditions and unique reactivity, a useful synthetic alternative compared to the conventional methods of synthesis has been provided to the researchers by the cross-coupling reactions [3]. Scheme **1** represents a general catalytic cycle for cross-coupling reactions.

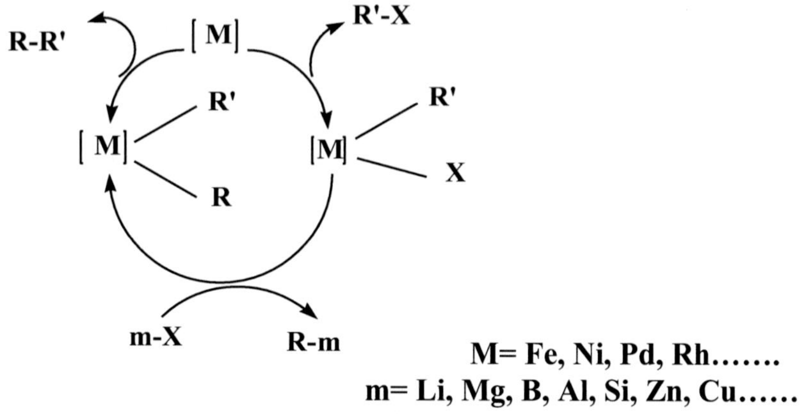

Scheme (1). General catalytic cycle for cross-coupling reactions [4].

The reaction involves the reductive elimination of the organic substituents R and R' on a metal complex. The intermediate MR(R') is generally formed in a step two-step process. In the first step, oxidative addition of the organic halide (RX) to M gives MR(X). Subsequently, the second partner undergoes transmetallation with a source of R'. The final step is reductive elimination of the two coupling fragments to regenerate the catalyst and produce the organic product [5].

In cross-coupling reactions, two dissimilar chemical species are joined together to afford a single product. Some notable examples of cross-coupling reactions include:

- The Grignard reaction [6].
- Heck reaction [7].
- Sonogashira cross-coupling [8].
- Suzuki coupling [9].
- The Buchwald-Hartwig reaction [10].
- Kumada coupling reaction [11].

The function of many organometallic compounds is often derived from the presence of heteroatoms, such as nitrogen, oxygen, and sulfur, which are held by C–heteroatom bonds [12]. Drugs and conductive polymers frequently contain amine C–N bonds, and practically all regular items contain ether, ketone, or ester C–O bonds [13]. Heterocyclic mixtures containing C–N, C–O, or C–S bonds present in the ring structure, are found in different fields of chemistry [14]. In addition, valuable intermediates in blend frequently contain C–B or C–Si bonds that are subsequently changed over into C–C, C–O, or C–N bonds in the last products [15].

The primary focus of the chapter is mainly on cross-coupling reactions involving C-N, C-O, and C-S bond formation reactions supported by transition metal-based catalyst.

C–N BOND-FORMING CROSS-COUPLING REACTIONS

Because of their interesting and diverse biological activities, nitrogen-containing compounds often find important implications in many different fields and also as a synthetic precursor [16]. The construction of the C–N bonds of aromatic compounds is particularly important and proved to be challenging to medicinal chemists due to the involvement of harsh reaction conditions or the use of expensive catalysts in many cases [17]. Recently, some significant achievements have been made in C–N coupling reactions using metal complexes with new ligands under mild conditions in a cost-effective manner [18].

Herein, we have selected some prime literature reports for an elaborate discussion on the scope of C-N bond-forming reactions.

Copper (Cu)-Catalysed C–N Bond Formation

In recent years, considerable advancements have been achieved in C–N bond formation using various copper salts [19]. These reactions utilize a catalytic amount of copper in the presence of various ligands [20]. Some recent and interesting reports where various copper salts were used for C–N bond formation have been discussed in this section.

• Bao and his co-workers reported a one-pot sequential coupling reaction for the synthesis of a series of 6H-isoindolo [2, 1-a] indol-6-ones (Scheme **2**). This sequence includes a Cu-catalyzed C–N coupling cyclization followed by a Pd-catalyzed C–H activation [21].

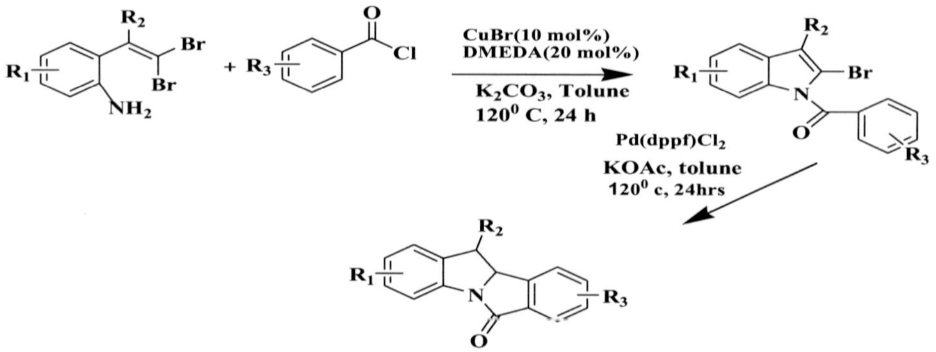

Scheme (2). Synthesis of 6H-isoindolo [2, 1-a] indol-6-ones [21].

• In yet another important reaction, Quach et. Al. introduced ligand and base-free copper (II)-catalyzed C–N bond-forming cross-coupling reactions of organoboron compounds with aliphatic amines and anilines (Scheme **3**). They utilized catalytic copper (II) acetate monohydrate and molecular sieves (4Å) in dichloromethane at slightly elevated temperatures under an atmosphere of oxygen. A broad range of functional groups tolerance was achieved for both the cases of cross-coupling partners [22].

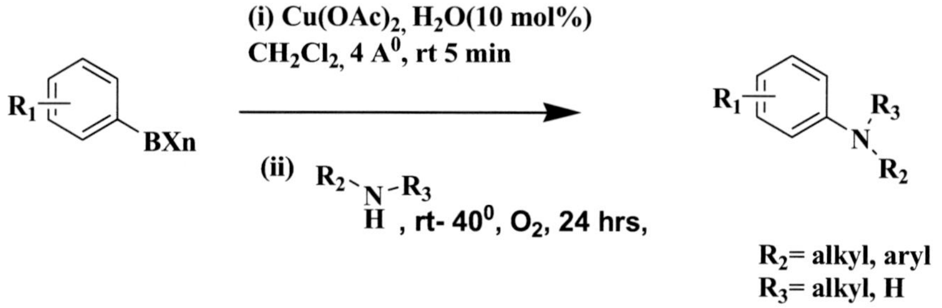

Scheme (3). Copper (II)-Catalyzed C–N Bond formation [22].

• Intramolecular N-arylation reactions are often employed for the convergent total synthesis of important organic precursors and numerous examples are available in the literature. Peng and his collaborators built up a clear, effective, and more reasonable copper-catalyzed strategy for intramolecular N-arylation giving the benzimidazole ring framework (Scheme **4**). They have utilized Cu_2O (5 mol %)

as the impetus, DMEDA (10 mol %) as the ligand, and K_2CO_3 as the base. The response was only completed in water, delivering the procedure profoundly important from both ecological and efficient places of view [23].

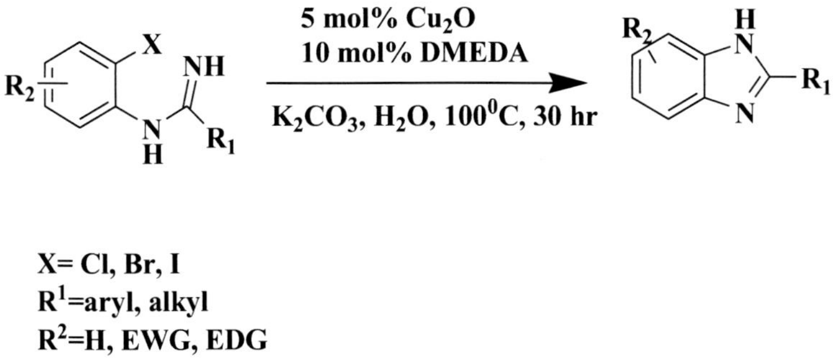

Scheme (4). Copper-catalyzed intramolecular C-N bond formation [23].

- Cu(II)-Catalyzed C-N Coupling of (Hetero)aryl Halides and N-Nucleophiles Promoted by α-Benzoin Oxime was reported by Yuan and his coworkers (Scheme **5**). They initially detailed the new utilization of an interprets metal chelating ligand α-Benzoin Oxime for improving Cu-catalyzed C-N coupling responses. The framework could catalyze coupling responses of (hetero)aryl halides with a wide of nucleophiles (*e.g.*, azoles, piperidine, pyrrolidine, and amino acids) in moderate to astounding yields. The convention permits quick admittance to the most well-known frameworks found in FDA-affirmed drugs [24].

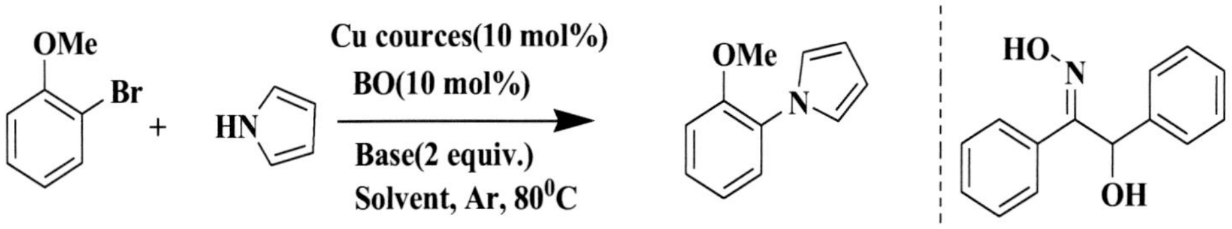

Scheme (5). C-N Coupling of (Hetero)aryl Halides and N-Nucleophiles Promoted by α-Benzoin Oxime [24].

- Ambient-Light-Mediated Copper-Catalyzed C-N Bond Formation was reported by Majek and his group. They developed a catalytic procedure (Scheme **6**). In the presence of 10 mol% of CuI, lithium carbazolide underwent phenylation in 65% yield. The mild conditions of this light-mediated amination (RT) will certainly fuel endeavors into new synthetic applications [25].

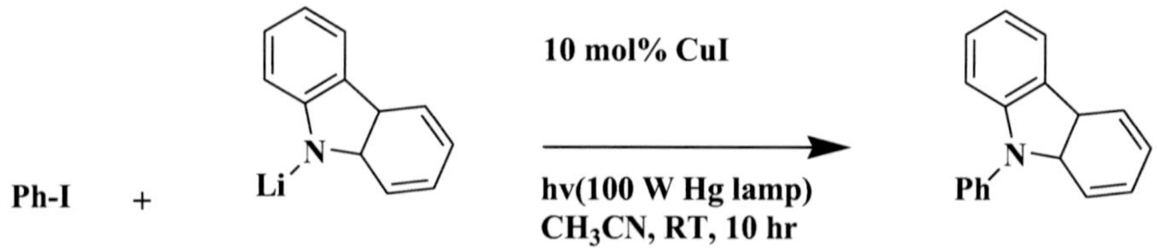

Scheme (6). Light-Mediated Copper-Catalyzed CN Bond Formation [25].

Palladium (Pd)-Catalysed C–N Bond Formation

Palladium-catalyzed C–N and C–O bond formation between 4-substituted 7-azaindoles and amides, amines, amino acid esters, or phenols have as of late acquired fame among established researchers for various medication revelation and improvement programs. Hence, some intriguing reports where different 'Pd' based compounds were utilized for C–N bond arrangement have been talked about in this part.

- Migita and co-workers reported the first palladium-catalyzed formation of aryl C-N bonds in 1983. The reaction of electronically neutral aryl bromides and amino tin compounds in the presence of catalytic $[(o\text{-tol})_3P]_2PdCl_2$ resulted in the efficient preparation of the corresponding aniline in moderate to good yield [26] (Scheme **7**).

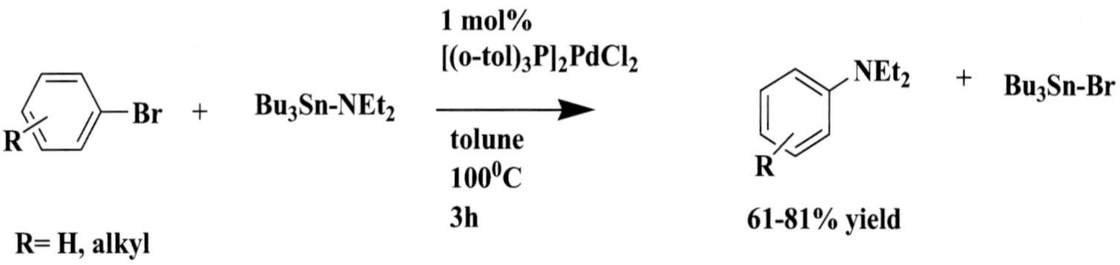

Scheme (7). $[(o\text{-tol})_3P]_2PdCl_2$ catalyzed formation of aryl C-N [26].

- Hartwig and his group also reported that $[(o\text{-tol})_3P]_2PdCl_2$ and $[(o\text{-tol})_3P]_2Pd$ effectively catalyzes the amine arylation reaction as demonstrated in Scheme **8** [27].

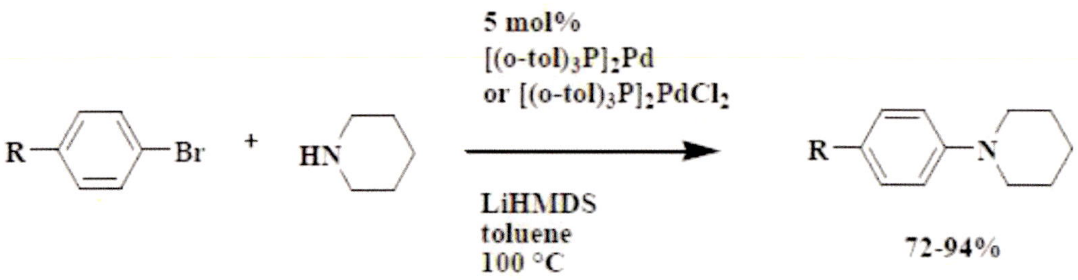

Scheme (8). $[(o\text{-}tol)_3P]_2PdCl_2$ and $[(o\text{-}tol)_3P]_2Pd$ catalyzed amine arylation reaction [27].

- The illustration of the synthesis of anti-psychotic agent arylpiperazines, which are metabolites of Aripiprazole, was reported by Morita and his co-workers utilizing the BINAP/Pd-catalyst constitutes another significant Pd-catalysed C-N cross-coupling reaction [28] (Scheme **9**).

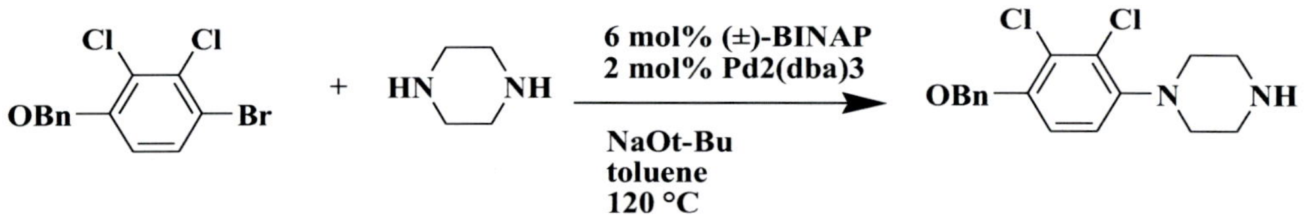

Scheme (9). BINAP/Pd-catalysed synthesis of arylpiperazines [28].

- Mann and his associates also successfully demonstrated Palladium-catalyzed C−N (sp^2) bond Formation reaction through *N*-arylation of aromatic and unsaturated nitrogen and the reductive elimination of Palladium Azolyl and Methyleneamido Complexes (Scheme **10**). This work includes data on imine arylation using DPPF-ligated palladium to demonstrate the differences in rates on the arylation of different types of sp2- hybridized nitrogen using the same catalyst [29].

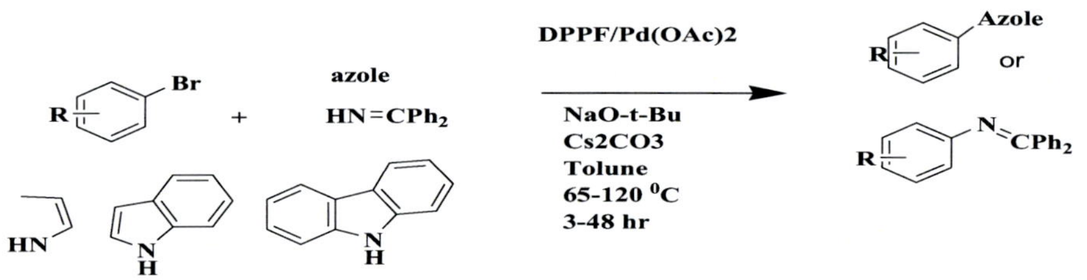

Scheme (10). Palladium-Catalyzed C−N (sp^2) Bond Formation: *N*-Arylation of Aromatic and Unsaturated Nitrogen [29].

Cobalt (Co) Catalysed C–N Bond Formation

Although significant progress has been achieved in cobalt catalyzed organic transformations, when we consider the case of cross-coupling reactions, a number of literature reports are very few. Therefore, sufficient scope remains for exploring the Co-catalysed C-N bond formation reactions and hence this constitutes an important research objective.

- Toma *et al.* reported cobalt-catalyzed C–N bond-forming reaction involving aromatic chlorides and cyclic secondary amines (Scheme **11**). The reaction was *ortho*- and *para*-selective; *meta*-substituted halides were unreactive [30].

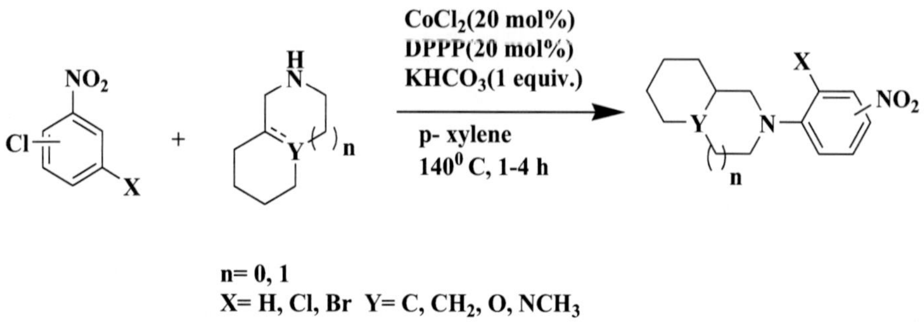

Scheme (11). cobalt-catalyzed C–N bond-forming reaction involving aromatic chlorides and cyclic secondary amines [30].

- N-bridged heterocyclic compounds like imidazopyridines are considered among the bioactive compounds [31]. Especially substituted Imizazopy- ridines display important pharmacological activities and hence these molecules are of significant pharmaceutical importance [32]. These molecules are involved in the synthesis of several important drugs. Rasheed and his co-workers reported a novel protocol for cobalt (II)-catalyzed synthesis of 2-arylimidazo [1, 2-a] pyridines and 3-arylimidazo [1, 2-a] pyridines by direct C-N bond formation (Scheme **12**). The process was regio-selective and had good functional group tolerability. This annulation reaction probably involves a Co (II)/Co (III) catalytic system and uses molecular oxygen (O_2) as a terminal oxidant [33].

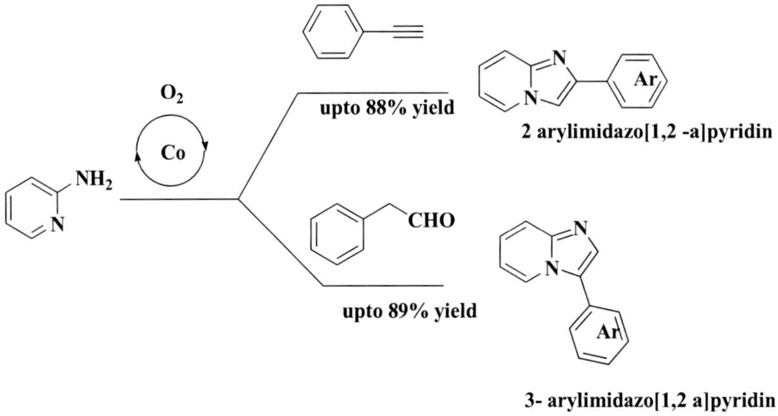

Scheme (12). cobalt (II)-catalyzed synthesis of 2-arylimidazo [1, 2-a] pyridines and 3-arylimidazo [1, 2-a] pyridines [33].

- In another report, Affo and his co-workers synthesized indolines using trimethylsilylmethylmagnesium-promoted and cobalt-catalyzed radical alkylation of alkyl halides to produce 3-methylenepyrrolidine from iodoalkanes [34] (Scheme **13**).

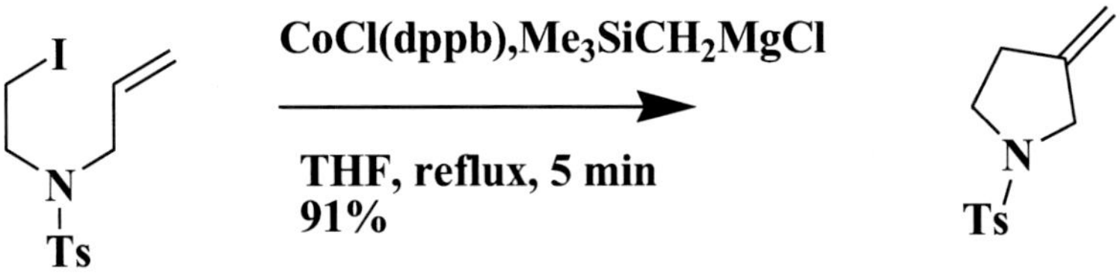

Scheme (13). Synthesis of 3-methylenepyrrolidine [34].

Nickel (Ni) Catalyzed C-N Bond Formation

Nickel complexes can elegantly orchestrate diverse amination reactions depending on specific reaction conditions. Many types of research are going on in recent times, which are making far-reaching impacts across chemistry, materials science, medicine, and biology. Some of the literature references are discussed below:

- An inexpensive nickel source ($NiBr_2.3H_2O$) was used as an efficient catalyst by Lim *et al.* in C-N cross-coupling reaction, which proceeds through direct irradiation of the nickel-amine complex [35] (Scheme **14**).

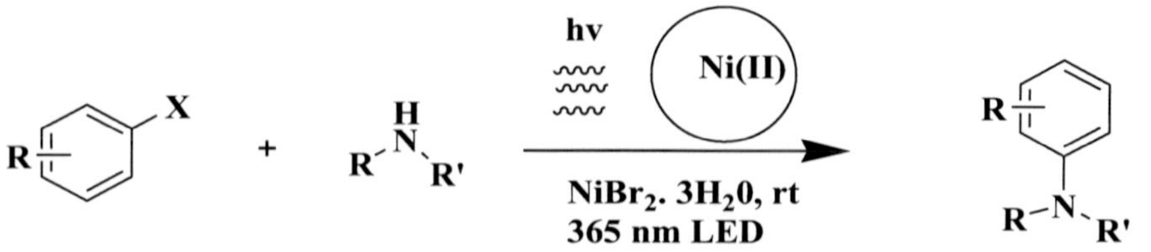

Scheme (14). C-N cross-coupling methodology using an inexpensive nickel source ($NiBr_2.3H_2O$) [35].

- Christopher *et al.* chipped away at nickel-catalyzed amine arylations empowered by customized auxiliary ligand configuration (Scheme **15**). They have detailed an operationally straightforward and air-stable ligand/nickel(II) pre-catalyst, which obliges the broadest blend of C(sp^2)– N coupling accomplices for any single nickel catalyst, without the requirement for a valuable metal co-catalyst [36].

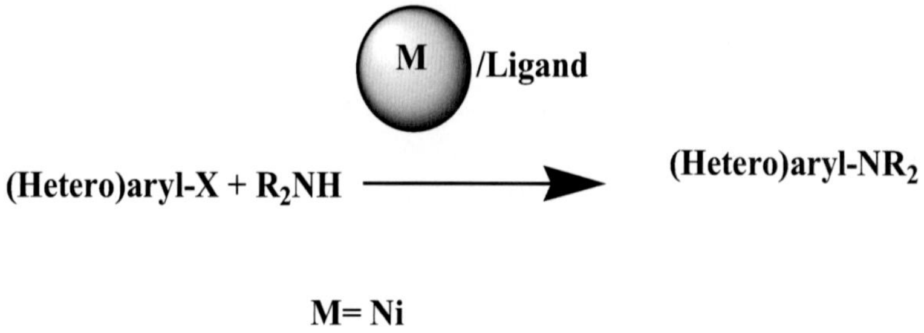

Scheme (15). Nickel-catalysed amine arylations [36].

- Nickel catalyzed protocol was employed by Tomoya and his group for the amination of aryl fluorides with primary amines (Scheme **16**). They have shown a cross-coupling reaction between aryl fluorides and primary amines, which was enabled by the 1,2 bis(dicyclohexylphosphino)benzene (DCYPBz) or 1,2-bis(dicyclohexylphosphino) ethane (DCYPE) ligands [37].

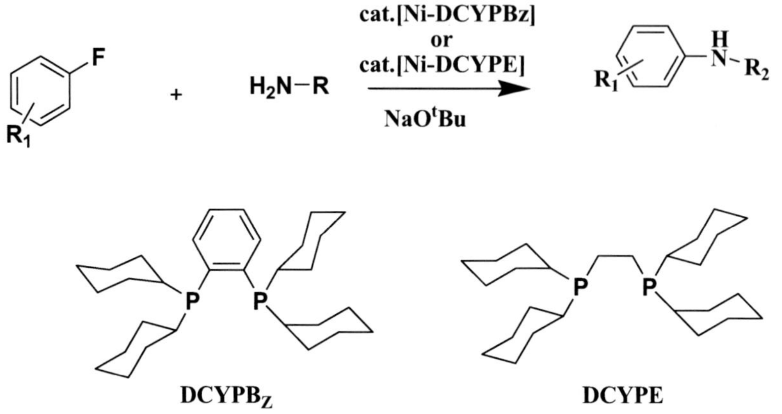

Scheme (16). Nickel-catalyzed amination of aryl fluorides with primary amines [37].

C–O BOND-FORMING CROSS-COUPLING REACTIONS

Carbon-oxygen bond formation reactions are considered among the most important organic transformations having tremendous industrial and pharmaceutical significance. In recent years some efficient oxidations using molecular oxygen have been reported. Here we have included a discussion on catalytic methods that employ copper or palladium complexes and metal-free radicals for installing oxygen functionality.

- Kuwabe and his collaborators depicted about Palladium-Catalyzed Intramolecular C-O bond Formation. They have combined various oxygen heterocycles utilizing the palladium-catalyzed intramolecular etherification of aryl halides by utilizing di-tert-butylphosphinobiaryl ligands. The response continues under gentle conditions utilizing weak bases like Cs_2CO_3 or K_3PO_4 (Scheme **17**). This technique was utilized as the critical advance in the combination of MKC-242, an antidepressant. The synthesis of MKC-242 was accomplished in 40% generally yield from industrially accessible sesamol and acrylonitrile [38].

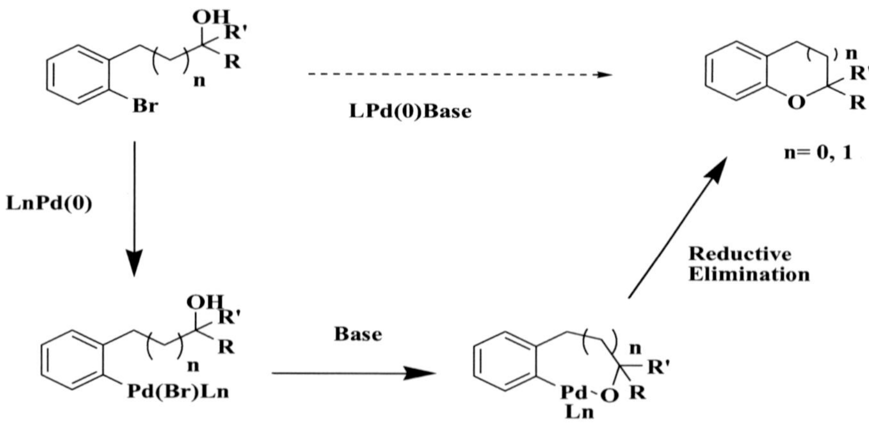

Scheme (17). Palladium-Catalyzed Intramolecular C-O Bond Formation [38].

- Chen *et al.* reported palladium-catalyzed C–O bond formation reaction through the direct synthesis of phenols and aryl/alkyl ethers from activated aryl halides. The palladium-tri-*tert*-butylphosphine complex was very much effective in converting haloarenes to corresponding substituted phenols (Scheme **18**). This methodology offers a direct transformation of aryl halides to phenols, just as the clear application to produce a wide assortment of diaryl or alkyl/aryl ethers [39].

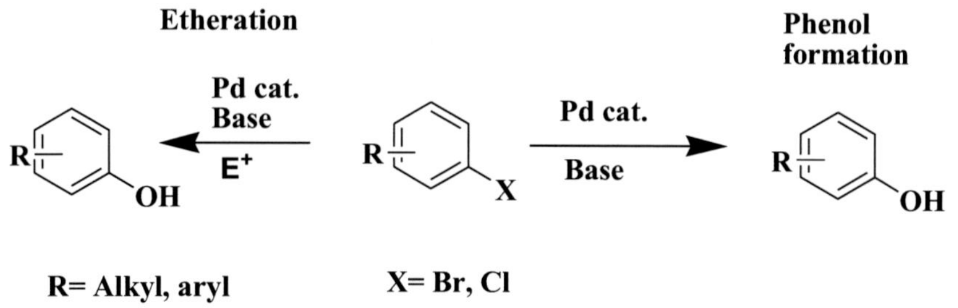

Scheme (18). Palladium-catalyzed C–O bond formation: direct synthesis of phenols and aryl/alkyl ethers from activated aryl halides [39].

- Palladium-Catalyzed C−O Cross-Coupling of Primary Alcohols was also employed by Zang and his group. They portrayed around two catalytic frameworks, which together give gentle and general conditions to the Pd-catalyzed C-O cross-coupling of essential alcohols (Scheme **19**). For actuated substrates, for example, electron-insufficient aryl halides, the industrially accessible ligand L2 advances proficient coupling for an assortment of liquor nucleophiles. On account of unactivated electrophiles, for example, electron-rich aryl halides, the new ligand L8 was created to improve these difficult C-O bond-forming reactions [40].

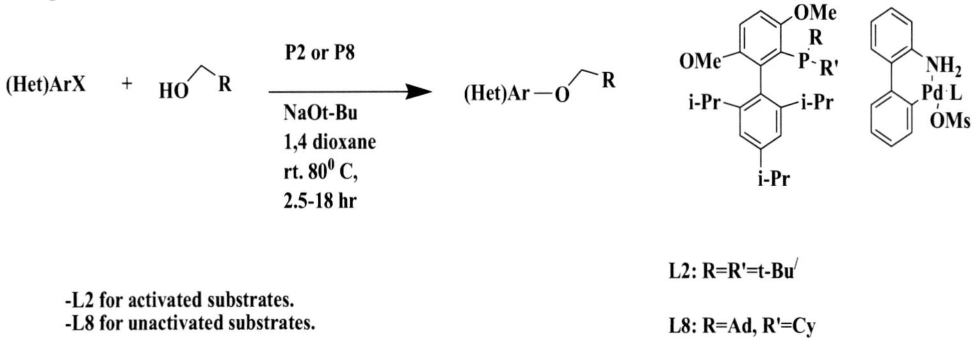

Scheme (19). Palladium-Catalyzed C–O Cross-Coupling of Primary auxiliarycohols [40].

- Niu and his associates described an efficient O-arylation of phenols and aliphatic alcohols with aryl halides utilizing an air-stable copper (I)−bipyridyl complex as the catalyst. This arylation reaction can be performed in good yield in the absence of Cs_2CO_3 (Scheme **20**). A variety of functional groups is compatible with these reaction conditions with low catalyst loading levels [41].

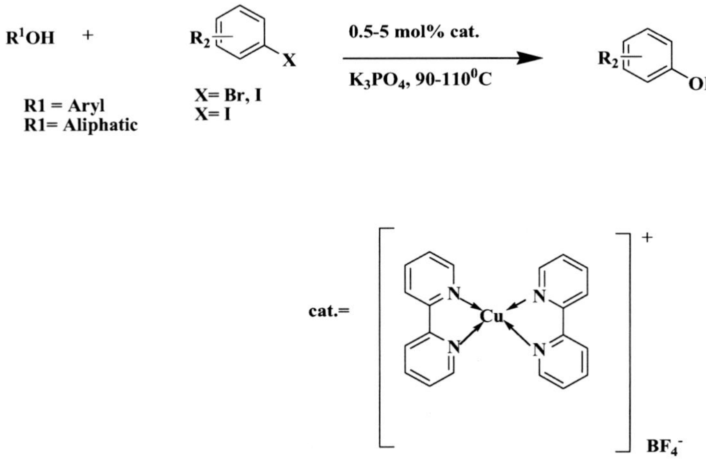

Scheme (20). Ullmann-type C–O bond formation catalyzed by an air-stable copper (I)−bipyridyl complex [41].

- An efficient Copper-catalyzed one-pot highly regioselective synthesis of furans from (2-furyl)carbene complexes was reported by Cao and his coworkers. It was a convenient one-pot Cu (I)-catalyzed strategy for the regioselective synthesis of trisubstituted furan derivatives. The scheme of synthesis is illustrated in Scheme **21**. This process has opened a new synthetic route to a variety of R-carbonyl furans using air as the oxidant good yields [42].

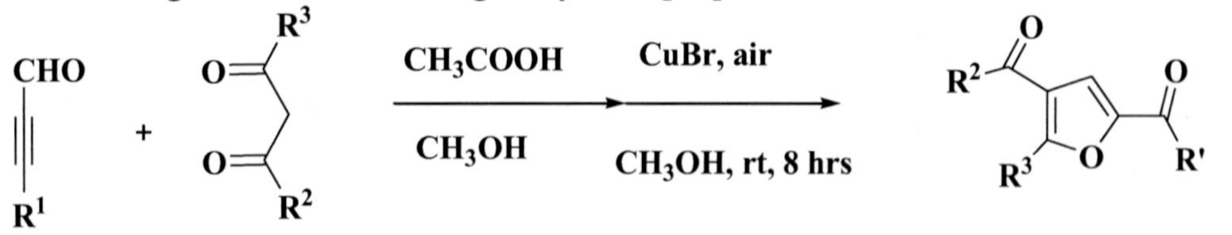

Scheme (21). Cu (I)-catalyzed strategy for the regioselective synthesis of trisubstituted furan derivatives [42].

- Wang and his group showed a Copper-catalyzed C-O bond formation reaction *via* oxidative cross-coupling reaction of aldehydes and ethers. In this reaction, practical and productive development of C-O bonds using oxidative cross-coupling response of aldehydes and ethers has been realized under open air (Scheme **22**). At the point when 2 mol % copper was utilized as the catalyst, different α-acyloxy ethers were obtained with up to 93% confined yield [43].

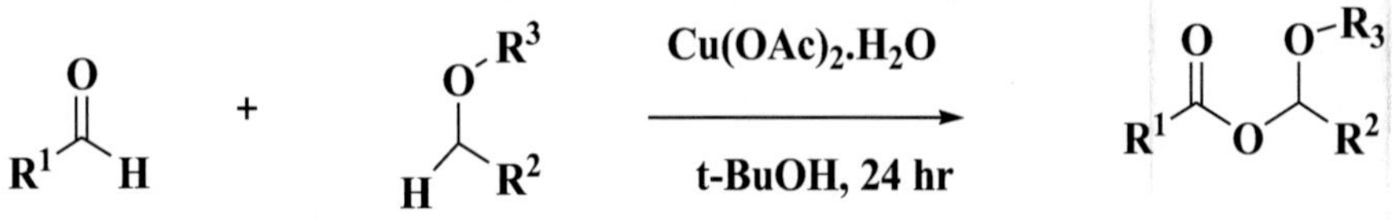

Scheme (22). Copper-catalyzed C-O bond formation reaction *via* oxidative cross-coupling reaction of aldehydes and others [43].

C-S BOND-FORMING CROSS-COUPLING REACTION

As the C-S bond is fundamental in various synthetically significant compounds, the headway of delicate and general procedures for C-S bond formation has received great attention of the researchers. Late substance approaches for C-S bond advancement, considering direct C-H bond functionalisation not simply give new pieces of information into the negligent cognizance of C-S coupling reactions yet likewise license the association of sulfur-containing compounds from more *via*ble synthetic routes with high atom economy. Usage of carbon-Sulfur bond development through C-H bond actuation using change metals establishes an incredible asset in therapeutic just as drug science.

- Sangeetha and his coworkers developed an efficient domino process for the synthesis of 2-acyldihydrobenzo[*b*]thiophenes and 2-acylbenzo[*b*] thiophenes using a Cu-catalyst and easily available xanthate as a sulfur surrogate in good yields. Treatment of xanthate with the copper (II)-catalyst reduced it to a Cu (I)-catalyst, which initiates the catalytic cycle [44] (Scheme **23**).

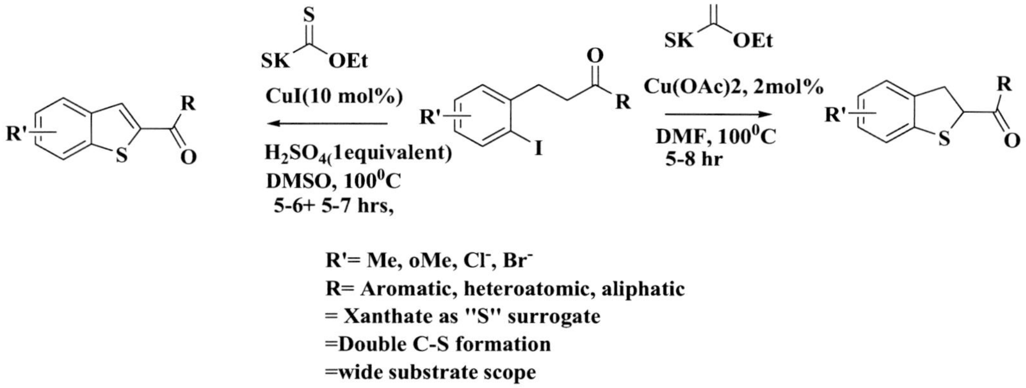

Scheme (23). Synthesis of 2-acyldihydrobenzo[*b*]thiophenes and 2-acylbenzo[*b*]thiophenes using a Cu-catalyst [44].

- Copper-catalyzed C-S bond arrangement through C-O bond cleavage in the presence of S8 as the sulfur source was accounted for by Rostami and his partners. They have first presented another proficient technique for the union of unsymmetrical sulfide utilizing cross-coupling response of phenolic ester like acetic acid derivations, tosylates, and triflates and with arylboronic corrosive or triphenyltin chloride as the coupling accomplices is reported (Scheme **24**). Contingent upon the response, S8/KF or S8/NaOt-Bu framework is discovered to be a *via*ble wellspring of sulfur within the sight of copper salts and in poly (ethylene glycol) as a green solvent [45].

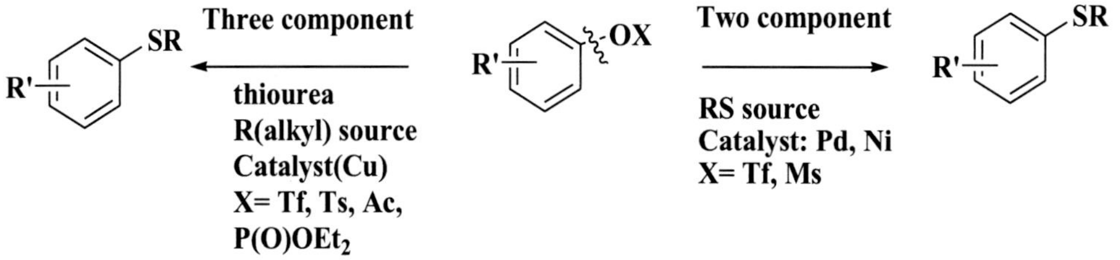

Scheme (24). copper-catalyzed C-S bond formation through C-O bond cleavage in presence of S8 as the sulfur source [45].

- Another work of copper-catalyzed cross-coupling response of thiols with aryl halides by Lee and his group revealed the utilization of CuI with oxalic diamide as an amazing synergist framework for cross-coupling response of thiols with aryl bromides and chlorides (Scheme **25**). It is essential to take note that unactivated aryl chlorides are coupled easily with aryl-and alkyl thiols to give the comparing thioethers in great to brilliant yields [46].

$$\text{ArX} \quad + \quad \text{R-SH} \xrightarrow[\text{Solvent, heat}]{\text{CuI(5 mol\%)}\ \ \text{L(10 mol\%)}} \text{Ar-SR}$$

X= Br, Cl: R= Aryl, Alkyl
L= Oxalic diamide

Scheme (25). Copper-catalyzed cross-coupling reaction of thiols with aryl halides [46].

APPLICATIONS OF CARBON-HETEROATOM BOND-FORMING CROSS-COUPLING REACTION

The carbon-carbon and carbon-heteroatom bond formation reactions perform an essential position in artificial natural chemistry. They had been used correctly inside the synthesis of medicinally critical molecules, agrochemicals, and precious compounds in cloth sciences. The cross-coupling technique is a dependable and flexible method that lets in for the bond-formation of an sp^2 -hybridized fragrant halide (performing as an electrophile with an organometallic (nucleophile) the use of a steel catalyst. In particular, palladium-catalyzed pass-coupling reactions had been confirmed as an extremely good resource and strong magnificence of reactions for the advent of carbon-carbon and carbon-heteroatom bonds withinside the pharmaceutical industry. Some important examples of cross-coupling reactions are as follows:

- In DNA-Encoded Libraries (DELs) [47].
- In macrocycles and cyclopeptides [48].
- In Allosteric modulator [49].
- In Proteolysis targeting chimeras (PROTAC) [50].

Furthermore, the utility of the cross-coupling method has ended up extra common for the big scale-up synthesis inside the improvement phase.

CONCLUDING REMARKS

The improvement of metal-catalyzed cross-coupling reactions during the last 30 years has revolutionized the way carbon-carbon bonds among sp and sp^2 carbon atoms are formed. These techniques have profoundly modified the protocols for the development of herbal products, constructing blocks for supramolecular chemistry and self-assembly, natural substances and polymers, and lead compounds in medicinal chemistry from less difficult entities. Given the considerable popularity, there is a need for complete know-how of the diverse greener techniques delivered for this essential response encompassing various catalytic structures primarily based totally on palladium, nickel, gold, ferrites, copper, silver, and cobalt.

CONSENT FOR PUBLICATION

Not applicable.

CONFLICT OF INTEREST

The author declares no conflict of interest, financial or otherwise.

ACKNOWLEDGEMENTS

The authors acknowledge the Department of Science and Humanities, NIT Nagaland.

REFERENCES

[1] Miyaura, N. *Cross-Coupling Reactions: A Practical Guide, Topics in Current Chemistry 219, 2002,*

[2] Rana, S.; Biswas, J.P.; Paul, S.; Paik, A.; Maiti, D. Organic synthesis with the most abundant transition metal-iron: from rust to multitasking catalysts. *Chem. Soc. Rev.,* **2021**, *50*(1), 243-472. [http://dx.doi.org/10.1039/D0CS00688B] [PMID: 33399140]

[3] Liu, W.; Li, J.; Querard, P.; Li, C.J. Transition-Metal-Free C-C, C-O, and C-N Cross-Couplings Enabled by Light. *J. Am. Chem. Soc.,* **2019**, *141*(16), 6755-6764. [http://dx.doi.org/10.1021/jacs.9b02684] [PMID: 30932483]

[4] Miyaura, N.; Yamamoto, Y. Main Group Compounds in Organic Synthesis, in Comprehensive Organometallic Chemistry III. **2007**.

[5] Molnár, Á. Efficient, selective, and recyclable palladium catalysts in carbon-carbon coupling reactions. *Chem. Rev.,* **2011**, *111*(3), 2251-2320. [http://dx.doi.org/10.1021/cr100355b] [PMID: 21391571]

[6] Peltzer, R.M.; Gauss, J.; Eisenstein, O.; Cascella, M. The Grignard Reaction - Unraveling a Chemical Puzzle. *J. Am. Chem. Soc.,* **2020**, *142*(6), 2984-2994. [http://dx.doi.org/10.1021/jacs.9b11829] [PMID: 31951398]

[7] Li, Y.; Gong, J.F.; Song, M.P. Palladium-catalyzed δ-selective reductive Heck reaction of alkenyl carbonyl compounds with aryl iodides and bromides. *Org. Chem. Front.,* **2020**, *7*(16), 2216-2223. [http://dx.doi.org/10.1039/D0QO00428F]

[8] Schilz, M.; Plenio, H. A guide to Sonogashira cross-coupling reactions: the influence of substituents in aryl bromides, acetylenes, and phosphines. *J. Org. Chem.,* **2012**, *77*(6), 2798-2807. [http://dx.doi.org/10.1021/jo202644g] [PMID: 22390837]

[9] Barder, T.E.; Walker, S.D.; Martinelli, J.R.; Buchwald, S.L. Catalysts for Suzuki-Miyaura coupling processes: scope and studies of the effect of ligand structure. *J. Am. Chem. Soc.,* **2005**, *127*(13), 4685-4696. [http://dx.doi.org/10.1021/ja042491j] [PMID: 15796535]

[10] Forero-Cortés, P.A.; Haydl, A.M. The 25th Anniversary of the Buchwald–Hartwig Amination: Development, Applications, and Outlook. *Org. Process Res. Dev.,* **2019**, *23*(8), 1478-1483. [http://dx.doi.org/10.1021/acs.oprd.9b00161]

[11] Heravi, M.A.; Zadsirjan, V.; Hajiabbasi, P.; Hamidi, H. Advances in Kumada–Tamao–Corriu cross-

coupling reaction: an update. *Monatsh. Chem.,* **2019**, *150*(4), 535-591.
[http://dx.doi.org/10.1007/s00706-019-2364-6]

[12] Williams, K.A.; Boydston, A.J.; Bielawski, C.W. Main-chain organometallic polymers: synthetic strategies, applications, and perspectives. *Chem. Soc. Rev.,* **2007**, *36*(5), 729-744.
[http://dx.doi.org/10.1039/b601574n] [PMID: 17471398]

[13] Hartwig, J.F. Carbon-heteroatom bond formation catalysed by organometallic complexes. *Nature,* **2008**, *455*(7211), 314-322.
[http://dx.doi.org/10.1038/nature07369] [PMID: 18800130]

[14] Kerru, N.; Gummidi, L.; Maddila, S.; Gangu, K.K.; Jonnalagadda, S.B. A review on recent advances in nitrogen-containing molecules and their biological applications. *Molecules,* **2020**, *25*(8), 1909.
[http://dx.doi.org/10.3390/molecules25081909] [PMID: 32326131]

[15] Boit, T.B.; Bulger, A.S.; Dander, J.E.; Garg, N.K. Activation of C-O and C-N Bonds Using Non-Precious-Metal Catalysis. *ACS Catal.,* **2020**, *10*(20), 12109-12126.
[http://dx.doi.org/10.1021/acscatal.0c03334] [PMID: 33868770]

[16] Saraswat, P.; Jeyabalan, G.; Hassan, M.J.; Rahman, M.U.; Nyola, N.K. Review of synthesis and various biological activities of spiro heterocyclic compounds comprising oxindole and pyrrolidine moieties. *Synth. Commun.,* **2016**, *20*(20), 1643-1664.
[http://dx.doi.org/10.1080/00397911.2016.1211704]

[17] Campeau, L.C.; Hazari, N. Cross-Coupling and Related Reactions: Connecting Past Success to the Development of New Reactions for the Future. *Organometallics,* **2019**, *38*(1), 3-35.
[http://dx.doi.org/10.1021/acs.organomet.8b00720] [PMID: 31741548]

[18] Bariwal, J.; Van der Eycken, E. C-N bond forming cross-coupling reactions: an overview. *Chem. Soc. Rev.,* **2013**, *42*(24), 9283-9303.
[http://dx.doi.org/10.1039/c3cs60228a] [PMID: 24077333]

[19] Ahmad, U.S.; Pulidindi, I.S.; Li, C. Recent advances in C–CN and C–H bond activation of green nitrile (MeCN) for organo-complexation, cyanation and cyanomethylation. *New J. Chem.,* **2020**, *44*(40), 17177-17197.
[http://dx.doi.org/10.1039/D0NJ01996H]

[20] Surry, D.S.; Buchwald, S.L. Diamine Ligands in Copper-Catalyzed Reactions. *Chem. Sci. (Camb.),* **2010**, *1*(1), 13-31.
[http://dx.doi.org/10.1039/c0sc00107d] [PMID: 22384310]

[21] He, H.; Dong, S.; Chen, Y.; Yang, Y.; Le, Y.; Bao, W. Synthesis of 6H-isoindolo[2,1-α] indol-6-ones through a sequential copper-catalyzed C-N cupling and palladium catalysed C-H activation reaction. *Tetrahedron,* **2012**, *68*(14), 3112-3116.
[http://dx.doi.org/10.1016/j.tet.2012.01.087]

[22] Quach, T.D.; Batey, R.A. Ligand- and base-free copper(II)-catalyzed C-N bond formation: cross-coupling reactions of organoboron compounds with aliphatic amines and anilines. *Org. Lett.,* **2003**, *5*(23), 4397-4400.
[http://dx.doi.org/10.1021/ol035681s] [PMID: 14602009]

[23] Peng, J.; Ye, M.; Zong, C.; Hu, F.; Feng, L.; Wang, X.; Wang, Y.; Chen, C. Copper-catalyzed intramolecular C-N bond formation: a straightforward synthesis of benzimidazole derivatives in water. *J. Org. Chem.,* **2011**, *76*(2), 716-719.
[http://dx.doi.org/10.1021/jo1021426] [PMID: 21175149]

[24] Yuan, C.; Zhang, L.; Zhao, Y. Cu(II)-Catalyzed C-N Coupling of (Hetero)aryl Halides and *N*-Nucleophiles Promoted by *α*-Benzoin Oxime. *Molecules,* **2019**, *24*(22), 4177.
[http://dx.doi.org/10.3390/molecules24224177] [PMID: 31752140]

[25] Ambient-Light-Mediated Copper-Catalyzed CC and CN Bond Formation. *Angew. Chem. Int. Ed.,* **2013**, *52*, 2-5.

[26] Kosugi, M.; Kameyama, M.; Migita, T. Cheminform abstract: Palladium-catalyzed aromatic amination of aryl bromides with N,N-diethylaminotributyltin. *Chem. Lett.,* **1983**, *19*, 927-928. [http://dx.doi.org/10.1246/cl.1983.927]

[27] Hartwig, J.F.; Richards, S.; Barañano, D.; Paul, F. Influences on the Relative Rates for C−N Bond-Forming Reductive Elimination and β-Hydrogen Elimination of Amides. A Case Study on the Origins of Competing Reduction in the Palladium-Catalyzed Amination of Aryl Halides. *J. Am. Chem. Soc.,* **1996**, *118*(15), 3626-3633. [http://dx.doi.org/10.1021/ja954121o]

[28] Romero, M.; Harrak, Y.; Basset, J.; Ginet, L.; Constans, P.; Pujol, M.D. Preparation of N-arylpiperazines and other N-aryl compounds from aryl bromides as scaffolds of bioactive compounds. *Tetrahedron,* **2006**, *62*(62), 9010-9016. [http://dx.doi.org/10.1016/j.tet.2006.07.011]

[29] Mann, G.; Hartwig, J.F.; Driver, M.S.; Fernández-Rivas, C. Palladium-Catalyzed C−N(sp^2) Bond Formation: *N*-Arylation of Aromatic and Unsaturated Nitrogen and the Reductive Elimination Chemistry of Palladium Azolyl and Methyleneamido Complexes. *J. Am. Chem. Soc.,* **1998**, *120*(4), 827-828. [http://dx.doi.org/10.1021/ja973524g]

[30] Toma, G.; Fujita, K.; Yamaguch, R. Cobalt-Catalyzed C–N Bond-Forming Reaction between N-Aromatic 2-Chlorides and Secondary Amines. *Eur. J. Org. Chem.,* **2009**, *27*(27), 4586-4588. [http://dx.doi.org/10.1002/ejoc.200900597]

[31] Enguehard-Gueiffier, C.; Gueiffier, A. Recent progress in the pharmacology of imidazo[1,2-a]pyridines. *Mini Rev. Med. Chem.,* **2007**, *7*(9), 888-899. [http://dx.doi.org/10.2174/138955707781662645] [PMID: 17897079]

[32] Dymińska, L. Imidazopyridines as a source of biological activity and their pharmacological potentials-Infrared and Raman spectroscopic evidence of their content in pharmaceuticals and plant materials. *Bioorg. Med. Chem.,* **2015**, *23*(18), 6087-6099. [http://dx.doi.org/10.1016/j.bmc.2015.07.045] [PMID: 26314922]

[33] Rasheed, S.; Rao, D.N.; Das, P. Cobalt(II)-Catalyzed C−N Bond Formation: Synthesis of Aryl-Substituted Imidazo[1,2-*a*]pyridines. *Asian J. Org. Chem.,* **2016**, *5*(10), 1213-1218. [http://dx.doi.org/10.1002/ajoc.201600326]

[34] Affo, W.; Ohmiya, H.; Fujioka, T.; Ikeda, Y.; Nakamura, T.; Yorimitsu, H.; Oshima, K.; Imamura, Y.; Mizuta, T.; Miyoshi, K. Cobalt-catalyzed trimethylsilylmethylmagnesium-promoted radical alkenylation of alkyl halides: a complement to the Heck reaction. *J. Am. Chem. Soc.,* **2006**, *128*(24), 8068-8077. [http://dx.doi.org/10.1021/ja061417t] [PMID: 16771523]

[35] Lim, C.H.; Kudisch, M.; Liu, B.; Miyake, G.M. C-N Cross-Coupling *via* Photoexcitation of Nickel-Amine Complexes 2018. *J. Am. Chem. Soc.,* **2018**, 7667-7673.

[36] Lavoie, C.M.; MacQueen, P.M.; Rotta-Loria, N.L.; Sawatzky, R.S.; Borzenko, A.; Chisholm, A.J.; Hargreaves, B.K.V.; McDonald, R.; Ferguson, M.J.; Stradiotto, M. Challenging nickel-catalysed amine arylations enabled by tailored ancillary ligand design. *Nat. Commun.,* **2016**, *7*(1), 11073. [http://dx.doi.org/10.1038/ncomms11073] [PMID: 27004442]

[37] Harada, T.; Ueda, Y.; Iwai, T.; Sawamura, M. Nickel-catalyzed amination of aryl fluorides with primary amines. *Chem. Commun. (Camb.),* **2018**, *54*(14), 1718-1721. [http://dx.doi.org/10.1039/C7CC08181B] [PMID: 29250630]

[38] Shin-itsu, K.; Torraca, K.; Buchwald, S.L. Palladium-Catalyzed Intramolecular C−O Bond Formation, January 2002. *J. Am. Chem. Soc.,* **2002**, *123*, 12202-12206.

[39] Chen, G.; Chan, A.S.C.; Kwong, F.Y. Palladium-catalyzed C–O bond formation: direct synthesis of phenols and aryl/alkyl ethers from activated aryl halides. *Tetrahedron Lett.,* **2007**, *48*(3), 473-476.

[http://dx.doi.org/10.1016/j.tetlet.2006.11.036]

[40] Zhang, H.; Ruiz-Castillo, P.; Buchwald, S.L. Palladium-Catalyzed C-O Cross-Coupling of Primary Alcohols. *Org. Lett.,* **2018**, *20*(6), 1580-1583. [http://dx.doi.org/10.1021/acs.orglett.8b00325] [PMID: 29474078]

[41] Niu, J.; Zhou, H.; Li, Z.; Xu, J.; Hu, S. An efficient Ullmann-type C-O bond formation catalyzed by an air-stable copper(I)-bipyridyl complex. *J. Org. Chem.,* **2008**, *73*(19), 7814-7817. [http://dx.doi.org/10.1021/jo801002c] [PMID: 18771324]

[42] Cao, H.; Zhan, H.; Cen, J.; Lin, J.; Lin, Y.; Zhu, Q.; Fu, M.; Jiang, H. Copper-catalyzed C-O bond formation: an efficient one-pot highly regioselective synthesis of furans from (2-furyl)carbene complexes. *Org. Lett.,* **2013**, *15*(5), 1080-1083. [http://dx.doi.org/10.1021/ol400080e] [PMID: 23413974]

[43] Wang, Q.; Zheng, H.; Chai, W.; Chen, D.; Zeng, X.; Fu, R.; Yuan, R. Copper catalyzed C-O bond formation *via* oxidative cross-coupling reaction of aldehydes and ethers. *Org. Biomol. Chem.,* **2014**, *12*(34), 6549-6553. [http://dx.doi.org/10.1039/C4OB00862F] [PMID: 25033411]

[44] Sangeetha, S.; Sekar, G. Copper-catalyzed double C-S bond formation for the synthesis of 2-acyldihydrobenzo[*b*]thiophenes and 2-acylbenzo[*b*]thiophenes. *Chem. Commun. (Camb.),* **2020**, *56*(74), 10906-10909. [http://dx.doi.org/10.1039/D0CC04647G] [PMID: 32789379]

[45] Rostami, A.; Rostami, A.; Ghaderi, A.; Gholinejad, M.; Gheisarzadeh, S. Copper-Catalyzed C-S Bond Formation *via* the Cleavage of C-O Bonds in the Presence of S_8 as the Sulfur Source. *Synthesis,* **2017**, *49*(22), 5025-5038. [http://dx.doi.org/10.1055/s-0036-1588508]

[46] Lee, C.F. Copper-catalyzed cross-coupling reaction of thiols with aryl halides, Phosphorus, Sulfur, and Silicon and the Related Elements, 7: ISOCS-28. **2019**, *194*, 13-17.

[47] de Pedro Beato, E.; Priego, J.; Gironda-Martínez, A.; González, F.; Benavides, J.; Blas, J.; Martín-Ortega, M.D.; Toledo, M.A.; Ezquerra, J.; Torrado, A. Mild and Efficient Palladium-Mediated C-N Cross-Coupling Reaction between DNA-Conjugated Aryl Bromides and Aromatic Amines. *ACS Comb. Sci.,* **2019**, *21*(2), 69-74. [http://dx.doi.org/10.1021/acscombsci.8b00142] [PMID: 30615417]

[48] Kemker, I.; Schnepel, C.; Schröder, D.C.; Marion, A.; Sewald, N. Cyclization of RGD Peptides by Suzuki-Miyaura Cross-Coupling. *J. Med. Chem.,* **2019**, *62*(16), 7417-7430. [http://dx.doi.org/10.1021/acs.jmedchem.9b00360] [PMID: 31306009]

[49] Popovici-Muller, J.; Lemieux, R.M.; Artin, E.; Saunders, J.O.; Salituro, F.G.; Travins, J.; Cianchetta, G.; Cai, Z.; Zhou, D.; Cui, D.; Chen, P.; Straley, K.; Tobin, E.; Wang, F.; David, M.D.; Penard-Lacronique, V.; Quivoron, C.; Saada, V.; de Botton, S.; Gross, S.; Dang, L.; Yang, H.; Utley, L.; Chen, Y.; Kim, H.; Jin, S.; Gu, Z.; Yao, G.; Luo, Z.; Lv, X.; Fang, C.; Yan, L.; Olaharski, A.; Silverman, L.; Biller, S.; Su, S.M.; Yen, K. Discovery of AG-120 (Ivosidenib): A first-in-class mutant IDH1 inhibitor for the treatment of IDH1 mutant cancers. *ACS Med. Chem. Lett.,* **2018**, *9*(4), 300-305. [http://dx.doi.org/10.1021/acsmedchemlett.7b00421] [PMID: 29670690]

[50] Wang, M.; Lu, J.; Wang, M.; Yang, C.Y.; Wang, S. Discovery of SHP2-D26 as a First, Potent, and Effective PROTAC Degrader of SHP2 Protein. *J. Med. Chem.,* **2020**, *63*(14), 7510-7528. [http://dx.doi.org/10.1021/acs.jmedchem.0c00471] [PMID: 32437146]

CHAPTER 5

Oxidation Reactions Using Nanomaterials as Heterogeneous Catalyst

Priyanka Gogoi[1], **Diganta Sarma**[1] and **Kalyanjyoti Deori**[1,*]

[1] *Department of Chemistry, Dibrugarh University, Dibrugarh, Assam, India*

Abstract: Oxidation reactions play a pivotal role in chemical synthesis. Its wide range of industrial applications and processing, however, make use of corrosive chemicals and harsh reaction conditions, which create threats to the ecosystem, environment, and public health. Also, the traditional homogeneous catalysts used for this purpose get easily deactivated. Heterogeneous catalysts with various supports have been developed to achieve more environmentally benign conditions as well as to defeat other disadvantages of homogeneous catalysts. Nanomaterials, due to their large surface available to reactants and ease of separation, show promising applications in heterogeneous catalysis and lead towards a greener way of synthesis. The current chapter enlists the applications of nanocatalysts in the oxidation reactions of various organic molecules.

Keywords: Composite/hybrid nanoparticles, Heterogeneous catalyst, Metal-oxide, Nanomaterials, Oxidation.

INTRODUCTION

Oxidation reactions, being the heart of chemical synthesis, cover a large part of the area in the field of production of many industrially important compounds [1]. The oxidation process carries around 30% of the total production in chemical industries. For example, selective oxidation of alcohols and olefins produces epoxides, aldehydes, ketones, and carboxylic acids, which have significant applications and mechanistic interests in modern organic synthesis [2]. Acetophenone and 1-phenylethanol, which are one of the oxidation products of ethylbenzene, are used in the synthesis of a wide variety of drugs [3, 4]. Earlier, oxidation of organic compounds was done by adding stoichiometric amounts of strong oxidants such as permanganates, selenium oxide (SeO_2), chromium (IV) reagents, ruthenium (VIII) oxide, nitric acid, oxygen, and hydrogen peroxide, *etc.*, however, the traditional homogeneous catalysts are not very useful as their

* **Corresponding author Kalyanjyoti Deori**: Department of Chemistry, Dibrugarh University, Assam, India; Email: kalchemdu@gmail.com, kalchemdu@dibru.ac.in

recirculation is a fussy and time-consuming process [5]. Moreover, most of these chemicals are corrosive, toxic, unstable to the environment, and produce many undesired side products which result in less atom economy and create heavy metal waste [6].

Heterogeneous catalysis, on the other hand, in which reactants are in a different phase from the catalyst, solves most of the drawbacks of homogeneous catalysts. In most cases, the majority of heterogeneous catalysts are solid and reactants are either gas or liquid [7]. The new era of chemistry is more inclined towards finding new innovative techniques that concern the environment, safety, and human health [8 - 12]. Therefore, chemical synthesis and reactions are being explored based on eco-friendly concepts without affecting the quantity of products and quality of the reaction. Supported metal catalysts are one of the most widely used catalysts in various industrially important reactions which include oxidation, hydrogenation, dehydrogenation, Fischer-Tropsch, and ammonia synthesis [13]. Over the last decade, a considerable effort has been devoted to the field of nanoscience and nanotechnology due to its many specific characteristics especially high surface to volume ratio [14 - 18]. Nanomaterials, due to their special properties and novel characteristics in comparison to single-molecule and bulk materials, have gained tremendous attention since the last decade, especially in the field of catalysis [1]. They are highly selective, reactive, and stable; hence, extensively used in many oxidation reactions in recent times. The purpose of the chapter is to provide a brief overview of recent developments in oxidation reactions using various types of nanomaterials as heterogeneous catalysts.

Fig. (**1**) depicts the generalized representation of nanomaterials as a heterogeneous catalyst for different oxidation reactions.

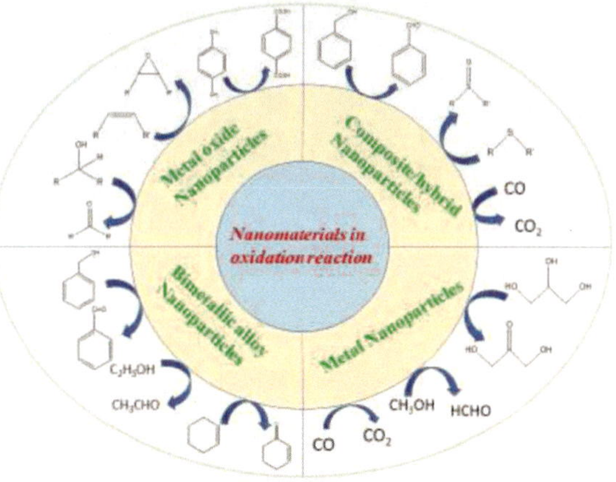

Fig. (1). Various types of nanomaterials as heterogeneous catalysts for oxidation reactions.

METAL OXIDE NANOPARTICLES IN THE OXIDATION REACTION

The application of metal oxide nanoparticles (NPs) in the field of catalysis has become a substantial research area during the last decades. Transition metal oxide nanoparticles have been extensively used as a catalyst for the oxidation of alcohols, alkenes, sulfides, and olefins towards the synthesis of various organic compounds. Traditionally, permanganate and dichromate were used as oxidizing agents, but they are costly, toxic, and produce a large quantity of waste [19 - 21]. A considerable effort has been devoted to the field of nanoscience and nanotechnology to reduce the disadvantages of traditional methods making the use of metal and metal oxide nanoparticles [22, 23]. Molybdenum oxide (MoO_x), amongst the transition metal oxides, has drawn a significant research interest. MoO_2 NPs can be used as an efficient catalyst in the epoxidation of olefins [24]. Also, it has been utilized in cracking and reforming hydrocarbon compounds as a catalyst [25, 26]. In this regard, to enhance the overall efficiency of the catalyst, the oxide nanoparticles were structurally or electronically modified by doping them with transition metals. As shown in Fig. (**2a**), hydrothermally synthesized Fe doped MoO_2 for epoxidation of olefins was developed [27]. Generally, most of the molybdenum (VI) complexes having organic ligands act as homogeneous catalysts [28, 29], besides a few other reports [30]. Partial oxidation of liquid hydrocarbons e.g., n-dodecane at 850 °C and 1 atm by MoO_2 NPs (Fig. **2b, c**), has been developed using molybdenum trioxide as starting material [31]. Another interesting piece of work was carried out by Shokouhimehr and co-workers. They developed magnetically separable MoO_2 NPs which show excellent performance for epoxidation reaction in carbon tetrachloride (CCl_4) [32].

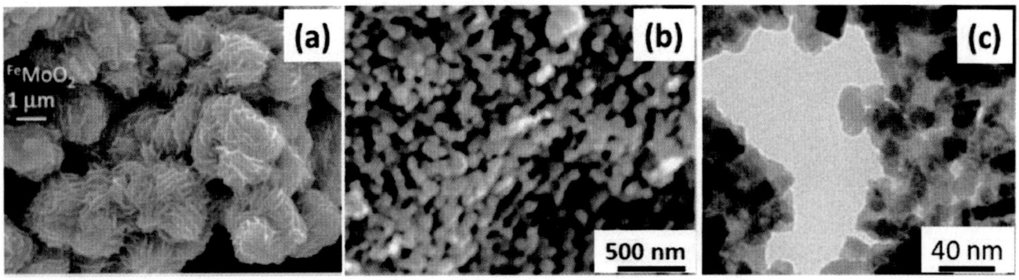

Fig. (2). SEM image of **(a)** hydrothermally synthesized Fe doped MoO_2. Reprinted with permission from ref [27]. Copyright 2016 Springer. **(b)** SEM, **(c)** TEM images of MoO_2 NPs used in partial oxidation of liquid hydrocarbons Reprinted with permission from ref [31]. Copyright 2010 American Institute of Aeronautics and Astronautics.

TiO_2 NPs, a well-known photocatalyst has also attracted worldwide interest due to their potential use in solar cell, photodegradation of inorganic and organic wastes *etc.* [33 - 35]. Selective photocatalytic oxidation of benzyl alcohol and its derivatives to their carbonyl products with 99% selectivity by TiO_2 NPs under visible light irradiation were reported [36 - 38]. In some cases, the oxidation of secondary alcohols to corresponding ketones was carried out in polyethylene glycol (PEG) as green solvent and H_2O_2 as oxidizing agent [39]. Fig. (**3**) shows the mechanism for alcohol oxidation over TiO_2 nanoparticles.

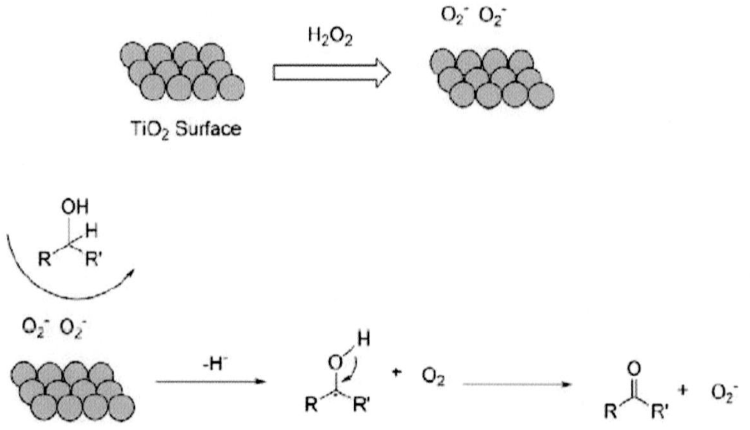

Fig. (3). Plausible reaction mechanism for alcohol oxidation over TiO_2 nanoparticles [39].

On applying H_2O_2, superoxide species O_2^- generates on the surface of TiO_2 NPs. These reactive superoxide species generates free radical and finally lead to the desired product *i.e.*, to corresponding ketones. On the other hand, the single crystalline rutile phase of TiO_2 nanorods was found to be efficient in the selective oxidation of benzyl alcohol into benzaldehyde [40]. When compared with the polyhedral type of nanoparticles, the catalytic activity of TiO_2 nanorods showed better results due to the high surface-to-volume ratio of later. It was found that the large surface area and porosity of metal oxide nanoparticles are the major criteria for determining the number of active sites. Therefore, mesoporous nano-materials with large accessible pore surfaces and three-dimensional networks, have gained significant attention in recent time [41, 42]. Surfactant-assisted synthesis of mesoporous binary nanoparticles was carried out by interconnecting CdS quantum dots (~5-6 nm) with the anatase phase of TiO_2. The mesoporous structure of as-synthesized nanoparticles exhibits a large internal surface area due to which, CdS-TiO_2 NPs shows an excellent photocatalytic activity for oxidation of para-substituted aromatic alcohols to corresponding carbonyl compounds using O_2 as oxidant [43, 44]. Similarly, photo-oxidation of various alcohols under UV irradiation shows excellent catalytic performance in the presence of single-crystal TiO_2 having 2D and 3D ordered mesoporous structures [45]. As shown in Fig. (**4**), TiO_2 semiconductor produces an electron (e^-) and a hole (h^+) after absorbing UV radiation, which is responsible for the activation of alcohols to alcohol radical cation and also O_2 to form $\cdot O_2^-$ which finally combines with alcohol radical cation to give the final desired product. Selective oxidation of hydrocarbons under mild and free solvent-free conditions can also be achieved by TiO_2 nanomaterials with different surface morphologies and structures e.g., TiO_2 nanotubes (TiO_2 NTs), TiO_2 nanorods (TiO_2 NRs), TiO_2 nanofibers (TiO_2 NFs), TiO_2 nanoparticles (TiO_2 NPs).

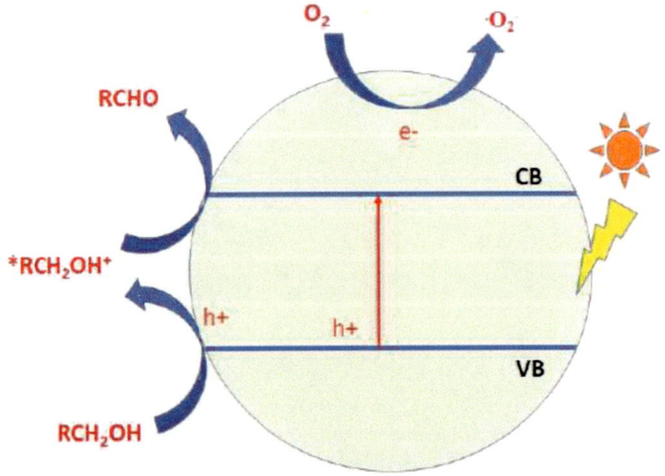

Fig. (4). A plausible mechanism for photocatalytic selective oxidation [45].

Vanadium oxide nanoparticles can be used as a catalyst in the oxidation process of cyclohexane. Scientists have put a tremendous effort into the selective oxidation of cyclohexane into cyclohexanone, which is an intermediate of various industrially important polymer products [46]. To improve the catalytic performance, vanadium oxide NPs are supported on the surface of other metal oxides. In this regard, MgO supports are used for vanadium oxide NPs by grafting and impregnation method which later on applied for the liquid phase oxidation of cyclohexane [47, 48]. Vanadium oxide NPs mixed with WO_3 was also used in cyclohexane oxidation reaction [49].

Moreover, tungsten (VI) oxide NPs, uniformly dispersed throughout the surface of mesoporous support MCM-48 (Mobil Composition of Matter) show outstanding catalytic results towards oxidation of olefins, sulfides, and cyclic ketones using hydrogen peroxide and peracetic acid as oxidants. It is noteworthy to mention here that the immobilization of nanoparticles in the high surface area supports boosting the catalytic activity and therefore, mesoporous materials have been extensively used as support [50]. In addition to all the above-mentioned metal oxide nanoparticles, research has been tremendously focused on several other oxides like manganese oxide, iron oxide, copper oxide, cerium oxide, *etc*., for various catalytic oxidation of industrially important organic moieties. For example, Mn_3O_4 NPs show exceptional activity towards the epoxidation of aromatic olefins, aerobic oxidation of hydrocarbons, *etc.* [50, 51].

Iron oxide NPs, due to their unique properties, has drawn considerable attention in recent years. Application of free iron oxide NPs is generally rare as they are prone to aggregation. Hence, to prevent aggregation, these are shielded by capping agents in most cases. Mesoporous silica, a praiseworthy example that can be used as a support system for the stabilization of iron oxide NPs. SBA-15: Santa Barbara Amorphous-15, amongst one of those, has high surface area and good thermal stability, which can be synthesized by the hydrothermal method. Supports other than silica are also sometimes used [52, 53]. The catalyst system Fe_2O_3/SBA-15 performs an excellent catalytic activity towards the oxidation of alcohols and sulfides (Fig. **5**).

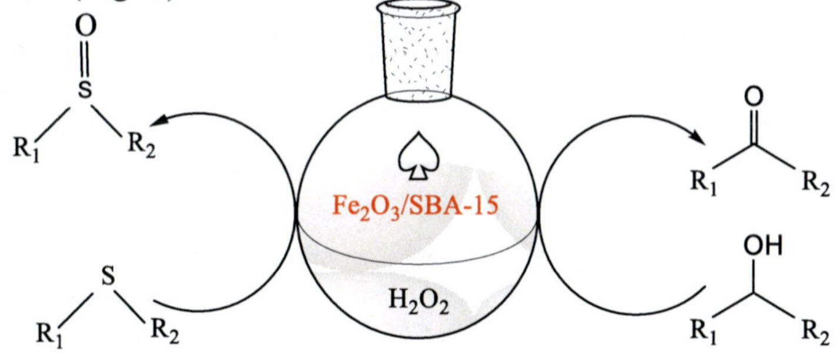

Fig. (5). Schematic reaction for the oxidation of sulfides and alcohols by Fe_2O_3/SBA-15 [52].

Supported CuO NPs can also be used for similar oxidation reactions under mild reaction conditions for the synthesis of various carbonyl compounds [54, 55]. Cerium dioxide, CeO_2, one of the most abundant rare earth oxides, has been successfully employed as an environmentally benign catalyst for aerobic oxidation of p-xylene to terephthalic acid by Deori and co-workers [56]. The same author synthesized three-dimensional self-assembled CeO_2 nanocubes of size 44-46 nm (Fig. **6**), which shows a breakthrough in catalytic performance during oxidation of p-xylene to terephthalic acid in an environmentally benign reaction condition (at 70 °C in water) with 100% yield of terephthalic acid [57]. Due to its great redox reversibility between Ce^{4+} and Ce^{3+} and enhanced oxygen storage capacity (OSC). Cerium dioxide or ceria is one of the most promising heterogeneous catalysts in lots of extensive applications such as CO oxidation, as electrocatalyst for methanol oxidation and oxidation of alcohol, toluene, *etc.* [56, 57].

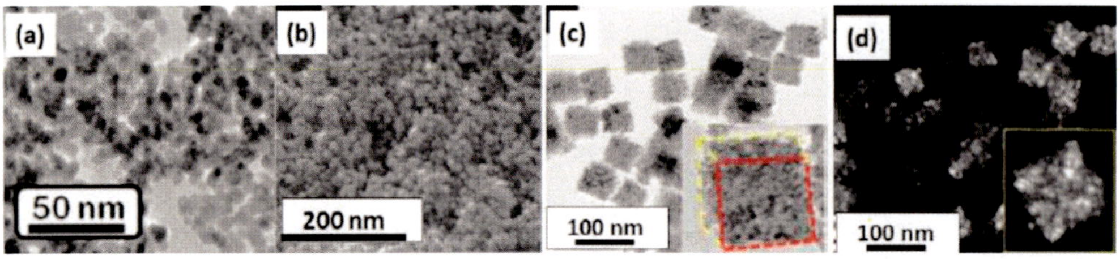

Fig. (6). **(a)** Low magnification TEM image **(b)** FESEM image of CeO_2 NCs of particle size 15nm. Reprinted with permission from ref [56]. Copyright 2013 Royal Society of Chemistry. **(c)**, **(d)** 44-46 nm size of CeO_2 nanocubes were used as a catalyst for the p-xylene oxidation reaction. Reprinted with permission from ref [57]. Copyright 2014 American Chemical Society.

BIMETALLIC ALLOY NANOPARTICLES FOR OXIDATION REACTION

Pd/Cu bimetallic nanoparticles (BNPs) supported on graphene have been prepared *via* the facile synthesis method, which can be utilized as a novel electrocatalyst for ethanol oxidation in the basic medium [58]. Electrooxidation of alcohols by this as-synthesized bimetallic nanomaterial has been investigated with the help of cyclic voltammograms in a basic medium. The experimental result provides a good electrocatalytic activity of the material. Modified Hummers' method followed by ultrasonication has been employed to synthesize the support material *i.e.*, graphene oxide then further mixed with copper source as shown in Fig. (**7**). Similar work has been reported by Wang *et al.* in which graphene supported Pd/Co bimetallic nanoparticles shows excellent catalytic activity in electrocatalytic oxidation reaction [59]. The electrooxidation of methanol and ethanol in an acidic medium was explored on carbon-ceramic supported Pt-Ni bimetallic nanoparticles as an electrocatalyst [60]. Mukherjee and co-workers carried out a unique piece of work for anodic oxidation of ethanol at room temperature using Pt-Cu alloy nanoparticles [61]. In another work, selective oxidation of ethanol to acetaldehyde has been carried out by Au-Ir bimetallic catalyst prepared by co-impregnation of Au chlorides and Ir on SiO_2 followed by inserting bimetallic nanoparticles with varying concentrations [62]. Bimetallic nanoparticles are multifunctional nanomaterials successfully applied in different fields especially in oxidation reactions. One of the basic reasons bimetallic nanomaterials get better importance as compared to monometallic nanomaterials is the synergic effect that exists between two atoms [63]. Wang and his group successfully synthesized Au@Pd core-shell BNPs which oxidize benzyl alcohols selectively [64]. To increase the effectiveness of bimetallic nanoparticles, bimetallic nanocomposites come into play which can also be used as sensors to detect early diagnosis of disease. Thus, it emerges as a very promising field that is being explored further by researchers.

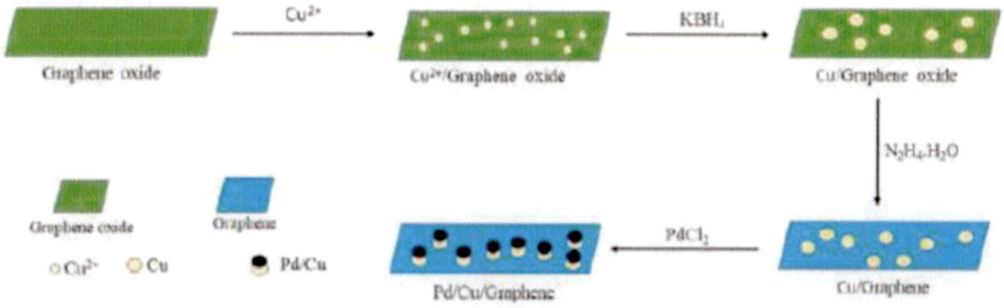

Fig. (7). Pictorial representation of synthetic route for Pd/Cu bimetallic nanoparticles at graphene [58].

METAL NANOPARTICLES FOR OXIDATION REACTION

Colloidal Au NPs are one of the earliest nanomaterials to be studied and applied in the field of technology and development. Au catalyzed CO oxidation reactions have been explored extensively by researchers [65]. Though isolated Au NPs are found capable of oxidizing CO by theoretical studies, the most widely accepted mechanism for oxidation occurs either on support or metal-support interface [66]. Many researchers have worked extensively on aerobic oxidation of alcohols over supported gold nanoparticles. The acidic and basic nature of support plays a key role in the activity and selectivity of alcohol oxidation. Au NPs supported on hydrotalcite (HT) has successfully been synthesized by Kaneda and co-workers for aerobic oxidation of alcohols [67]. The result shows catalyst with MgO and HT gives the highest yield of product. Another group, Wang *et al.* designed a catalyst system of Au NPs supported on Ni-Al layered double hydroxides (LDH) [68, 69]. Besides monohydric alcohols, supported Au NPs defend as a very good catalyst for oxidation of polyhydric alcohols. Recently, Pan and co-workers reported ZnO supported Au NPs for base-free oxidation of glycerol to form dihydroxyacetone [70]. Rossi's group reported functional magnetic nanomaterials (MNP) supported Au NPs to catalyze aerobic oxidation of benzyl alcohols [71]. MNP is one of the excellent classes of materials that has tremendous potential for a variety of transformation reactions including oxidation.

COMPOSITE/HYBRID NANOPARTICLES FOR OXIDATION REACTIONS

Hybrid nanocrystals (HNCs) are one of the most promising multi-component systems due to their unexpected enhanced chemical and physical responses in comparison to single materials. In addition to this, new properties and behaviors are expected to observe because of the coupling between two materials. TiO_2 NPs embedded into different carbon allotropes such as Graphene (GR), carbon nanotubes (CNT), and fullerene (C_{60}) are found to be very active photocatalysts for selective oxidation of benzyl alcohol to benzaldehyde under visible light irradiation [72]. In another report, RuO_2 NPs were fabricated to TiO_2 nanobelt surface to form RuO_2/TiO_2 nano heterostructure that acts as an effective photocatalyst for oxidation of benzyl alcohol as compared to RuO_2 NPs and TiO_2 nanobelt alone [73]. SBA-15, one of the very good mesoporous silica supports, as discussed in an earlier section, fabricated with tungsten oxide NPs to form WO_x-SBA-15 nanocomposites showed high activity in selective oxidation of sulfides at room temperature [74]. Similarly, the MnO_2/graphene oxide (GO) nanocomposites are used in alcohol oxidation to get 100% corresponding aldehyde product selectively under mild reaction conditions [75]. In olefin epoxidation reaction CuO colloidal nanocrystal clusters (CNCs) were found to be active. As shown in Fig. (**8**), the incorporation of mesoporous SiO_2 as a protective core-shell to improve the stability which results in the formation of CNCs@meso-SiO_2 nanocomposites enhances the catalytic activity [76].

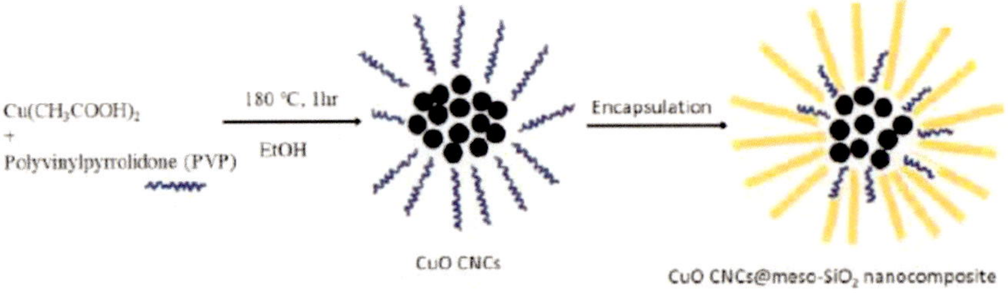

Fig. (8). Schematic representation of the synthesis pathways to design CuO CNCs@meso-SiO_2 nanocomposites for olefin epoxidation reaction [76].

Immobilizing iron aminobis (phenol) ligand complex over silica-coated magnetic nanoparticles (MNPs) to form Fe_3O_4@SiO_2-APTES-FeL magnetic nanocomposite performs very good activity during oxidation of alcohols and sulfides (see Fig. **9 a,b** for FETEM images of as-synthesized particles) [77].

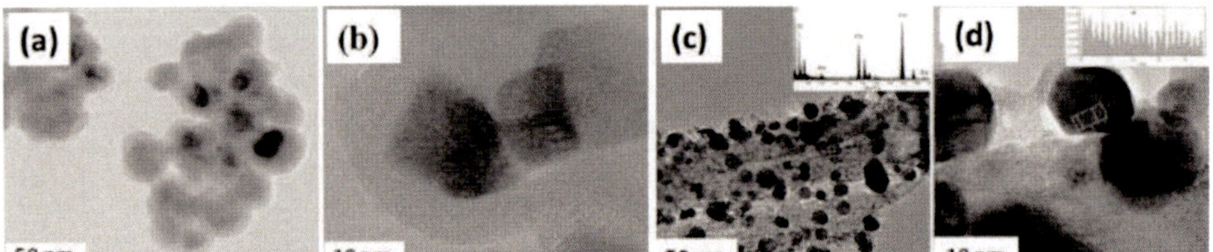

Fig. (9). **(a)** FETEM image **(b)** high-resolution image of Fe_3O_4@SiO_2-APTES-FeL. Reprinted with permission from ref [77]. Copyright 2018 Wiley-VCH Verlag GmbH & Co. KGaA, Weinheim **(c)** TEM and **(d)** HRTEM images of gold supported on CeO_2 nanocrystals by chemical precipitation method, effectively used for CO oxidation. Reprinted with permission from ref [78]. Copyright 2010 American Chemical Society.

CeO_2 nanostructure-supported gold NPs have been produced by microwave synthesis process which later found to be a very effective catalyst in CO oxidation [78] (Fig. **9c, d**). The high catalytic activity of metal-CeO_2 HNCs towards CO oxidation is mainly because of pure electronic interaction involving metal atoms to CeO_2 nanocrystals.

CONCLUDING REMARKS

Nanocatalysis, a very fascinating field of modern research and development, can be successfully employed for a wide variety of oxidation reactions. In recent years, increasing awareness towards environmentally benign and sustainable pathway have driven catalytic research with nanomaterials as a heterogeneous catalyst towards various oxidation reactions. In this chapter, we have discussed various oxidation reactions utilizing a variety of nanomaterials that can provide information to chemists about the different synthetic pathways of the materials and their effective catalytic activities. As discussed in this chapter; surface area, porosity, exposed crystal plane, and morphology are the most important parameters of the as-synthesized nanomaterials which profoundly affect catalytic performance towards oxidation reaction. This morphological dependence becomes more significant with decreasing size along specific dimensions as distinct crystallographic facets exhibit a unique coordination environment and hence distinct catalytic properties. This implies that the surface reactions may show distinct activity and selectivity for different facets and that the catalytic performance may be fine-tuned through selective exposure of the desired facets by controlling the shape of the catalyst nanoparticle. Although numerous techniques for oxidation reactions have been improved in recent research, however, there are still several challenges in the development of heterogeneous nanocatalysts and design for more sustainable catalytic schemes for oxidation reactions that are compatible at the industrial level.

CONSENT FOR PUBLICATION

Not applicable.

CONFLICT OF INTEREST

The author declares no conflict of interest, financial or otherwise.

ACKNOWLEDGEMENTS

Declared none.

REFERENCES

[1] Hemalatha, K.; Madhumitha, G.; Kajbafvala, A.; Anupama, N.; Sompalle, R.; Mohana, S. Function of nanocatalyst in chemistry of organic compounds revolution: An overview. *J. Nanomater.*, **2013**, *2013*, 1-23.
[http://dx.doi.org/10.1155/2013/341015]

[2] Mori, K.; Hara, T.; Mizugaki, T.; Ebitani, K.; Kaneda, K. Hydroxyapatite-supported palladium nanoclusters: a highly active heterogeneous catalyst for selective oxidation of alcohols by use of molecular oxygen. *J. Am. Chem. Soc.*, **2004**, *126*(34), 10657-10666.
[http://dx.doi.org/10.1021/ja0488683] [PMID: 15327324]

[3] Hasirci, V.N. PVNO-DVB hydrogels: Synthesis and characterization. *J. Appl. Polym. Sci.*, **1982**, *27*(1), 33-41.
[http://dx.doi.org/10.1002/app.1982.070270104]

[4] Blickenstaff, R.T.; Hanson, W.R.; Reddy, S.; Witt, R. Potential radioprotective agents--VI. Chalcones, benzophenones, acid hydrazides, nitro amines and chloro compounds. Radioprotection of murine intestinal stem cells. *Bioorg. Med. Chem.*, **1995**, *3*(7), 917-922.
[http://dx.doi.org/10.1016/0968-0896(95)00076-S] [PMID: 7582968]

[5] Zhang, Q.; Yang, X.; Guan, J. Applications of magnetic nanomaterials in heterogeneous catalysis. *ACS Appl. Nano Mater.*, **2019**, *2*(8), 4681-4697.
[http://dx.doi.org/10.1021/acsanm.9b00976]

[6] Gallezot, P. Selective oxidation with air on metal catalysts. *Catal. Today*, **1997**, *37*(4), 405-418.
[http://dx.doi.org/10.1016/S0920-5861(97)00024-2]

[7] Hamilton, D. C.; Tooze, R. Homogeneous catalysis-advantages and problems. *Catalyst separation, recovery and recycling*, **2006**, 1-6.

[8] Roopan, S.M.; Khan, F.R.N. ZnO nanoparticles in the synthesis of AB ring core of camptothecin. *Chem. Pap.*, **2010**, *64*(6), 812-817.
[http://dx.doi.org/10.2478/s11696-010-0058-y]

[9] Madhumitha, G.; Roopan, S.M. Devastated crops: multifunctional efficacy for the production of nanoparticles. *J. Nanomater.*, **2013**, *2013*, 1-12.
[http://dx.doi.org/10.1155/2013/951858]

[10] Newkome, G.; Fishel, D. Preparation of hydrazones: acetophenone hydrazone. *Org. Synth.*, **1988**, *50*, 102-102.

[11] Corma, A.; García, H.; Llabrés i Xamena, F.X. Engineering metal organic frameworks for heterogeneous catalysis. *Chem. Rev.*, **2010**, *110*(8), 4606-4655.
[http://dx.doi.org/10.1021/cr9003924] [PMID: 20359232]

[12] Kajbafvala, A.; Zanganeh, S.; Kajbafvala, E.; Zargar, H.R.; Bayati, M.R.; Sadrnezhaad, S.K. Microwave-assisted synthesis of narcis-like zinc oxide nanostructures. *J. Alloys Compd.*, **2010**, *497*(1-2), 325-329.
[http://dx.doi.org/10.1016/j.jallcom.2010.03.057]

[13] Kumar, A.; Kumar, V.P.; Kumar, B.P.; Vishwanathan, K.V.R.; Chary, K.V.R. Vapor phase oxidation of benzyl alcohol over gold nanoparticles supported on mesoporous TiO_2. *Catal. Lett.*, **2014**, *144*(8), 1450-1459.
[http://dx.doi.org/10.1007/s10562-014-1285-6]

[14] Roychoudhury, S.; Lyubovsky, M.; Walsh, D.; Chu, D.; Kallio, E. Design and development of a diesel and JP-8 logistic fuel processor. *J. Power Sources*, **2006**, *160*(1), 510-513.
[http://dx.doi.org/10.1016/j.jpowsour.2005.12.101]

[15] Wang, L.; Yamauchi, Y. Block copolymer mediated synthesis of dendritic platinum nanoparticles. *J. Am. Chem. Soc.*, **2009**, *131*(26), 9152-9153.
[http://dx.doi.org/10.1021/ja902485x] [PMID: 19534469]

[16] Wang, L.; Yamauchi, Y. Metallic nanocages: synthesis of bimetallic Pt-Pd hollow nanoparticles with dendritic shells by selective chemical *etc*hing. *J. Am. Chem. Soc.*, **2013**, *135*(45), 16762-16765. [http://dx.doi.org/10.1021/ja407773x] [PMID: 24171407]

[17] Wang, L.; Yamauchi, Y. Autoprogrammed synthesis of triple-layered Au@Pd@Pt core-shell nanoparticles consisting of a Au@Pd bimetallic core and nanoporous Pt shell. *J. Am. Chem. Soc.*, **2010**, *132*(39), 13636-13638. [http://dx.doi.org/10.1021/ja105640p] [PMID: 20831169]

[18] Tsolekile, N.; Parani, S.; Matoetoe, M.C.; Songca, S.P.; Oluwafemi, O.S. Evolution of ternary I-III-VI QDs: Synthesis, characterization and application. Nano-Struct. *Nano-Objects*, **2017**, *12*, 46-56. [http://dx.doi.org/10.1016/j.nanoso.2017.08.012]

[19] Badri, R.; Haddad, A.A.; Adlu, M.; Mohammadi, M.K. Facile, mild and selective silica sulfuric acid catalyzed oxidation of benzyl alcohols to benzaldehyde derivatives by potassium peroxosulfate. *Bull. Chem. Soc. Ethiop.*, **2013**, *27*, 131-136.

[20] Mallat, T.; Baiker, A. Oxidation of alcohols with molecular oxygen on solid catalysts. *Chem. Rev.*, **2004**, *104*(6), 3037-3058. [http://dx.doi.org/10.1021/cr0200116] [PMID: 15186187]

[21] Matsumoto, T.; Ueno, M.; Wang, N.; Kobayashi, S. Recent advances in immobilized metal catalysts for environmentally benign oxidation of alcohols. *Chem. Asian J.*, **2008**, *3*(2), 196-214. [http://dx.doi.org/10.1002/asia.200700359] [PMID: 18232022]

[22] Esfahani, H.A.; Wang, L.; Nemoto, Y.; Yamauchi, Y. Synthesis of Bimetallic Au@Pt Nanoparticles with Au Core and Nanostructured Pt Shell toward Highly Active Electrocatalysts. *Chem. Mater.*, **2010**, *22*(23), 6310-6318. [http://dx.doi.org/10.1021/cm102074w]

[23] Arsalani, N.; Akbari, A.; Amini, M.; Jabbari, E.; Gautam, S.; Chae, K.H. POSS-Based covalent networks: supporting and stabilizing Pd for Heck reaction in aqueous media. *Catal. Lett.*, **2017**, *147*(4), 1086-1094. [http://dx.doi.org/10.1007/s10562-016-1954-8]

[24] Bento, A.; Sanches, A.; Medina, E.; Nunes, C.D.; Vaz, P.D. MoO_2 nanoparticles as highly efficient olefin epoxidation catalysts. *Appl. Catal. A Gen.*, **2015**, *504*, 399-407. [http://dx.doi.org/10.1016/j.apcata.2015.03.024]

[25] Song, J.H.; Chen, P.; Kim, S.H.; Somorjai, G.; Gartside, R.J.; Dautzenberg, F.M. Catalytic cracking of *n*-hexane over MoO_2. *J. Mol. Catal. Chem.*, **2002**, *184*(1-2), 197-202. [http://dx.doi.org/10.1016/S1381-1169(01)00522-2]

[26] Katrib, A.; Mey, D.; Maire, G. Molybdenum and tungsten dioxides, XO_2 (X=Mo,W), as reforming catalysts for hydrocarbon compounds. *Catal. Today*, **2001**, *65*(2-4), 179-183. [http://dx.doi.org/10.1016/S0920-5861(00)00580-0]

[27] Bento, A.; Sanches, A.; Vaz, P.D.; Nunes, C.D. Catalytic Application of Fe-doped MoO_2 Tremella-Like Nanosheets. *Top. Catal.*, **2016**, *59*(13-14), 1123-1131. [http://dx.doi.org/10.1007/s11244-016-0631-x]

[28] Bruno, S.M.; Balula, S.S.; Valente, A.A.; Paz, F.A.A.; Pillinger, M.; Sousa, C.; Klinowski, J.; Freire, C.; Ribeiro-Claro, P.; Gonçalves, I.S. Synthesis and catalytic properties in olefin epoxidation of dioxomolybdenum (VI) complexes bearing a bidentate or tetradentate salen-type ligand. *J. Mol. Catal. Chem.*, **2007**, *270*(1-2), 185-194. [http://dx.doi.org/10.1016/j.molcata.2007.01.050]

[29] Freund, C.; Herrmann, W.; Kühn, F. E. Organorhenium and organomolybdenum oxides as heterogenised catalysts. *organometallic oxidation catalysis,* **2006**, *22*, 39-77.

[30] Castro, A.; Alonso, J.C.; Valente, A.A.; Neves, P.; Brandão, P.; Felix, V.; Ferreira, P. Nanostructured Dioxomolybdenum (VI) Catalyst for the Liquid□Phase Epoxidation of Olefins. *Eur. J. Inorg. Chem.*,

2010, *2010*(9), 1405-1412.
[http://dx.doi.org/10.1002/ejic.200901093]

[31] Flores, O.M.; Turba, T.; Ellefson, C.; Wang, K.; Breit, J.; Ahn, J.; Norton, M.G.; Ha, S. Catalytic combustion of VOCs over a series of manganese oxide catalysts. *Appl. Catal. B,* **2010**, *98*, 186-192.

[32] Shokouhimehr, M.; Piao, Y.; Kim, J.; Jang, Y.; Hyeon, T. A magnetically recyclable nanocomposite catalyst for olefin epoxidation. *Angew. Chem. Int. Ed.,* **2007**, *46*(37), 7039-7043.
[http://dx.doi.org/10.1002/anie.200702386] [PMID: 17685372]

[33] Oftadeh, M.; Aghtar, A.; Esfahani, M.N.; Niasari, M.S.; Mir, N. Fabrication of highly efficient dye-sensitized solar cell and CO_2 reduction photocatalyst using TiO_2 nanoparticles prepared by spin coating-assisted sol–gel method. *J. Iran. Chem. Soc.,* **2012**, *9*(2), 143-149.
[http://dx.doi.org/10.1007/s13738-011-0017-8]

[34] Rajamanickam, N.; Jayakumar, K.; Ramachandran, K. Influence of Mn ion on flower shaped perovskite $BaTiO_3$ nanostructures-based dye-sensitized solar cell. Nano-Struct. *Nano-Objects,* **2017**, *9*, 19-25.
[http://dx.doi.org/10.1016/j.nanoso.2016.12.003]

[35] Challagulla, S.; Nagarjuna, R.; Ganesan, R.; Roy, S. TiO_2 synthesized by various routes and its role on environmental remediation and alternate energy production. Nano-Struct. *Nano-Objects,* **2017**, *12*, 147-156.
[http://dx.doi.org/10.1016/j.nanoso.2017.10.002]

[36] Higashimoto, S.; Kitao, N.; Yoshida, N.; Sakura, T.; Azuma, M.; Ohue, H.; Sakata, Y. Selective photocatalytic oxidation of benzyl alcohol and its derivatives into corresponding aldehydes by molecular oxygen on titanium dioxide under visible light irradiation. *J. Catal.,* **2009**, *266*(2), 279-285.
[http://dx.doi.org/10.1016/j.jcat.2009.06.018]

[37] Higashimoto, S.; Suetsugu, N.; Azuma, M.; Ohue, H.; Sakata, Y. Efficient and selective oxidation of benzylic alcohol by O_2 into corresponding aldehydes on a TiO_2 photocatalyst under visible light irradiation: Effect of phenyl-ring substitution on the photocatalytic activity. *J. Catal.,* **2010**, *274*(1), 76-83.
[http://dx.doi.org/10.1016/j.jcat.2010.06.006]

[38] Higashimoto, S.; Okada, K.; Morisugi, T.; Azuma, M.; Ohue, H.; Kim, T.H.; Matsuoka, M.; Anpo, M. Effect of surface treatment on the selective photocatalytic oxidation of benzyl alcohol into benzaldehyde by O_2 on TiO_2 under visible light. *Top. Catal.,* **2010**, *53*(7-10), 578-583.
[http://dx.doi.org/10.1007/s11244-010-9490-z]

[39] Kidwai, M.; Bhardwaj, S.; Jain, A. A green oxidation protocol for the conversion of secondary alcohols into ketones using heterogeneous nanocrystalline titanium (IV) oxide in polyethylene glycol. *Green Chem. Lett. Rev.,* **2012**, *5*(2), 195-202.
[http://dx.doi.org/10.1080/17518253.2011.611824]

[40] Li, C.J.; Xu, G.R.; Zhang, B.; Gong, J.R. Gong, High selectivity in visible-light-driven partial photocatalytic oxidation of benzyl alcohol into benzaldehyde over single-crystalline rutile TiO_2 nanorods. *Appl. Catal. B,* **2012**, *115*, 201-208.
[http://dx.doi.org/10.1016/j.apcatb.2011.12.003]

[41] Wei, J.; Sun, Z.; Luo, W.; Li, Y.; Elzatahry, A.A.; Al-Enizi, A.M.; Deng, Y.; Zhao, D. New insight into the synthesis of large-pore ordered mesoporous materials. *J. Am. Chem. Soc.,* **2017**, *139*(5), 1706-1713.
[http://dx.doi.org/10.1021/jacs.6b11411] [PMID: 28085258]

[42] Dimitrov, M.; Ivanova, R.; Velinov, N.; Henych, J.; Slusna, M.; Stengl, V.; Tolasz, J.; Mitov, I.; Tsoncheva, T. Mesoporous TiO_2 powders as host matrices for iron nanoparticles. Effect of the preparation procedure and doping with Hf. Nano-Struct. *Nano-Objects,* **2016**, *7*, 56-63.
[http://dx.doi.org/10.1016/j.nanoso.2016.06.001]

[43] Tamiolakis, I.; Lykakis, I.N.; Armatas, G.S. Mesoporous CdS-sensitized TiO_2 nanoparticle assemblies

with enhanced photocatalytic properties: Selective aerobic oxidation of benzyl alcohols. *Catal. Today,* **2015**, *250*, 180-186.
[http://dx.doi.org/10.1016/j.cattod.2014.03.047]

[44] Tamiolakis, I.; Lykakis, I.N.; Katsoulidis, A.P.; Armatas, G.S. One-pot synthesis of highly crystalline mesoporous TiO_2 nanoparticle assemblies with enhanced photocatalytic activity. *Chem. Commun. (Camb.),* **2012**, *48*(53), 6687-6689.
[http://dx.doi.org/10.1039/c2cc33122e] [PMID: 22643333]

[45] Wang, J.; Bian, Z.; Zhu, J.; Li, H. Ordered mesoporous TiO_2 with exposed (001) facets and enhanced activity in photocatalytic selective oxidation of alcohols. *J. Mater. Chem. A Mater. Energy Sustain.,* **2013**, *1*(4), 1296-1302.
[http://dx.doi.org/10.1039/C2TA00035K]

[46] Qadir, M.I.; Scholten, J.D.; Dupont, J. TiO_2 nanomaterials: Highly active catalysts for the oxidation of hydrocarbons. *J. Mol. Catal. Chem.,* **2014**, *383*, 225-230.
[http://dx.doi.org/10.1016/j.molcata.2013.12.012]

[47] Aboelfetoh, E.F.; Pietschnig, R. Preparation and catalytic performance of Al_2O_3, TiO_2 and SiO_2 supported vanadium based-catalysts for C–H activation. *Catal. Lett.,* **2009**, *127*(1-2), 83-94.
[http://dx.doi.org/10.1007/s10562-008-9634-y]

[48] Aboelfetoh, E.F.; Fechtelkord, M.; Pietschnig, R. Structure and catalytic properties of MgO-supported vanadium oxide in the selective oxidation of cyclohexane. *J. Mol. Catal. Chem.,* **2010**, *318*(1-2), 51-59.
[http://dx.doi.org/10.1016/j.molcata.2009.11.007]

[49] Makgwane, P.; Ray, S.S. Efficient room temperature oxidation of cyclohexane over highly active hetero-mixed WO_3/V_2O_5 oxide catalyst. *Catal. Commun.,* **2014**, *54*, 118-123.
[http://dx.doi.org/10.1016/j.catcom.2014.05.031]

[50] Koo, D.H.; Kim, M.; Chang, S. WO_3 nanoparticles on MCM-48 as a highly selective and versatile heterogeneous catalyst for the oxidation of olefins, sulfides, and cyclic ketones. *Org. Lett.,* **2005**, *7*(22), 5015-5018.
[http://dx.doi.org/10.1021/ol052019i] [PMID: 16235946]

[51] Qi, B.; Lou, L.L.; Yu, K.; Bian, W.; Liu, S. Selective epoxidation of alkenes with hydrogen peroxide over efficient and recyclable manganese oxides. *Catal. Commun.,* **2011**, *15*(1), 52-55.
[http://dx.doi.org/10.1016/j.catcom.2011.08.012]

[52] Atashin, H.; Malakooti, R. Magnetic iron oxide nanoparticles embedded in SBA-15 silica wall as a green and recoverable catalyst for the oxidation of alcohols and sulfides. *J. Saudi Chem. Soc.,* **2017**, *21*, 17-24.
[http://dx.doi.org/10.1016/j.jscs.2013.09.007]

[53] Tamoradi, T.; Moeini, N.; Ghadermazi, M.; Choghamarani, A.G. Fe_3O_4-AMPD-Pd: A noble and efficient magnetic nanocatalyst for synthesis of sulfides and oxidation reactions. *Polyhedron,* **2018**, *153*, 104-109.
[http://dx.doi.org/10.1016/j.poly.2018.07.002]

[54] Soleimani, E.; Taheri, R. Synthesis and surface modification of CuO nanoparticles: Evaluation of dispersion and lipophilic properties. Nano-Struct. *Nano-Objects,* **2017**, *10*, 167-175.
[http://dx.doi.org/10.1016/j.nanoso.2017.05.002]

[55] Ye, Z.; Hu, L.; Jiang, J.; Tang, J.; Cao, X.; Gu, H. Jiang. J.; Tang, J.; Cao, X.; Gu, H. Cuo@Ag as a highly active catalyst for the selective oxidation of trans-stilbene and alcohols. *Catal. Sci. Technol.,* **2012**, *2*(6), 1146-1149.
[http://dx.doi.org/10.1039/c2cy20138k]

[56] Deori, K.; Gupta, D.; Saha, B.; Awasthi, S.K.; Deka, S. Introducing nanocrystalline CeO_2 as heterogeneous environmentally friendly catalyst for the aerobic oxidation of para-xylene to terephthalic acid in water. *J. Mater. Chem. A Mater. Energy Sustain.,* **2013**, *1*(24), 7091-7099.

[http://dx.doi.org/10.1039/c3ta01590d]

[57] Deori, K.; Gupta, D.; Saha, B.; Deka, S. S. Design of 3-dimensionally self-assembled CeO_2 nanocube as a breakthrough catalyst for efficient alkylarene oxidation in water. *ACS Catal.,* **2014**, *4*(9), 3169-3179.
[http://dx.doi.org/10.1021/cs500644j]

[58] Dong, Y.Z.; Han, X. Wang. Y.; Liu, M.; Li. Y. Pd/Cu bimetallic nanoparticles supported on graphene nanosheets: Facile synthesis and application as novel electrocatalyst for ethanol oxidation in alkaline media. *Int. J. Hydrogen Energy,* **2014**, *39*(27), 14669-14679.
[http://dx.doi.org/10.1016/j.ijhydene.2014.06.139]

[59] Wang, Y.; Zhao, Y.; Yin, J.; Liu, M.; Dong, Q.; Su, Y. Synthesis and electrocatalytic alcohol oxidation performance of Pd–Co bimetallic nanoparticles supported on graphene. *Int. J. Hydrogen Energy,* **2014**, *39*(3), 1325-1335.
[http://dx.doi.org/10.1016/j.ijhydene.2013.11.002]

[60] Habibi, B.; Dadashpour, E. Carbon-ceramic supported bimetallic Pt–Ni nanoparticles as an electrocatalyst for electrooxidation of methanol and ethanol in acidic media. *Int. J. Hydrogen Energy,* **2013**, *38*(13), 5425-5434.
[http://dx.doi.org/10.1016/j.ijhydene.2012.06.045]

[61] Mukherjee, P.; Sarathi, P.; Mandal, K.; Bhattacharjee, D.; Dasgupta, S.; Kumar, S. Improved catalysis of room temperature synthesized Pd-Cu alloy nanoparticles for anodic oxidation of ethanol in alkaline media. *Electrochim. Acta,* **2015**, *184*, 447-455.
[http://dx.doi.org/10.1016/j.electacta.2014.12.054]

[62] Guen, Y.; Hensen, E.J.M. Selective oxidation of ethanol to acetaldehyde by Au–Ir catalysts. *J. Catal.,* **2013**, *305*, 135-145.
[http://dx.doi.org/10.1016/j.jcat.2013.04.023]

[63] Sharma, G.; Kumar, A.; Sharma, S.; Naushad, M.; Dwivedi, R.P. ALOthman, Z. A.; Tessema, G. Novel development of nanoparticles to bimetallic nanoparticles and their composites: A review. *Science,* **2019**, *31*, 257-269.

[64] Wang, Y.; Zhao, H.; Zhao, G. Iron-copper bimetallic nanoparticles embedded within ordered mesoporous carbon as effective and stable heterogeneous Fenton catalyst for the degradation of organic contaminants. *Appl. Catal. B,* **2015**, *164*, 396-406.
[http://dx.doi.org/10.1016/j.apcatb.2014.09.047]

[65] Kung, M.C.; Davis, R.J.; Kung, H.H. Understanding Au- catalyzed low-temperature CO oxidation. *J. Phys. Chem. C,* **2007**, *111*(32), 11767-11775.
[http://dx.doi.org/10.1021/jp072102i]

[66] Sun, K. Theoretical investigations on CO oxidation reaction catalyzed by gold nanoparticles. *Chin. J. Catal.,* **2016**, *37*(10), 1608-1618.
[http://dx.doi.org/10.1016/S1872-2067(16)62476-2]

[67] Kaneda, K.; Mitsudome, T. Metal–support cooperative catalysts for environmentally benign molecular transformations. *Chem. Rec.,* **2017**, *17*(1), 4-26.
[http://dx.doi.org/10.1002/tcr.201600036] [PMID: 27455891]

[68] Wang, J.; Lang, X.; Zhaorigetu, B.; Jia, M.; Wang, J.; Guo, X.; Zhao, J. Aerobic oxidation of alcohols on Au nanocatalyst: insight to the roles of the Ni-Al layered double hydroxides support. *ChemCatChem,* **2014**, *6*(6), 1737-1747.
[http://dx.doi.org/10.1002/cctc.201400046]

[69] Wang, S.; Yin, S.; Chen, G.; Li, L.; Zhang, H. Nearly atomic precise gold nanoclusters on nickel-based layered double hydroxides for extraordinarily efficient aerobic oxidation of alcohols. *Catal. Sci. Technol.,* **2016**, *6*(12), 4090-4104.
[http://dx.doi.org/10.1039/C6CY00186F]

[70] Pan, Y.; Wu, G.; He, Y.; Feng, J.; Li, D. Identification of the Au/ZnO interface as the specific active site for the selective oxidation of the secondary alcohol group in glycerol. *J. Catal.,* **2019**, *369*, 222-232.
[http://dx.doi.org/10.1016/j.jcat.2018.10.030]

[71] Oliveira, R.L.; Kiyohara, P.K.; Rossi, L.M. High performance magnetic separation of gold nanoparticles for catalytic oxidation of alcohols. *Green Chem.,* **2010**, *12*(1), 144-149.
[http://dx.doi.org/10.1039/B916825G]

[72] Yang, M.Q.; Zhang, N.; Xu, Y.J. Synthesis of fullerene-, carbon nanotube-, and graphene-TiO_2 nanocomposite photocatalysts for selective oxidation: a comparative study. *ACS Appl. Mater. Interfaces,* **2013**, *5*(3), 1156-1164.
[http://dx.doi.org/10.1021/am3029798] [PMID: 23324025]

[73] Tian, J.; Hu, X.; Wei, N.; Zhou, Y.; Xu, X.; Cui, H.; Liu, H. RuO_2/TiO_2 nanobelt heterostructures with enhanced photocatalytic activity and gas-phase selective oxidation of benzyl alcohol. *Sol. Energy Mater. Sol. Cells,* **2016**, *151*, 7-13.
[http://dx.doi.org/10.1016/j.solmat.2016.02.017]

[74] Bordoloi, A.; Halligudi, S. Studies in structural characterization and correlation with the activity of an efficient and stable WOx/SBA-15 nanocomposite catalyst. *J. Catal.,* **2008**, *257*(2), 257-283.
[http://dx.doi.org/10.1016/j.jcat.2008.05.015]

[75] Hu, Z.; Zhao, Y.; Liu, J.; Wang, J.; Zhang, B.; Xiang, X. Ultrafine MnO_2 nanoparticles decorated on graphene oxide as a highly efficient and recyclable catalyst for aerobic oxidation of benzyl alcohol. *J. Colloid Interface Sci.,* **2016**, *483*, 26-33.
[http://dx.doi.org/10.1016/j.jcis.2016.08.010] [PMID: 27544446]

[76] Chen, C.; Qu, J.; Cao, C.; Niu, F.; Song, W. CuO nanoclusters coated with mesoporous SiO_2 as highly active and stable catalysts for olefin epoxidation. *J. Mater. Chem.,* **2011**, *21*(15), 5774-5779.
[http://dx.doi.org/10.1039/c0jm04568c]

[77] Karimpour, T.; Safaei, E.; Karimi, B.; Lee, Y.I. Iron (III) amine bis(phenolate) complex immobilized on silica-coated magnetic nanoparticles: A highly efficient catalyst for the oxidation of alcohols and sulfides. *ChemCatChem,* **2018**, *10*(8), 1889-1899.
[http://dx.doi.org/10.1002/cctc.201701217]

[78] Anumol, E.A.; Kundu, P.; Deshpande, P.A.; Madras, G.; Ravishankar, N. Seebeck tuning in chalcogenide nanoplate assemblies by nanoscale heterostructuring. *ACS Nano,* **2011**, *5*, 8049-8061.
[http://dx.doi.org/10.1021/nn202639f] [PMID: 21888416]

CHAPTER 6

Nanocatalysis for Reduction/Hydrogenation Reactions

Leena Khanna[1,*], **Mansi**[1] and **Pankaj Khanna**[2]

[1] *University School of Basic and Applied Sciences, Guru Gobind Singh Indraprastha University, Dwarka, New Delhi-110078, India*

[2] *Department of Chemistry, Acharya Narendra Dev College, University of Delhi, Kalkaji, New Delhi-110019, India*

Abstract: Heterogeneous nanocatalyst demonstrate excellent catalytic activity for the hydrogenation/reduction of nitro-aromatics, carbonyls, alkenes/alkynes in the presence of different reductants such as $NaBH_4$ and H_2 using various solvents such as ethanol, methanol, dioxane, THF, and water, as green solvents. Earth-abundant coordinating elements such as Pd, Pt, Fe, Cu, Co, Ag, Au, and Ni, elementary synthesis, short time reactions, high selectivity, mild reaction conditions, and reusability of nanocatalyst for at least 4-5 cycles without any loss in catalytic activity, are some priorities for the hydrogenation reactions using nanocatalyst.

Keywords: Alkenes, Alkynes, Bimetallic, Carbonyls, Hydrogenation, Nano-catalyst, Nitro aromatics, Reducing agents, Reduction.

INTRODUCTION

Hydrogenation and reduction of compounds are important chemical reactions in organic synthesis as well as industries. The use of heterogeneous metal catalysts has been the most versatile and dynamic process for these reactions. It was about 100 years ago when Paul Sabatier hydrogenated alkenes efficiently over Ni metal catalysts [1]. Since then, the use of these solid catalysts has multiplied thousand times. The heterogeneous metal catalysis owns a broad scope, besides Ni, Pd, and Pt, we now have several cheap metals to catalyse the hydrogen-ation/reduction reactions. A wide variety of multifunctional molecules are reduced by this method, and numerous value-added products can be obtained in high yield in a short time, with chemo and/or regioselective control.

* **Corresponding author Leena Khanna**: University School of Basic and Applied Sciences, Guru Gobind Singh Indraprastha University, Dwarka, New Delhi-110078, India; E-mail: leenakhanna@ipu.ac.in

Recently, the emphasis is on the development of sustainable, eco-friendly methods, utilizing metallic nanoparticles as a catalyst in these reactions, which can curb chemical waste generation with more safety, recyclability, and high-atom efficiency [2]. Moreover, the utility of monometallic catalysts has been surpassed by bimetallic and multimetallic nanocatalysts. The addition of another metal might induce surface-electronic structure reconfiguration, which generates diverse interaction between hydrogen and reactants in turnover cycles, leading to high catalytic activity [3].

In this chapter, we have featured numerous methods for hydrogenation/ reduction reactions using several reusable nanocatalysts [4, 5]. The simplest method of preparation of nanocatalyst follows the scheme, where metal precursors are mixed with polymer stabilizers and reduced by using chemical agents such as H_2, $NaBH_4$, glycerol, amines, *etc*., to achieve different surface morphologies [6]. The novel structured mono, bi, or mixed nanoparticles, mostly obtained from Pt, Ru, Au, Pd, Fe, Cu, Co, and Ag, are mainly employed for reduction and hydrogenation catalysis of olefins, nitroaromatics, and carbonyls [7].

Material scientists are continuously designing different types of routes to obtain different nanoparticles with distinguishing specifications. Novel structured catalysts, such as nanometal clusters, particles with uniform but particular designs (*e.g*., octahedral, tetrahedral, cubic, rhombic), as well as bimetallic or multimetallic nanoparticles with disciplined structures (*e.g*., alloy, core--shell, cluster-in-cluster, layer-by-layer),core-shell are generally found with improved catalytic activities as compared to traditional catalysts [3]. Different shapes of nanocatalysts can be obtained with different types of polymer and control over the rate of kinetic reduction of metal in the solvent [8].

REDUCTION/ HYDROGENATION OF NITRO COMPOUNDS

The reduction of nitro compounds is the most simplistic route to produce amino derivatives, which are important intermediates for producing nitrogen compounds, useful in pharmaceuticals, agriculture, polymers, dyes, cosmetic industries, *etc.* Some nitroaromatics are a serious threat to the environment. However, after reduction into amino aryls, they become harmless (Fig. **1**). The previously used catalytic methods had harsh conditions and expensive reagents to obtain products in high yield. The modern use of nano catalytic systems has minimized the use of catalyst, is eco-friendly with low cost and providing excellent yield and a high rate of conversion. We have tabulated Table **1**, the efficient reduction of nitro compounds to anilines using a variety of nanocatalysts. These catalysts can be recycled many times with high stability and no leaching up to many cycles. Suitably designing the catalyst structure and their wise selection has been successful in improving the reduction power, sustainability, and recoverability of the nanocatalyst. The use of cheap transition metals as nanocatalysts has enhanced the use of the catalytic system for the reduction of nitro compounds.

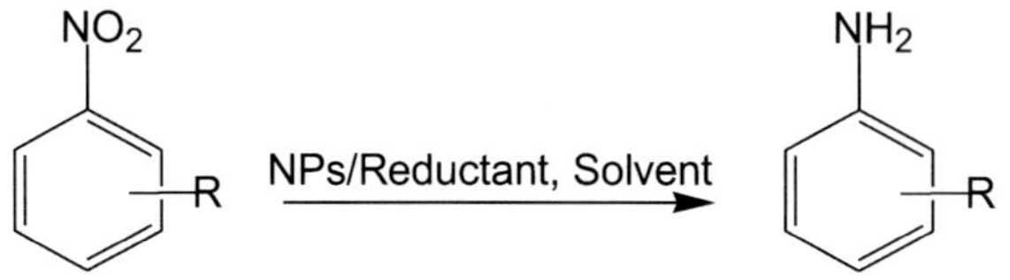

Fig. (1). Reduction of nitro compounds.

Table 1. Reduction/hydrogenation of nitro compounds.

Catalyst	Morphological Features	Solvent	Reductant	Reaction Condition	Yield	Refs.
Cu(0)NP	Metallic copper in the cubic phase.	Alcohols, ethers, and amines as Proton donor	Glycerol	130 °C for 1h 10 mol % of catalyst and 2 eq of KOH	93%	[9]
PdCu@ MWCNT	Face-centered cubic (fcc) 4.78 ± 0.43 nm.	1 ml of water: methanol mixture (7:3),	$NaBH_4$	r.t. 10 min.	99%	[10]
CuNPs	Mean size 10.2 ± 3.0 nm, roundish shape	Glycerol, KOH,	Glycerol	130 °C, 5 h.	99%	[11]
Co/N-SiC				r.t., 1 atm		[12]
Fe-Ni catalyst (Fe-Ag-ATP-CA beads)	Mean size 1.6 nm	Methanol	$NaBH_4$	r.t.	99%	[13]
Pd@MOF-1	Mean size 1–2 nm	Ethanol	H_2	2 h, r.t.	98%	[14]
(Co-DABCO-TPA@C-800)	Mean size from <5 nm to 30 nm.	t-BuOH, THF/H_2O	Aldehyde	40 bar H_2, 120 °C	80%- 90%	[15]
Pd@TPAPOP-1	External Surface Area (164 m^2/g)	EtOH	H_2	r.t.	85%- 95%	[16]
Pd@CTF	Mean size of Pd 5.34 nm	EtOH/ H_2O	NH_4COOH	25 °C	85%- 99%	[17]
Fe_3O_4-MWCNTs@PEI-Ag.	Mean size is 60 nm	Water	$NaBH_4$	70 °C-90 °C	70%-90%	[18]
Fe_3O_4/NH_2/Ru	Diameter 2.1 nm	Ethanol	$NaBH_4$	30 °C, 40 min.	90%	[19]

(Table 1) cont.....

Catalyst	Morphological Features	Solvent	Reductant	Reaction Condition	Yield	Refs.
MWCNTs/MA@PC/Pt(II)	Specific surface area 32.65 cm^2/g	Water	$NaBH_4$	15 min., 25 °C	>99%	[20]
Rh@PS NPs.		1,4-dioxane	NaOtBu	100 °C for 8 h	91%	[21]
Pd-PDA	Round shaped particles of 150–250 nm Mean size.	EtOH	HCOONa	85 °C, 60 min.	75%-90%	[22]
Pd@C12mimBr (Pd@βCDMOD14)	Mean size 4.2–4.6 nm	Water	$NaBH_4$	-		[23]
Clin@Py-Tzl@Pd		THF/ H_2O	$NaBH_4$	r.t., 8 h	80%-95%	[24]
Ag-rGO/g-C_3N_4	BET surface area (SBET) is 85.46±2 M^2/g	Methanol	Visible light, $NaBH_4$	r.t.	90%- 98%	[25]
$BiFeO_3$	Surface area 6.4 m^2/g	EtOH-H_2O	$NaBH_4$	r.t., 800 rpm	80%- 95%	[26]
$CuFe_2O_4$	Crystal mean size 9 nm. Cubic unit cell	Water	$NaBH_4$	Reflux, 10-80 min.	83%- 98%	[27]
$CuFe_2O_4$@SiO_2@g-C_3N_4/Cu		Water	$NaBH_4$	55 °C, 20 min.	98%	[28]
CoOx@NC-800		Toluene	Na_2CO_3	120 °C, 15 h	60%- 90%	[29]
Fe_3O_4@SiO_2-NH_2–Au	BET surface area (53m^2/g), 6.7 nm	Ethanol, H_2O	$NaBH_4$	3.7 min.	80%	[30]
Pd-Pt/m-GO	Specific surface area 31 m^2/g	Methanol	H_2	50 °C, 15 min.	98%- 100%	[31]
Ru/ReS_2		Water	$NaBH_4$			[32]
Ag/HMS	Mean size between 15 and 30 nm	THF/ MeOH	$NaBH_4$	1 h, r.t.	90% - 95%	[33]
Co/$Co(OH)_2$	Pore radius of 15.3 Å	MeOH	$NaBH_4$	3 h, r.t.	90%- 95%	[34]
Fe_3O_4/SiO_2/CdS	BET specific surface area 11.26 m^2/g	i-PrOH	Blue LED irradiation	20 h.	<95%	[35]
Fe-Cu @MCC		Water	$NaBH_4$	r.t. 120 min., 70 °C.	90%	[36]

(Table 1) cont.....

Catalyst	Morphological Features	Solvent	Reductant	Reaction Condition	Yield	Refs.
Pd–Au/TiO_2	Particles mostly have spherical shape	Water	$NH_2NH_2.H_2O$	5 min. r.t.	96%	[37]
$CuCl_2.2H_2O$	Mean size range 4–7 nm, mean particle diameter was about 5.6 nm.	Urea, distilled water	$NaBH_4$	>100 min.	>90%	[38]
CuNPs	Mean size 8 nm	THF/ water	$NaBH_4$	50 °C, 2 h.	65%-95%	[39]
(CuNPs)	30–85 nm	Water	($NaBH_4$), polyacrylic acid (PAA)	r.t.		[39]
CuO–GO nanocomposite	Particle Mean size 10 nm	DeionizeH_2O	Aqueous $NaBH_4$	r.t., 30 min.	98%	[40]
$CoFe_2O_4$@AA-Co (30)	50 nm, TEM	H_2O	$NaBH_4$	r.t, 5 min.	98%	[41]
Fe/Ni@RH-SiO_2	BET surface area of 129.42 m^2/g	H_2O	$NaBH_4$	50 °C, 10 min.	97%	[42]
Ir-PVP	Crystallite Mean size 1.22 nm	Methanol	H_2	30 °C, 24 h	95%	[43]
$Fe_2O_3/BiVO_4$	Surface Area 19.25 m^2/g	Methanol	$NH_2NH_2.H_2O$	r.t., hv, 12 h	98%	[44]
Pt/SBA-15	Cubic, octahedral, SBA-15 support	Water	EG	25 °C		[45]
Pt/ Pd APTS	Amino functionalized nay support	Water	H_2	110 °C, 40 min.		[46]
Fe_3O_4@-C/Pt	Amino functionalized carbon support	Water	H_2	30 °C, 2 h		[47]
Ru- NP's	3-6 nm, supported on clay	EG	EG	12 h	97%	[48]
Ni- NP's	15-20 nm, TiO_2 support	Water	H_2	2 h	80%- 95%	[49]
Co_3O_4-NGr@C	20-80 nm, carbon support	Water	H_2	11 h, 100 °C	89%- 97%	[50]
Co/CN	Mesoporous carbon nitride	Water	NH_3BH_3	30 °C, 5 min.	>99%	[51]
Pt-Ru/SnO_2	2.3 nm	-	H_2	50 °C	>99%	[52]
Pd-100	4–15 nm	-	H_2	5 h	>99%	[53]

(Table 1) cont.....

Catalyst	Morphological Features	Solvent	Reductant	Reaction Condition	Yield	Refs.
Ni–B(3)	Mixed metal oxide particles	KOH	KBH_4	80 min., 120 °C	90%	[54, 55]
Ni–B	Mixed metal oxide particles	Methanol	H_2	70 °C	90%	[56]

NPs: Nanoparticles; THF: Tetrahydrofuran.

HYDROGENATION OF ALKENES/ALKYNES

Alkenes and alkynes can be readily hydrogenated to alkanes or alkenes, respectively, using several nanocatalysts consisting of Ni, Pd, Pt, Ru, and Rh under mild reaction conditions (Fig. **2**, Table **2**) even under stereoselective way. Many precious nano metal catalysts like Ru, Pd, and Au have also been acknowledged based on the semi-hydrogenation process of alkynes. In recent years, Fe, Co, Ni, Cu, and V-based nanocatalysts have also been developed for this process with high reusability and very selective hydrogenation power. Several examples of double and triple bond hydrogenation reactions have been reviewed by us, demonstrating detailed reaction conditions, morphological features of nanocatalyst, and variable reducing agent used.

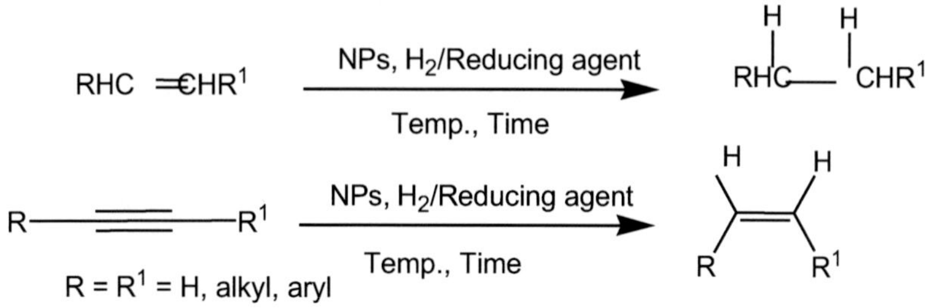

Fig. (2). Hydrogenation of alkenes/alkynes.

Table 2. Hydrogenation of alkenes and alkynes.

Catalyst	Morphological Features	Solvent	Reducing Agent	Reaction Condition	References
Pt/Al_2O_3	4–18 nm, cubic, truncated octahedral, Al_2O_3 support	Water	H_2	210 °C, 2 h	[57]
Pt- NP's	Cuboctahedral, cubic	Water	Ascorbic, $NaBH_4$	50 °C	[58, 59]
Pt@hmC	2 nm, mesoporous carbon support	Water	Ethanol	2 h	[60]
Pt(0)	Spherical, $Fe_3O_4@SiO_2$ support	Water	H_2	1 h, 75 °C	[61]

(Table 2) cont.....

Catalyst	Morphological Features	Solvent	Reducing Agent	Reaction Condition	References
Pt/RGO	3 nm, graphene support	OLA toluene	OLA	40 °C	[62]
Pd/SiO_2	Embedded in SiO_2 support	H_2 THF	H_2	25 °C, 2 h	[63]
$Pd(0)\text{-}SiO_2$	Polymer coated on Pd particle surface	H_2, water	H_2	40 °C	[64]
Pd NC's	Deposited on Fe_3O_4	Methanol	H_2	1 h	[65]
Pd NPs	2-6 nm, spherical,	Ethanol, water	$NaBH_4$	30 min.	[66]
$Pd_{1289}L_{164}$	2-5 nm	Water, toluene	$NaBH_4$	4 h	[67]
$PdMgF_2$-100	6-10 nm, MgF_2 support	Methanol	H_2	r.t., 3 h	[68]
Pd@Ag NP's	3 nm, encapsulated in metal organic Framework	DMF	DMF	r.t.	[69]
Fe_3O_4–SiO_2 NH_2-Rh	3 nm, Fe_3O_4–SiO_2 support	Water	$NaBH_4$	75 °C	[70]
Ir(0) NP's	2.7 nm, embedded in polymer Membrane	BMI.PF6	H_2	75 °C	[71]
Pt-FeCo/GC@MSS	6 nm	Acetone	H_2	25 °C, 1 h	[72]
$Pd_{34}Rh_{39}P_{27}$	Spherical, 6 nm	Dioxane	H_2	35 °C, 4 h	[73]
Pd–Ag	3 nm on dendrimer surface	-	H_2	80 °C, 1 h	[74]
Pd@Ag	3 nm mean size	-	H_2	r.t., 1 h	[69]
$Ru_{0.56}Ni_{0.44}/C$	4.4 nm mean size	-	H_2	60 °C, 30 min.	[75]
Co/CN	Mesoporous carbon nitride	Water	NH_3BH_3	30 °C, 5 min.	[51]
Pd@TPAPOP-1	External Surface Area (164 m^2/g)	EtOH	H_2	r.t.	[16]
Fe-Ni catalyst (Fe-Ag-ATP-CA beads)	Mean size 1.6 nm	THF	H_2	30 °C	[13]

NPs: Nanoparticles; DMF: Dimethylformamide; THF: Tetrahydrofuran.

REDUCTION/ HYDROGENATION OF CARBONYL COMPOUNDS

The reduction of carbonyl compounds into the corresponding alcohols is the most common organic transformation in synthetic organic chemistry. Distinguished methods to carry out this conversion involve the use of metal hydrides, catalytic hydrogenation, dissolving metals, and transfer hydrogenation. The hydrogen donor method is cheap and easy to handle, and eco-friendly with few possible hazards. The ketones can be reduced *via* H-transfer from isopropanol as hydrogen donor under homogeneous catalytic conditions. However, rare interest has been shown in the reduction of non-aromatic carbonyl compounds by hydrogen transfer. However, the use of heterogeneous catalysts, mainly nano metals, has shown many advantages like easy handling, high recovery, and recyclability, and enhanced stability over homogeneous catalysts (Fig. **3**, Table **3**).

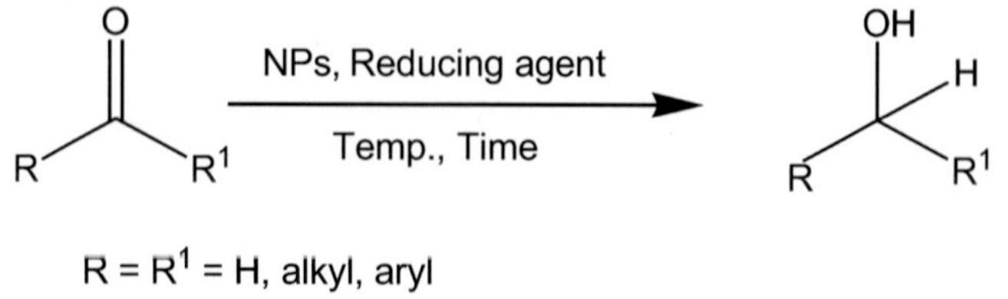

Fig. (3). Hydrogenation/Reduction of carbonyl compounds.

Table 3. Reduction of carbonyl compounds.

Catalyst	Morphological Features	Solvent	Reducing Agent	Reaction Condition	Yield	References
Pt NPs	Spherical, Fe_3O_4@SiO_2 support	Water	H_2	1.5 h, 75 °C	>80%	[61]
PdOnps/ GO	-	Ethanol	Et_3SiH	10-15 min., r.t.	-	[5]
Cu-BTC-MOF	Surface area BET 758 m²/g	Ethanol	$N_2H_4.H_2O$	r.t., 25-27 °C, 6 h-12 h	-	[76]
Au/TiO_2	0.20 mm thick, mean size 25 μm	Dry benzene	Et_2SiH_2	25 °C, ~1 min.	85%- 90%	[33]
PdCu/CβC	-	Ethanol or water	H_2	100 °C, 2 h,	60%- 80%	[77]
Pd/SiO_2	Embedded in SiO_2 support	THF	H_2	25 °C, 2 h		[63]
Pd/$C_{0.5}$–LDH	9 nm particles on layered double OH-support	NaOH, water	H_2	100 °C, 1 h	100%	[78]
u-GNPs-RuO_2NRs	<1 nm	Physically blended with Grapheme	$(CH_3)_3COK$	82 °C	>99%	[79]

(Table 3) cont.....

Catalyst	Morphological Features	Solvent	Reducing Agent	Reaction Condition	Yield	References
Ir-4	2-4 nm, N-doped carbon Support	Ethanol	H_2	80 °C	>99%	80
5-Al_2O_3 -Ni-o-t-ANTs	5 nm, embedded in Al_2O_3 Tube	Gas deposition	H_2	80 °C, 2 h		[81]
Pd@SiO_2	Embedded in SiO_2 support	Water	HCHO	1 h		[82]
PtFe/C	Nanoporous structure	Acetone	H_2	2 h	>80%	[83]
Pt NPs	Mean size control		H_2	30 °C, 20 h	>99%	[84]
γ-Fe_2O_3@HAP	Mixed metal oxide particles		iso-Propanol	180 °C, 10 h	91%	[85]
Pt/Al_2O_3	~2 nm in TiO_2 nanotube	Water	H_2	25 °C,	>96%	[86]
Co-ZIF-350	Mean size of Co-ZIF-350 is 9.4nm	Ethanol	H_2	50 °C, 4 h.	>99%	[87]
Ni-P-1 H-5/1	BET surface area 23 m^2/g	Water	H_2	80 °C, 5 h	90%- 99%	[88]
Mpg-C_3N_4@Pd	BET surface area 190 m^2/g	H_2O/MeOH	Ammonia borane	5 min., r.t.	>99%	[89]
CuO/$CuCr_2O_4$			H_2	150 °C, 2 h		[90]
AuNPs	BET Surface area (5.0±0.5) m^2/g,	H_2O	$NaBH_4$	25 °C, 1.5 h	95%	[91]
Ag NPs	Surface plasmon band 400–420 nm.	Dioxane	H_2	24 h, 80 °C	40%-90%	[92]
Pd-PDA	Round shaped particles of 150–250 nm Mean size.	EtOH	HCOONa	85 °C, 60 min.	85%-90%	[22]
$Pd_3P_{0.95}$	Surface area 45 m^2/g	2-propanol	KOH	80 °C, 3 h;	>99%	[93]
Co/CN	Mesoporous carbon nitride	Water	NH_3BH_3	30 °C, 5 min.	>99%	[51]
Fe-Ni catalyst (Fe-Ag-ATP-CA beads)	Mean size 1.6 nm	KOH	NH_4Cl	r.t.	99%	[13]

NPs: Nanoparticles; THF: Tetrahydrofuran

MISCELLANEOUS REACTIONS USING NANO METAL CATALYST

Metallic nanocatalysts are also employed for a wide variety of other applications in organic synthesis/chemistry such as oxidation of alcohols, reductive amination, reduction of acid, hydration of nitrile, coupling reactions, oximation of aldehydes, cycloaddition reactions, and many more (Table **4**). More importantly, these days priority is given to sustainable development, ecofriendly nanocatalyst having higher catalytic efficiency as well as good control over chemoselectivity/regioselectivity. All these novel structured nanocatalysts constitute different types of coordinating metals, Pt, Pd, Fe, Ni, Co, Cu, Au, Ag, Ru. Some of them are supported on various surface supports such as TiO_2, SiO_2, and amino-functionalised carbon support resulting in disciplined structures with distinctive surface morphologies for improved catalytic activities.

Table 4. Miscellaneous reactions.

Nano Catalyst	Application Reaction	References
Au/TiO_2	Reductive amination	[7]
Cu/SiO_2	Reduction of acid	[90]
Cu/SiO_2 /$CuCr_2O_4$	Hydrogenation of ethyl acetate	[90]
MWCNTs/MA@PC/Pt(II)	Reduction of 4-pyridinecarbonitrile	[20]
$Pd_3P_{0.95}$ QDs	Hydration of nitrile	[93]
Rh@PS	Reduction reaction of sulfoxides	[21]
Pd-PDA	Heck reaction of aryl Halides with ethyl acrylate	[22]
$CoFe_2O_4$@AA-V	Oximation of aldehydes	[41]
Pd/SiO_2	Hydrogenation of Benzonitrile	[63]
Ir-PVP	Hydrogenation of nitrile, One-pot synthesis of pyrrolidone	[43]
$CuFe_2O_4$	Reduction of nitriles	[27]
Co/Co$(OH)_2$	Reduction of nitriles	[34]
Au/TiO_2	Reduction of amides	[33]
MWCNTs/MA@PC/Pt(II)	Reduction nitrile compounds	[20]
Fe_3O_4-MWCNTs@PEI-Ag	Reduction of nitrile	[18]

CONCLUDING REMARKS

This chapter has briefly reviewed the applications of mono, bi, and mixed nano metallic or non-metallic catalysts in hydrogenation and reduction reactions. In particular, the focus has been on the synthesis of alkanes, alkenes, amines, and alcohols in excellent yield *via* hydrogenation of alkynes, alkenes, nitro compounds, and carbonyl compounds respectively. Many examples using nanocatalyst of palladium, platinum, silver, carbon, ruthenium, copper, nickel, cobalt, gold, iridium *etc.* for the organic reduction have been successfully summarized. Specific examples under each category with main focus on the particle size of catalyst used, range of reaction conditions, major aspects in catalytic activity and selectivity control, and product yield are discussed. It has been observed that a recent trend is evolving in the designing of nano-size bimetallic catalysts for carrying out these reactions. In general, we can say that the nanocatalysts with different surface configurations have shown their key role in augmenting the catalytic activity and yield of hydrogenation reactions.

CONSENT FOR PUBLICATION

Not applicable.

CONFLICT OF INTEREST

The author declares no conflict of interest, financial or otherwise.

ACKNOWLEDGEMENTS

The authors are thankful to Guru Gobind Singh Indraprastha University, New Delhi for providing funding under Faculty Research Grant Scheme (FRGS).

REFERENCES

[1] Che, M. Nobel Prize in Chemistry 1912 to Sabatier: Organic Chemistry or Catalysis? *Catal. Today,* **2013**, *218–219*(April), 162-171. [http://dx.doi.org/10.1016/j.cattod.2013.07.006]

[2] S, K.; M, C.; H, S.; Sangeeta; CB, M. The uses of various nanoparticles in organic synthesis: A Review. *J. Nanomed. Nanotechnol.,* **2020**, *11*(2), 1-16. [http://dx.doi.org/10.35248/2157-7439.19.11.543]

[3] Kottappara, R.; Pillai, S. C.; Kizhakkekilikoodayil Vijayan, B. Copper-Based Nanocatalysts for Nitroarene Reduction-A Review of Recent Advances. *Inorganic Chemistry Communications,* **2020**, *108181* [http://dx.doi.org/10.1016/j.inoche.2020.108181]

[4] Zhang, R.; Hou, Z. *Soluble Pd Nanoparticles for Catalytic Hydrogenation*; Nanocatalysis Ion. Liq, **2016**, pp. 83-95. [http://dx.doi.org/10.1002/9783527693283.ch5]

[5] Mirza-Aghayan, M.; Kalantari, M.; Boukherroub, R. Palladium oxide nanoparticles supported on graphene oxide: a convenient heterogeneous catalyst for reduction of various carbonyl compounds using triethylsilane. *Appl. Organomet. Chem.,* **2019**, *33*(4), 1-11. [http://dx.doi.org/10.1002/aoc.4837]

[6] Zhang, K.; Suh, J.M.; Choi, J.W.; Jang, H.W.; Shokouhimehr, M.; Varma, R.S. Recent Advances in the Nanocatalysts-assisted $NaBH_4$ Reduction of Nitroaromatics in water. *ACS Omega,* **2019**, *4*(1), 483-495.

[http://dx.doi.org/10.1021/acsomega.8b03051] [PMID: 31032469]

[7] Louka, A.; Kidonakis, M.; Saridakis, I.; Zantioti-Chatzouda, E.M.; Stratakis, M. Diethylsilane as a Powerful Reagent in Au Nanoparticle-Catalyzed Reductive Transformations. *Eur. J. Org. Chem.*, **2020**, *2020*(23), 3508-3514.
[http://dx.doi.org/10.1002/ejoc.202000483]

[8] Qin, L.; Zeng, G.; Lai, C.; Huang, D.; Zhang, C.; Cheng, M.; Yi, H.; Liu, X.; Zhou, C.; Xiong, W.; Huang, F.; Cao, W. Synthetic strategies and application of gold-based nanocatalysts for nitroaromatics reduction. *Sci. Total Environ.*, **2019**, *652*, 93-116.
[http://dx.doi.org/10.1016/j.scitotenv.2018.10.215] [PMID: 30359806]

[9] Moran, M.J.; Martina, K.; Baricco, F.; Tagliapietra, S.; Manzoli, M.; Cravotto, G. Tuneable copper catalysed transfer hydrogenation of nitrobenzenes to aniline or azo derivatives. *Adv. Synth. Catal.*, **2020**, *362*(13), 2689-2700.
[http://dx.doi.org/10.1002/adsc.202000127]

[10] Göksu, H.; Zengin, N.; Burhan, H.; Cellat, K.; Şen, F. A Novel Hydrogenation of Nitroarene Compounds with Multi Wall Carbon Nanotube Supported Palladium/Copper Nanoparticles (PdCu@MWCNT NPs) in Aqueous Medium. *Sci. Rep.*, **2020**, *10*(1), 8043.
[http://dx.doi.org/10.1038/s41598-020-64988-0] [PMID: 32415148]

[11] Moran, M.J.; Martina, K.; Stefanidis, G.D.; Jordens, J.; Gerven, T.V.; Goovaerts, V.; Manzoli, M.; Groffils, C.; Cravotto, G. Glycerol: An Optimal Hydrogen Source for Microwave-Promoted Cu-Catalyzed Transfer Hydrogenation of Nitrobenzene to Aniline. *Front Chem.*, **2020**, *8*, 34.
[http://dx.doi.org/10.3389/fchem.2020.00034] [PMID: 32064251]

[12] Schönauer, T.; Thomä, S.L.J.; Kaiser, L.; Zobel, M.; Kempe, R. General Synthesis of Secondary Alkylamines by Reductive Alkylation of Nitriles by Aldehydes and Ketones. *Chemistry,* **2021**, *27*(5), 1609-1614.
[http://dx.doi.org/10.1002/chem.202004755] [PMID: 33236790]

[13] Farooqi, Z.H.; Begum, R.; Naseem, K.; Wu, W.; Irfan, A. Zero Valent Iron Nanoparticles as Sustainable Nanocatalysts for Reduction Reactions. *Catal. Rev., Sci. Eng.,* **2020**, *00*(00), 1-70.
[http://dx.doi.org/10.1080/01614940.2020.1807797]

[14] Bao, L.; Yu, Z.; Fei, T.; Yan, Z.; Li, J.; Sun, C.; Pang, S. Palladium supported on metal–organic framework as a catalyst for the hydrogenation of nitroarenes under mild conditions. *Appl. Organomet. Chem.*, **2020**, *34*(6), 1-8.
[http://dx.doi.org/10.1002/aoc.5607]

[15] Murugesan, K.; Chandrashekhar, V.G.; Senthamarai, T.; Jagadeesh, R.V.; Beller, M. Reductive amination using cobalt-based nanoparticles for synthesis of amines. *Nat. Protoc.*, **2020**, *15*(4), 1313-1337.
[http://dx.doi.org/10.1038/s41596-019-0258-z] [PMID: 32203487]

[16] Yadav, C.; Maka, V.K.; Payra, S.; Moorthy, J.N. Multifunctional Porous Organic Polymers (POPs): Inverse Adsorption of Hydrogen over Nitrogen, Stabilization of Pd(0) Nanoparticles, and Catalytic Cross-Coupling Reactions and Reductions. *J. Catal.*, **2020**, *384*, 61-71.
[http://dx.doi.org/10.1016/j.jcat.2020.02.002]

[17] Li, J.; Zhang, L.; Liu, X.; Shang, N.; Gao, S.; Feng, C.; Wang, C.; Wang, Z. Pd Nanoparticles Supported on a Covalent Triazine-Based Framework Material: An Efficient and Highly Chemoselective Catalyst for the Reduction of Nitroarenes. *New J. Chem.*, **2018**, *42*(12), 9684-9689.
[http://dx.doi.org/10.1039/C8NJ01404C]

[18] Ansari, S.; Khorshidi, A.; Shariati, S. Chemoselective Reduction of Nitro and Nitrile Compounds Using an Fe_3O_4-MWCNTs@PEI-Ag Nanocomposite as a Reusable Catalyst. *RSC Advances,* **2020**, *10*(6), 3554-3565.
[http://dx.doi.org/10.1039/C9RA09561F]

[19] Liu, Y.; Yin, D.; Xin, Q.; Lv, M.; Bian, B.; Li, L.; Xie, C.; Yu, S.; Liu, S. Preparation of Highly

Dispersed Ru Nanoparticles Supported on Amine-Functionalized Magnetic Nanoparticles: Efficient Catalysts for the Reduction of Nitro Compounds. *Solid State Sci,* **2020**.
[http://dx.doi.org/10.1016/j.solidstatesciences.2019.106100]

[20] Tabatabaei Rezaei, S.J.; Khorramabadi, H.; Hesami, A.; Ramazani, A.; Amani, V.; Ahmadi, R. Chemoselective reduction of nitro and nitrile compounds with magnetic carbon nanotubes-supported pt(ii) catalyst under mild conditions. *Ind. Eng. Chem. Res.,* **2017**, *56*(43), 12256-12266.
[http://dx.doi.org/10.1021/acs.iecr.7b02795]

[21] Sharma, S.; Bhattacherjee, D.; Das, P. Supported rhodium nanoparticles catalyzed reduction of nitroarenes, arylcarbonyls and aryl/benzyl sulfoxides using ethanol/methanol as *in situ* hydrogen source. *Adv. Synth. Catal.,* **2018**, *360*(11), 2131-2137.
[http://dx.doi.org/10.1002/adsc.201701609]

[22] Kunfi, A.; Szabó, V.; Mastalir, Á.; Bucsi, I.; Mohai, M.; Németh, P.; Bertóti, I.; London, G. Palladium on Polydopamine: Its True Potential in Catalytic Transfer Hydrogenations and Heck Coupling Reactions. *ChemCatChem,* **2017**, *9*(16), 3236-3244.
[http://dx.doi.org/10.1002/cctc.201700609]

[23] Bravo, M.V.; Silva, O.F.; Adam, C.; Granados, A.M. A Simple Way to Prepare Palladium Nanoparticles Decorated with Cyclodextrins and Ionic Liquid. The Effects of Coating on the Catalytic Activity and Colloidal Stability. *J. Mol. Liq.,* **2020**, *304*, 112725.
[http://dx.doi.org/10.1016/j.molliq.2020.112725]

[24] Gholinejad, M.; Oftadeh, E.; Shojafar, M.; Sansano, J.M.; Lipshutz, B.H. Synergistic Effects of ppm Levels of Palladium on Natural Clinochlore for Reduction of Nitroarenes. *ChemSusChem,* **2019**, *12*(18), 4240-4248.
[http://dx.doi.org/10.1002/cssc.201901535] [PMID: 31390483]

[25] Kumar, A.; Paul, B.; Boukherroub, R.; Jain, S.L. Highly efficient conversion of the nitroarenes to amines at the interface of a ternary hybrid containing silver nanoparticles doped reduced graphene oxide/ graphitic carbon nitride under visible light. *J. Hazard. Mater.,* **2020**, *387*, 121700.
[http://dx.doi.org/10.1016/j.jhazmat.2019.121700] [PMID: 31806437]

[26] Singh, H.; Rajput, J.K. Novel Perovskite Nanocatalyst ($BiFeO_3$) for the Photodegradation of Rhodamine B/Tartrazine and Swift Reduction of Nitro Compounds. *J. Iran. Chem. Soc.,* **2019**, *16*(11), 2409-2432.
[http://dx.doi.org/10.1007/s13738-019-01710-6]

[27] Zeynizadeh, B.; Mohammad Aminzadeh, F.; Mousavi, H. Green and Convenient Protocols for the Efficient Reduction of Nitriles and Nitro Compounds to Corresponding Amines with $NaBH_4$ in Water Catalyzed by Magnetically Retrievable $CuFe_2O_3$ Nanoparticles. *Res. Chem. Intermed.,* **2019**, *45*(0123456789), 3329-3357.
[http://dx.doi.org/10.1007/s11164-019-03794-4]

[28] Bagherzade, A.; Nemati, F.; Nahzomi, H.T.; Elhampour, A. Experimental and Quantum Chemical Study on Nano-Copper Immobilized on Magnetic Graphitic Carbon Nitride Core Shell Particles; a Reusable Heterogeneous Catalyst toward Reduction of Nitro Arenes. *J. Mol. Struct.,* **2019**, *1185*, 38-49.
[http://dx.doi.org/10.1016/j.molstruc.2019.02.065]

[29] Song, T.; Duan, Y.; Chen, X.; Yang, Y. Switchable access to amines and imines from reductive coupling of nitroarenes with alcohols catalyzed by biomass-derived cobalt nanoparticles. *Catalysts,* **2019**, *9*(2), 116.
[http://dx.doi.org/10.3390/catal9020116]

[30] Bhaduri, K.; Das, B.D.; Kumar, R.; Mondal, S.; Chatterjee, S.; Shah, S.; Bravo-Suárez, J.J.; Chowdhury, B. Recyclable Au/SiO_2-Shell/Fe_3O_4-Core Catalyst for the Reduction of Nitro Aromatic Compounds in Aqueous Solution. *ACS Omega,* **2019**, *4*(2), 4071-4081.
[http://dx.doi.org/10.1021/acsomega.8b03655] [PMID: 31459616]

[31] Salahshournia, H.; Ghiaci, M. Pd-Pt/Modified GO as an Efficient and Selective Heterogeneous Catalyst for the Reduction of Nitroaromatic Compounds to Amino Aromatic Compounds by the Hydrogen Source. *Appl. Organomet. Chem.,* **2019**, *33*(4), 1-13. [http://dx.doi.org/10.1002/aoc.4832]

[32] Dhenadhayalan, N.; Lin, T.W.; Veerakumar, P.; Lin, K.C. Metal nanoparticles anchored on rhenium disulfide nanosheets as catalysts for the reduction of aromatic nitro compounds. *ChemNanoMat,* **2018**, *4*(12), 1262-1269. [http://dx.doi.org/10.1002/cnma.201800353]

[33] Iordanidou, D.; Zarganes-Tzitzikas, T.; Neochoritis, C.G.; Dömling, A.; Lykakis, I.N. Application of Silver Nanoparticles in the Multicomponent Reaction Domain: A Combined Catalytic Reduction Methodology to Efficiently Access Potential Hypertension or Inflammation Inhibitors. *ACS Omega,* **2018**, *3*(11), 16005-16013. [http://dx.doi.org/10.1021/acsomega.8b02749] [PMID: 30533584]

[34] Mullangi, D.; Chakraborty, D.; Pradeep, A.; Koshti, V.; Vinod, C.P.; Panja, S.; Nair, S.; Vaidhyanathan, R. Highly Stable COF-Supported Co/Co(OH)$_2$ nanoparticles heterogeneous catalyst for reduction of nitrile/nitro compounds under mild conditions. *Small,* **2018**, *14*(37), e1801233. [http://dx.doi.org/10.1002/smll.201801233] [PMID: 30062759]

[35] Eskandari, P.; Kazemi, F. Preparation and Characterization of Fe_3O_4/SiO_2/CdS Nanocomposites as Efficient Magnetic Photocatalysts for the Reduction of Nitro Compounds under Visible LED Irradiation. *J. Photochem. Photobiol. Chem.,* **2018**, *364*, 233-239. [http://dx.doi.org/10.1016/j.jphotochem.2018.06.011]

[36] Karami, S.; Zeynizadeh, B.; Shokri, Z. Cellulose supported bimetallic Fe–Cu nanoparticles: a magnetically recoverable nanocatalyst for quick reduction of nitroarenes to amines in water. *Cellulose,* **2018**, *25*(6), 3295-3305. [http://dx.doi.org/10.1007/s10570-018-1809-0]

[37] Lakshminarayana, B.; Satyanarayana, G.; Subrahmanyam, C. Bimetallic Pd-Au/TiO_2 nanoparticles: an efficient and sustainable heterogeneous catalyst for rapid catalytic hydrogen transfer reduction of nitroarenes. *ACS Omega,* **2018**, *3*(10), 13065-13072. [http://dx.doi.org/10.1021/acsomega.8b02064] [PMID: 31458027]

[38] Deka, P.; Deka, R.C.; Bharali, P. *In situ* generated copper nanoparticle catalyzed reduction of 4-nitrophenol. *New J. Chem.,* **2014**, *38*(4), 1789-1793. [http://dx.doi.org/10.1039/c3nj01589k]

[39] Gawande, M.B.; Goswami, A.; Felpin, F.X.; Asefa, T.; Huang, X.; Silva, R.; Zou, X.; Zboril, R.; Varma, R.S. Cu and Cu-based nanoparticles: synthesis and applications in catalysis. *Chem. Rev.,* **2016**, *116*(6), 3722-3811. [http://dx.doi.org/10.1021/acs.chemrev.5b00482] [PMID: 26935812]

[40] Nasrollahzadeh, M.; Nezafat, Z.; Gorab, M. G.; Sajjadi, M. Recent progresses in graphene-based (photo)catalysts for reduction of nitro compounds. *Mol. Catal,* **2020**, *484* [http://dx.doi.org/10.1016/j.mcat.2019.110758]

[41] Sorkhabi, S; Ghadermazi, M.; Mozafari, R. Adhesive functionalized ascorbic acid on $CoFe_2O_4$: a core-shell nanomagnetic heterostructure for the synthesis of aldoximes and Amines. *RSC Advances,* **2020**, *10*(68), 41336-41352. [http://dx.doi.org/10.1039/D0RA08244A]

[42] Ghadermazi, M.; Moradi, S.; Mozafari, R. Rice Husk-SiO_2supported Bimetallic Fe-Ni Nanoparticles: As a New, Powerful Magnetic Nanocomposite for the Aqueous Reduction of Nitro Compounds to Amines. *RSC Advances,* **2020**, *10*(55), 33389-33400. [http://dx.doi.org/10.1039/D0RA05381C]

[43] Chaudhari, C.; Shiraishi, M.; Nishida, Y.; Sato, K.; Nagaoka, K. One-Pot Synthesis of Pyrrolidones from Levulinic Acid and Amines/Nitroarenes/Nitriles over the Ir-PVP Catalyst. *Green Chem.,* **2020**,

22(22), 7760-7764.
[http://dx.doi.org/10.1039/D0GC01725F]

[44] Liu, N.; Ouyang, D.; Cai, Y.; Li, Y. Heterostructured $Fe_2O_3/BiVO_4$ nano-photocatalyst for the reduction of nitroarenes into amines. *Ceram. Int.,* **2020**, *46*(15), 24534-24543.
[http://dx.doi.org/10.1016/j.ceramint.2020.06.240]

[45] Rioux, R.M.; Song, H.; Grass, M.; Habas, S.; Niesz, K.; Hoefelmeyer, J.D.; Yang, P.; Somorjai, G.A. Monodisperse platinum nanoparticles of well-defined shape: synthesis, characterization, catalytic properties and future prospects. *Top. Catal.,* **2006**, *39*(3–4), 167-174.
[http://dx.doi.org/10.1007/s11244-006-0053-2]

[46] Mandal, S.; Roy, D.; Chaudhari, R.V.; Sastry, M. Pt and Pd Nanoparticles Immobilized on Amine-Functionalized Zeolite: Excellent Catalysts for Hydrogenation and Heck Reactions. *Chem. Mater.,* **2004**, *16*(19), 3714-3724.
[http://dx.doi.org/10.1021/cm0352504]

[47] Sun, Q.; Guo, C.Z.; Wang, G.H.; Li, W.C.; Bongard, H.J.; Lu, A.H. Fabrication of magnetic yolk-shell nanocatalysts with spatially resolved functionalities and high activity for nitrobenzene hydrogenation. *Chemistry,* **2013**, *19*(20), 6217-6220.
[http://dx.doi.org/10.1002/chem.201300307] [PMID: 23536472]

[48] Sarmah, P.P.; Dutta, D.K. Chemoselective reduction of a nitro group through transfer hydrogenation catalysed by ru-nanoparticles stabilized on modified montmorillonite clay. *Green Chem.,* **2012**, *14*(4), 1086-1093.
[http://dx.doi.org/10.1039/c2gc16441h]

[49] Li, Y.; Yu, J.; Li, W.; Fan, G.; Yang, L.; Li, F. The promotional effect of surface defects on the catalytic performance of supported nickel-based catalysts. *Phys. Chem. Chem. Phys.,* **2016**, *18*(9), 6548-6558.
[http://dx.doi.org/10.1039/C5CP07371E] [PMID: 26864098]

[50] Jagadeesh, R.V.; Banerjee, D.; Arockiam, P.B.; Junge, H.; Junge, K.; Pohl, M.M.; Radnik, J.; Brückner, A.; Beller, M. Highly Selective Transfer Hydrogenation of Functionalised Nitroarenes Using Cobalt-Based Nanocatalysts. *Green Chem.,* **2015**, *17*(2), 898-902.
[http://dx.doi.org/10.1039/C4GC00731J]

[51] Zhao, T.J.; Zhang, Y.N.; Wang, K.X.; Su, J.; Wei, X.; Li, X.H. General Transfer Hydrogenation by Activating Ammonia-Borane over Cobalt Nanoparticles. *RSC Advances,* **2015**, *5*(124), 102736-102740.
[http://dx.doi.org/10.1039/C5RA19869K]

[52] Liu, M.; Bai, Q.; Xiao, H.; Liu, Y.; Zhao, J.; Yu, W.W. Selective Hydrogenation of O-Chloronitrobenzene over Tin Dioxide Supported Platinum-Ruthenium Bimetallic Nanocatalysts without Solvent. *Chem. Eng. J.,* **2013**, *232*, 89-95.
[http://dx.doi.org/10.1016/j.cej.2013.07.078]

[53] Corbos, E.C.; Ellis, P.R.; Cookson, J.; Briois, V.; Hyde, T.I.; Sankar, G.; Bishop, P.T. Tuning the Properties of PdAu Bimetallic Nanocatalysts for Selective Hydrogenation Reactions. *Catal. Sci. Technol.,* **2013**, *3*(11), 2934-2943.
[http://dx.doi.org/10.1039/c3cy00255a]

[54] Shanbogh, P.P.; Peter, S.C. Low cost nano materials crystallize in the NIAS structure type as an alternative to the noble metals in the hydrogenation process. *RSC Advances,* **2013**, *3*(45), 22887-22890.
[http://dx.doi.org/10.1039/c3ra44744h]

[55] Li, T.; Zhang, W.; Lee, R.Z.; Zhong, Q. Nickel-boron alloy catalysts reduce the formation of trans fatty acids in hydrogenated soybean oil. *Food Chem.,* **2009**, *114*(2), 447-452.
[http://dx.doi.org/10.1016/j.foodchem.2008.09.068]

[56] Chen, Y.C.; Tan, C.S. Hydrogenation of P-Chloronitrobenzene by Ni-B Nanocatalyst in CO_2-

Expanded Methanol. *J. Supercrit. Fluids,* **2007**, *41*(2), 272-278. [http://dx.doi.org/10.1016/j.supflu.2006.10.009]

[57] Yoo, J.W.; Hathcock, D.J.; El-Sayed, M.A. Propene Hydrogenation over Truncated Octahedral Pt Nanoparticles Supported on Alumina. *J. Catal.,* **2003**, *214*(1), 1-7. [http://dx.doi.org/10.1016/S0021-9517(02)00136-7]

[58] Lee, H.; Habas, S.E.; Kweskin, S.; Butcher, D.; Somorjai, G.A.; Yang, P. Morphological control of catalytically active platinum nanocrystals. *Angew. Chem. Int. Ed.,* **2006**, *45*(46), 7824-7828. [http://dx.doi.org/10.1002/anie.200603068] [PMID: 17089428]

[59] Wang, F.; Li, C.; Sun, L.D.; Xu, C.H.; Wang, J.; Yu, J.C.; Yan, C.H. Porous single-crystalline palladium nanoparticles with high catalytic activities. *Angew. Chem. Int. Ed. Engl.,* **2012**, *51*(20), 4872-4876. [http://dx.doi.org/10.1002/anie.201107376] [PMID: 22473999]

[60] Ikeda, S.; Ishino, S.; Harada, T.; Okamoto, N.; Sakata, T.; Mori, H.; Kuwabata, S.; Torimoto, T.; Matsumura, M. Ligand-free platinum nanoparticles encapsulated in a hollow porous carbon shell as a highly active heterogeneous hydrogenation catalyst. *Angew. Chem. Int. Ed.,* **2006**, *45*(42), 7063-7066. [http://dx.doi.org/10.1002/anie.200602700] [PMID: 17009288]

[61] Jacinto, M.J.; Landers, R.; Rossi, L.M. Preparation of Supported Pt(0) Nanoparticles as Efficient Recyclable Catalysts for Hydrogenation of Alkenes and Ketones. *Catal. Commun.,* **2009**, *10*(15), 1971-1974. [http://dx.doi.org/10.1016/j.catcom.2009.07.011]

[62] Sun, Z.; Rong, Z.; Wang, Y.; Xia, Y.; Du, W.; Wang, Y. Selective hydrogenation of cinnamaldehyde over pt nanoparticles deposited on reduced graphene oxide. *RSC Advances,* **2014**, *4*(4), 1874-1878. [http://dx.doi.org/10.1039/C3RA44962A]

[63] Dominguez-Quintero, O.; Martinez, S.; Henriquez, Y.; D'Ornelas, L.; Krentzien, H.; Osuna, J. Silica-supported palladium nanoparticles show remarkable hydrogenation catalytic activity. *J. Mol. Catal. Chem.,* **2003**, *197*(1–2), 185-191. [http://dx.doi.org/10.1016/S1381-1169(02)00583-6]

[64] Wu, T.; Jiang, T.; Hu, B.; Han, B.; He, J.; Zhou, X. Cross-Linked Polymer Coated Pd Nanocatalysts on SiO_2 Support: Very Selective and Stable Catalysts for Hydrogenation in Supercritical CO_2. *Green Chem.,* **2009**, *11*(6), 798-80. [http://dx.doi.org/10.1039/b822930a]

[65] Kim, Y.; Kim, M.J. Magnetically Recoverable Palladium Nanocatalyst for Chemoselective Olefin Hydrogenation. *Bull. Korean Chem. Soc.,* **2010**, *31*(5), 1368-1370. [http://dx.doi.org/10.5012/bkcs.2010.31.5.1368]

[66] Yarulin, A.; Yuranov, I.; Cárdenas-Lizana, F.; Abdulkin, P.; Kiwi-Minsker, L. Size-Effect of Pd-(Poly(N -Vinyl-2-Pyrrolidone)) Nanocatalysts on Selective Hydrogenation of Alkynols with Different Alkyl Chains. *J. Phys. Chem. C,* **2013**, *117*(26), 13424-13434. [http://dx.doi.org/10.1021/jp402258s]

[67] Gavia, D.J.; Koeppen, J.; Sadeghmoghaddam, E.; Shon, Y.S. Tandem semi-hydrogenation/isomerization of propargyl alcohols to saturated carbonyl analogues by dodecanethiolate-capped palladium nanoparticle catalysts. *RSC Advances,* **2013**, *3*(33), 13642-13645. [http://dx.doi.org/10.1039/c3ra42119h] [PMID: 27642511]

[68] Acham, V.R.; Biradar, A.V.; Dongare, M.K.; Kemnitz, E.; Umbarkar, S.B. Palladium Nanoparticles Supported on Magnesium Hydroxide Fluorides: A Selective Catalyst for Olefin Hydrogenation. *ChemCatChem,* **2014**, *6*(11), 3182-3191. [http://dx.doi.org/10.1002/cctc.201402469]

[69] Chen, L.; Huang, B.; Qiu, X.; Wang, X.; Luque, R.; Li, Y. Seed-mediated growth of MOF-encapsulated Pd@Ag core-shell nanoparticles: toward advanced room temperature nanocatalysts. *Chem. Sci. (Camb.),* **2016**, *7*(1), 228-233.

[http://dx.doi.org/10.1039/C5SC02925B] [PMID: 28758001]

[70] Pélisson, C.H.; Vono, L.L.R.; Hubert, C.; Denicourt-Nowicki, A.; Rossi, L.M.; Roucoux, A. Moving from surfactant-stabilized aqueous rhodium(0) colloidal suspension to heterogeneous magnetite-supported rhodium nanocatalysts: synthesis, characterization and catalytic performance in hydrogenation reactions. *Catal. Today,* **2012**, *183*(1), 124-129. [http://dx.doi.org/10.1016/j.cattod.2011.08.046]

[71] Faria, V.W.; Brunelli, M.F.; Scheeren, C.W. Iridium nanoparticles supported in polymeric membranes: a new material for hydrogenation reactions. *RSC Advances,* **2015**, *5*(103), 84920-84926. [http://dx.doi.org/10.1039/C5RA16426E]

[72] Kim, D.J.; Li, Y.; Kim, Y.J.; Hur, N.H.; Seo, W.S. A highly stable and magnetically recyclable nanocatalyst system: mesoporous silica spheres embedded with FECO/graphitic shell magnetic nanoparticles and pt nanocatalysts. *Chem. Asian J.,* **2015**, *10*(12), 2755-2761. [http://dx.doi.org/10.1002/asia.201500773] [PMID: 26312570]

[73] Ren, M.; Li, C.; Chen, J.; Wei, M.; Shi, S. Preparation of a ternary Pd-Rh-P amorphous alloy and its catalytic performance in selective hydrogenation of alkynes. *Catal. Sci. Technol.,* **2014**, *4*(7), 1920-1924. [http://dx.doi.org/10.1039/C4CY00338A]

[74] Karakhanov, E.A.; Maximov, A.L.; Zolotukhina, A.V.; Yatmanova, N.; Rosenberg, E. Alkyne Hydrogenation Using Pd-Ag Hybrid Nanocatalysts in Surface-Immobilized Dendrimers. *Appl. Organomet. Chem.,* **2015**, *29*(11), 777-784. [http://dx.doi.org/10.1002/aoc.3367]

[75] Zhu, L.; Sun, H.; Fu, H.; Zheng, J.; Zhang, N.; Li, Y.; Chen, B.H. Effect of Ruthenium Nickel Bimetallic Composition on the Catalytic Performance for Benzene Hydrogenation to Cyclohexane. *Appl. Catal. A Gen.,* **2015**, *499*, 124-132. [http://dx.doi.org/10.1016/j.apcata.2015.04.016]

[76] Kar, A.K.; Srivastava, R. An Efficient and Sustainable Catalytic Reduction of Carbon-Carbon Multiple Bonds, Aldehydes, and Ketones Using a Cu Nanoparticle Decorated Metal Organic Framework. *New J. Chem.,* **2018**, *42*(12), 9557-9567. [http://dx.doi.org/10.1039/C8NJ01704B]

[77] Emanuela Calcio Gaudino 1 , Elisa Acciardo 1, S. T. 2; Giancarlo Cravotto 1, and R. S. V. 3 M. M. Cross-Linked Cyclodextrins Bimetallic Nanocatalysts: Applications in Microwave-Assisted Reductive Aminations. **2015**.

[78] Han, R.; Nan, C.; Yang, L.; Fan, G.; Li, F. Direct synthesis of hybrid layered double hydroxide-carbon composites supported Pd nanocatalysts efficient in selective hydrogenation of citral. *RSC Advances,* **2015**, *5*(42), 33199-33207. [http://dx.doi.org/10.1039/C5RA03201F]

[79] Gopiraman, M.; Ganesh Babu, S.; Khatri, Z.; Kai, W.; Kim, Y.A.; Endo, M.; Karvembu, R.; Kim, I.S. Dry synthesis of easily tunable nano ruthenium supported on graphene: novel nanocatalysts for aerial oxidation of alcohols and transfer hydrogenation of ketones. *J. Phys. Chem. C,* **2013**, *117*(45), 23582-23596. [http://dx.doi.org/10.1021/jp402978q]

[80] Yang, Z.; Zhu, Z.; Luo, R.; Qiu, X.; Liu, J.T.; Yang, J.K.; Tang, W. Iridium-catalyzed highly efficient chemoselective reduction of aldehydes in water using formic acid as the hydrogen source. *Green Chem.,* **2017**, *19*(14), 3296-3301. [http://dx.doi.org/10.1039/C7GC01289F]

[81] Gao, Z.; Dong, M.; Wang, G.; Sheng, P.; Wu, Z.; Yang, H.; Zhang, B.; Wang, G.; Wang, J.; Qin, Y. Multiply confined nickel nanocatalysts produced by atomic layer deposition for hydrogenation reactions. *Angew. Chem. Int. Ed. Engl.,* **2015**, *54*(31), 9006-9010. [http://dx.doi.org/10.1002/anie.201503749] [PMID: 26150352]

[82] Li, K.T.; Hsu, M.H.; Wang, I. Palladium core-porous silica shell-nanoparticles for catalyzing the hydrogenation of 4-carboxybenzaldehyde. *Catal. Commun.*, **2008**, *9*(13), 2257-2260. [http://dx.doi.org/10.1016/j.catcom.2008.05.012]

[83] Ji, Y.; Wu, Y.; Zhao, G.; Wang, D.; Liu, L.; He, W.; Li, Y. Porous bimetallic Pt-Fe nanocatalysts for highly efficient hydrogenation of acetone. *Nano Res.*, **2015**, *8*(8), 2706-2713. [http://dx.doi.org/10.1007/s12274-015-0777-z]

[84] Tsang, S.C.; Cailuo, N.; Oduro, W.; Kong, A.T.S.; Clifton, L.; Yu, K.M.K.; Thiebaut, B.; Cookson, J.; Bishop, P. Engineering preformed cobalt-doped platinum nanocatalysts for ultraselective hydrogenation. *ACS Nano,* **2008**, *2*(12), 2547-2553. [http://dx.doi.org/10.1021/nn800400u] [PMID: 19206291]

[85] Wang, F.; Zhang, Z. Catalytic transfer hydrogenation of furfural into furfuryl alcohol over magnetic γ-Fe_2O_3@HAP catalyst. *ACS Sustain. Chem.& Eng.*, **2017**, *5*(1), 942-947. [http://dx.doi.org/10.1021/acssuschemeng.6b02272]

[86] Ruiz, D.; Mäki-Arvela, P.; Aho, A.; Chimentão, R.; Claver, C.; Godard, C.; Fierro, J.L.G.; Murzin, D.Y. Immobilized chiral rhodium nanoparticles stabilized by chiral P-Ligands as efficient catalysts for the enantioselective hydrogenation of 1-Phenyl-1,2-propanedione. *Mol. Catal.,* **2019**, *477*(June), 110551. [http://dx.doi.org/10.1016/j.mcat.2019.110551]

[87] Yue, F.; Hu, A.; Lu, X.; Wang, C.; Pan, H.; Zhang, H.; Yang, L.; Xia, Q.; Zhou, D.; Xia, Y. Mild-temperature hydrogenation of carbonyls over Co-ZIF-9 derived Co-ZIF-x nanoparticle catalyst. *Mol. Catal.,* **2020**, *495*(April), 111149. [http://dx.doi.org/10.1016/j.mcat.2020.111149]

[88] Liu, P.; Zhu, Y.L.; Zhou, L.; Zhang, W.H.; Li, Y.X. Amorphous nickel phosphide nanoparticles for selective hydrogenation of cinnamaldehyde. *Catal. Lett.*, **2020**, *150*(9), 2695-2702. [http://dx.doi.org/10.1007/s10562-019-03022-z]

[89] Nişancı, B.; Dağalan, Z. A facile and highly efficient transfer hydrogenation of ketones and aldehydes catalyzed by palladium nanoparticles supported on mesoporous graphitic carbon nitride. *J. Chem. Res.,* **2020**, *44*(1–2), 14-19. [http://dx.doi.org/10.1177/1747519819883792]

[90] Watari, R.; Kayaki, Y. Copper catalysts unleashing the potential for hydrogenation of carbon–oxygen bonds. *Asian J. Org. Chem.,* **2018**, *7*(10), 2005-2014. [http://dx.doi.org/10.1002/ajoc.201800436]

[91] Virgili, A.H.; Luza, L.; Fernandes, J.A.; Costa, T.M.H.; de Menezes, E.W.; Benvenutti, E.V. Self-supported gold/chitosan nanocatalyst for chemoselective hydrogenation in Π-conjugated C[Dbnd]C–C[Dbnd]O system. *Catal. Commun.,* **2018**, *116*, 32-37. [http://dx.doi.org/10.1016/j.catcom.2018.08.003]

[92] Landry, M.J.; Gellé, A.; Meng, B.Y.; Barrett, C.J.; Moores, A. Surface-plasmon-mediated hydrogenation of carbonyls catalyzed by silver nanocubes under visible light. *ACS Catal.,* **2017**, *7*(9), 6128-6133. [http://dx.doi.org/10.1021/acscatal.7b02128]

[93] Sharma, A.K.; Joshi, H.; Bhaskar, R.; Singh, A.K. Solvent-tailored $Pd_3P_{0.95}$ nano catalyst for amide-nitrile inter-conversion, the hydration of nitriles and transfer hydrogenation of the CO bond. *Dalton Trans.,* **2019**, *48*(29), 10962-10970. [http://dx.doi.org/10.1039/C8DT04667K] [PMID: 31180393]

CHAPTER 7

Recent Trends in Metal Nanoparticles (MNPs) Catalyzed Synthesis of Aza- and Oxa-Heterocycles

Tejas M. Dhameliya[1,*], **Rutvi J. Patel**[1], **Normi D. Gajjar**[1], **Rajvi H. Amin**[1], **Kunjan B. Bodiwala**[1] and **Dipen K. Sureja**[1]

[1] *L. M. College of Pharmacy, Navrangpura, Ahmedabad 380 009, Gujarat, India*

Abstract: Most of the currently marketed and approved drugs possess nitrogen and oxygen-containing heterocycles in their scaffolds. The importance of metal nanoparticles (MNPs) catalyzed synthesis of such aza- and oxa-heterocycles should not be underestimated due to their versatile applications. Previously, we reported the decennary update on MNPs catalyzed synthesis of *N*- and *O*-containing heterocycles reported during 2010-2019. In the present work, attempts have been made to summarize the prospects to highlight those promising MNPs of worthy catalytic potential reported during the year, 2020. The present compilation shall provide the new thought processes to address the unmet synthetic needs for the construction of nitrogen and oxygen-containing heterocyclic scaffolds.

Keywords: Catalysis, Heterocycles, Metal nanoparticles, Nanocatalysis.

INTRODUCTION

The medicinal importance of heterocyclic scaffolds should not be underestimated due to their wide spectrum of biological activities against pathogenic mycobacterial [1 - 7], bacterial [8] fungal [9], and viral strains [10]. Among the United States food and drug administration (USFDA) approved drugs, nitrogen [11] and oxygen [12] containing heterocycles ranked among the predominant heterocycles. Organic synthetic chemistry has been considered as the bottleneck for the transformation of drug discovery through new avenues such as C-C, C-heteroatom bond formation, and C-H activation [13]. To achieve the synthesis of these scaffolds, the enriched medicinal chemists' synthetic armory or toolbox [14] can be easily constructed using the green chemistry protocols such as reaction in water [15], ionic liquids [16 - 18], and nanoparticles (NPs) [19, 20].

* **Corresponding author Tejas M. Dhameliya:** L. M. College of Pharmacy, Navrangpura, Ahmedabad 380 009, Gujarat, India; E-mail: tejas.dhameliya@lmcp.ac.in; tmdhameliya@gmail.com Tel: +91 79 2630 2746.

SCOPE AND APPLICATIONS OF NPS IN SYNTHESIS OF HETEROCYCLES

Nowadays, the avenue for the growth of nanotechnology has opened several new vistas in synthetic chemistry, peculiarly in the arena of synthesis of novel biodynamic heterocycles [21]. To catalyze the organic transformations, the catalytic assistance of MNPs has emerged as sustainable alternatives to conventional materials, as robust, high surface-area containing heterogeneous catalysts with the potential of extendable recyclability [22]. MNPs have attained significant importance in green chemistry developments in benign aqueous reaction media for the construction of heterocycles [23]. Despite some potential hazards [24], MNPs still have greater triumph in the field of material chemistry with the positive spirit of versatile reactions to catalyze the synthetic transformations. Plasmodic gold NPs (AuNPs) have also emerged as important NPs in the field of advanced energy, environmental and biomedical applications [25].

Recently, we have reviewed the synthesis of *N*- and *O*-containing heterocyclic scaffolds catalyzed by MNPs for the decade of 2010 to the end of 2019 [26]. In the present work, we have presented the most recent trends of MNPs catalyzed synthesis of aza and oxa-heterocyclic scaffolds reported by several research groups during 2020 (Fig. **1**), which shall impart the inherent intuitions to the readers to design the synthetic strategy of desired heterocyclic scaffolds using NPs in catalytic amount along with the mechanistic insights of MNPs.

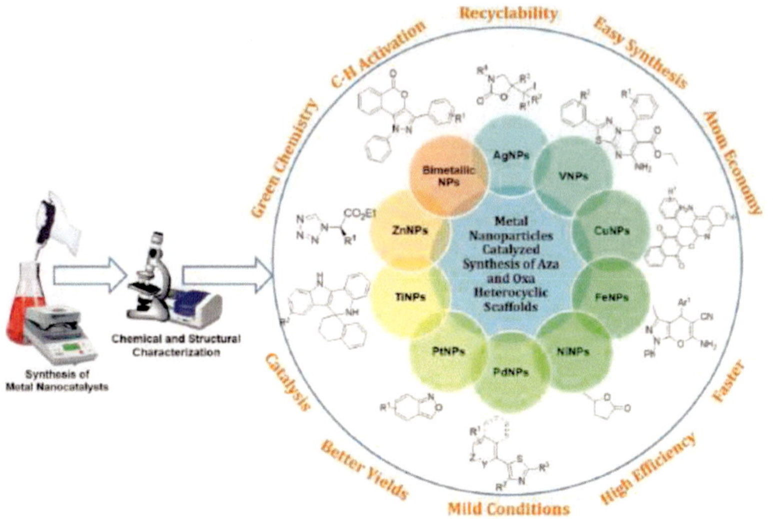

Fig. (1). Schematic representation of NPs with their catalytic applications in the synthesis of aza and oxa heterocycles.

AgNPs Catalyzed Synthesis of Heterocycles

Differently shaped silver nanoparticles (AgNPs) have been recognized in several forms such as cubic, rods, wires, bars, pyramidal, prisms, flower-shaped, and spherical due to their unique morphological properties [27]. Zhang *et al.* have performed carboxylative cyclization of propargylic alcohols (**1**) with CO_2 under atmospheric pressure using silver (0) nanoparticles [(Ag-rGO)] as catalyst and tetra butyl ammonium triazolide $[N_{4444}]$[Triz] (0.02 mmol) at 30 °C for 12 h (Scheme **1**) [28]. Ag-rGO has been prepared from the reduction of graphene oxide and $AgNO_3$ and has been characterized by X-ray diffraction (XRD), energy-dispersive XRD (EDX), scanning electron microscopy (SEM), transition electron microscopy (TEM), Fourier-transform infrared (FT-IR), and X-ray photoelectron spectroscopy (XPS). Ag-rGO catalysts show good stability and can be reused up to at least five times without obvious activity loss and integrity of the catalysts. An increase in the yield of **2** has been found while increasing the loading of AgNPs on rGO. However, a further increase in catalytic loading of AgNPs leads to severe aggregations of AgNPs.

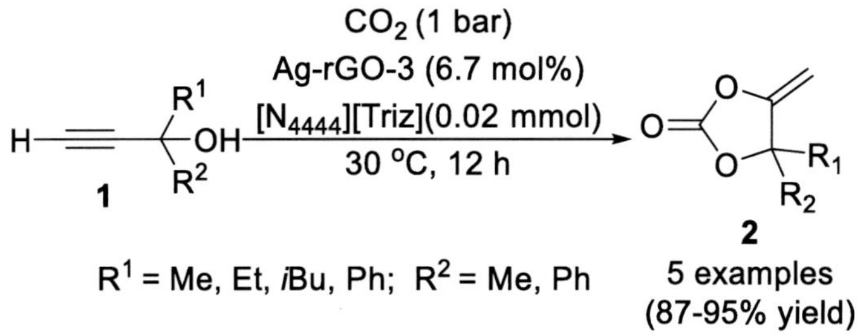

Scheme (1). Carboxylative cyclization of propargylic alcohols with CO_2 under atmospheric pressure catalyzed by silver (0) NPs (Ag-rGO).

Ghosh *et al.* have performed carbonylative cyclization of allylamines (**3**) to form cyclic carbamates (**4**) using highly reactive organic nanomaterials like AgNPs decorated covalent organic framework (COF) such as Ag@TpTta or Ag@TpPa-1 in the presence of *N*-iodosuccinimide (NIS) at 25 °C for 16 h without using any solvent (Scheme **2**) [29]. AgNPs architecture COFs (Ag@TpPa-1 and Ag@TpTta) were synthesized by decorating AgNPs on the exterior surfaces of TpPa-1 and TpTta, wherein TpTta has been prepared by the treatment of 4,4′,4″-(1,3,5-triazine-2,4,6-triyl)-trianiline with 1,3,5-triformyl phloroglucinol and TpPa-1 has been synthesized from trifluoroacetic acid and phloroglucinol. The nanoparticles have been characterized using field emission SEM (FE-SEM), high-resolution transition electron microscopy (UHR-TEM), thermogravimetry analysis (TGA), Brunner-Emmett-Teller analysis (BET), 1H and 13C NMR, transition electron microscopy (TEM), FT-IR, UV-visible spectroscopy (UV-Vis), XPS and powder XRD (PXRD). The catalysts Ag@TpTta and Ag@TpPa-1 exhibited excellent catalytic activity, stability, and selectivity towards the synthesis of **4**. The AgNPs (Ag@TpTta) could be easily recovered and reused for up to five cycles without any noticeable decay in substrate conversion.

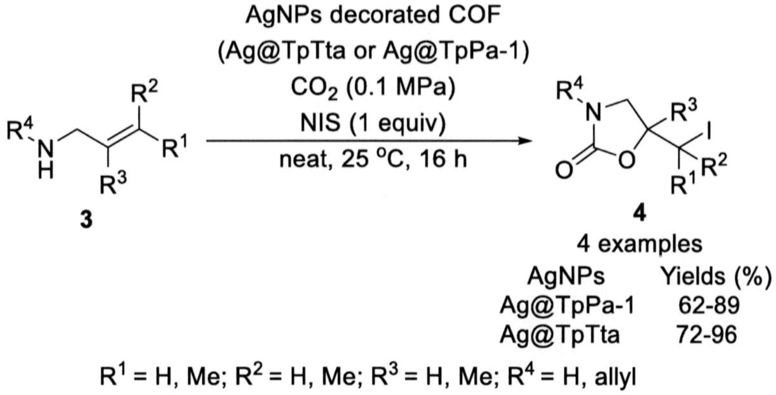

Scheme (2). Cyclization of allylamines (**3**) to cyclic carbamates (**4**) with CO_2 catalyzed by AgNPs architecture on COF.

Bamoharram *et al.* reported the catalytic use of AgNPs with Preyssler functionalized cellulose biocomposites (PC/AgNPs) for the one-pot, multicomponent synthesis of 2-amino-4*H*-pyrans (**8**/**10a**/**12**/**14a**) from malononitrile or ethyl cyanoacetate (**6a**), substituted benzaldehyde (**7a**) and various 1,3-diketones such as barbituric acid (**5a**)/dimedone (**9a**)/4-hydroxycoumarin (**11a**)/3-methyl 1*H*-pyrazol-5(4*H*)-one (**13a**) at reflux in EtOH/H_2O (1:1, Scheme **3**) [30]. The final MNPs have been synthesized by treatment of AgNPs with $AgNO_3$ and Preyssler functionalized cellulose biocomposites and characterized using SEM, TEM, UV-Vis, and FT-IR. The catalyst was recovered and reused three times without diminishing its efficiency.

CuNPs Catalyzed Synthesis of Heterocycles

Abiraman *et al.* performed the synthesis of 1,2,3-triazoles (**17**/**19**) from azides (**15**/**18**) and alkynes (**16**) with PAACC [poly(acrylic acid)-capped copper] NP as a catalyst, using sodium ascorbate as a mild base in the THF: water (1:1) at rt within 3 h (Scheme **4**) [31]. The PAACC NPs were synthesized using a high-intensity ultrasound sonication (30 kHz) method. The reduction of CuNPs from copper (II) salt by mild reducing agent L-ascorbic acid in an aqueous medium was achieved in the presence of poly(acrylic acid). The PAACC NPs were characterized by Differential reflectance spectroscopy (DRS) UV-Vis, XPS, FE-SEM, and high-resolution TEM (HRTEM). The PAACC NPs demonstrate good catalytic activity for the synthesis of 1,2,3-triazole compounds *via* click reaction with excellent yields under mild reaction conditions.

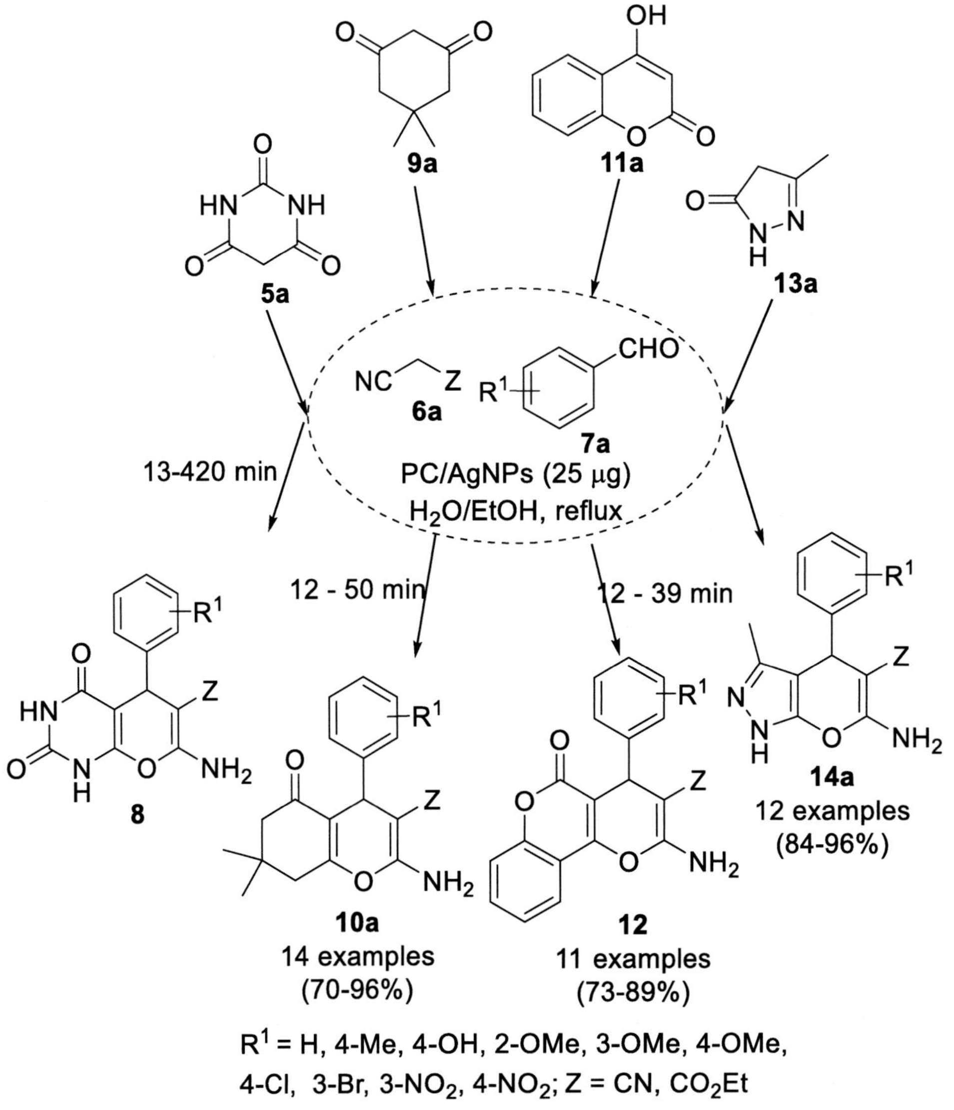

Scheme (3). Synthesis of 2-amino-4*H*-pyrans (**8**/**10a**/**12**/**14a**) catalyzed by PC/Ag NPs.

Taheri *et al.* reported the catalytic activity of Cu-metal organic frameworks (Cu-MOF) as nano-catalysts (NCs) for the one-pot synthesis of tacrine derivative (**22**) through a four-component reaction of benzaldehyde (**7a**), 2-hydroxynaphthalene-

1,4-dione (**20a**), malononitrile (**6b**), and cyclohexanone (**21**) in the presence of $AlCl_3$ as Lewis acid catalyst in dichloroethane in 2.5-5 h (Scheme **5**) [32]. The Cu-MOF turquoise powder has been synthesized by treating copper (II) nitrate trihydrate with 1,4-benzenedioic acid in *N,N'*-dimethylformamide (DMF) followed by consecutive autoclaving, cooling, and centrifugation. The characterization of the Cu-MOF has been achieved using FT-IR, PXRD, and SEM. The NCs have shown satisfactory performance up to five cycles in the recyclability experiments.

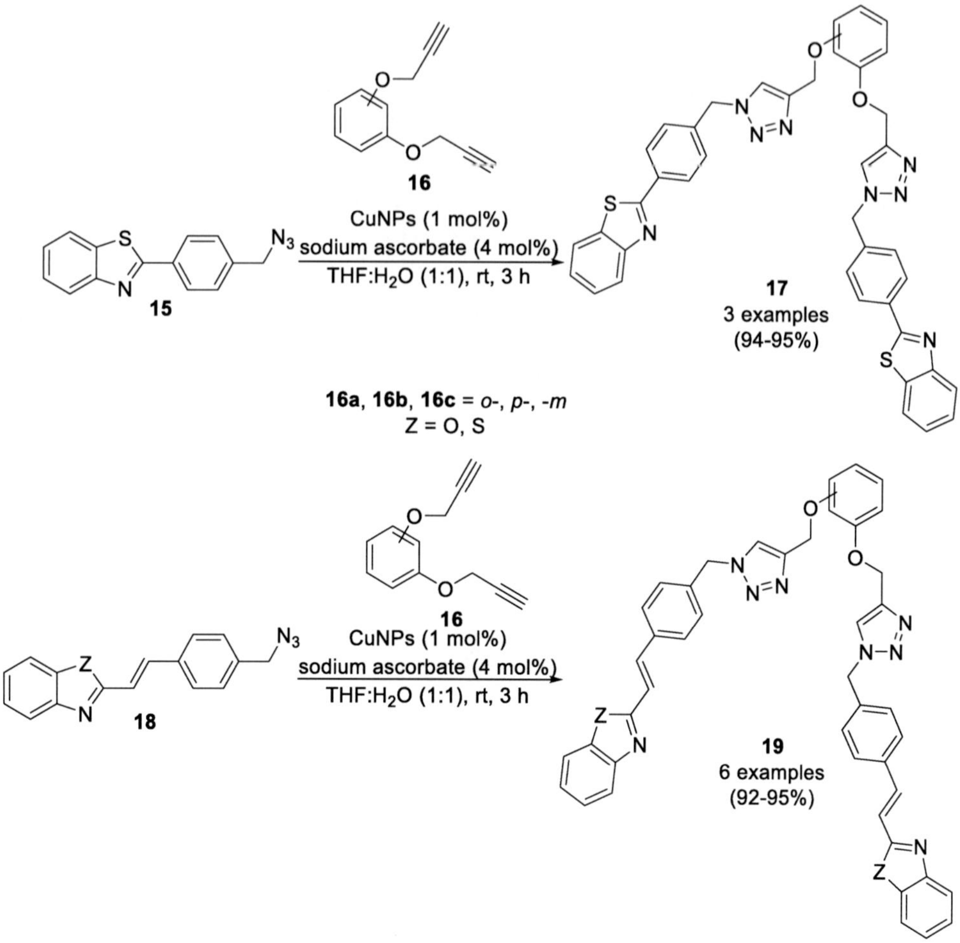

Scheme (4). Synthesis of 1,2,3-triazole using azides and alkynes in presence of PAACC [poly(acrylic acid)-capped copper] NPs as catalysts.

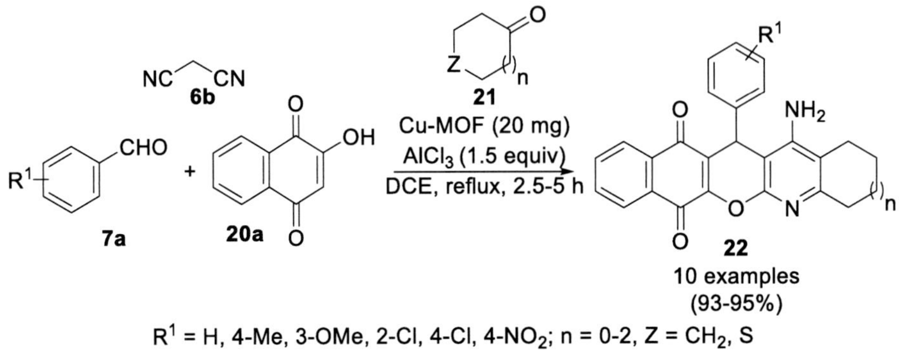

Scheme (5). One-pot synthesis of tacrine derivatives (**22**) catalyzed by Cu-MOF.

Ghorbani-Vaghei *et al.* have performed the synthesis of imidazo[1,2-*a*] pyridines (**25**) from the reaction of 2-aminopyridine (**23a**), phenylacetylene (**24**), and aldehydes (**7a**) catalyzed by the heterogeneous catalysts, immobilized CuI NPs on porous cross-linked poly(ethyleneamine)-polysulfonamide (PEA-PSA@CuI) in ethanol for 1.5-6 h (Scheme **6**) [33]. The final NCs have been constructed from benzene disulfonyl chloride, ethylenediamine, and tetrakis(2-aminoethyl)ammonium on SiO_2 NPs followed by treatment with HF and CuI NPs, which have been characterized using FT-IR and FE-SEM. The reusability of PEA-PSA@CuI NPs has been studied up to six runs with a nominal decrease in its activity.

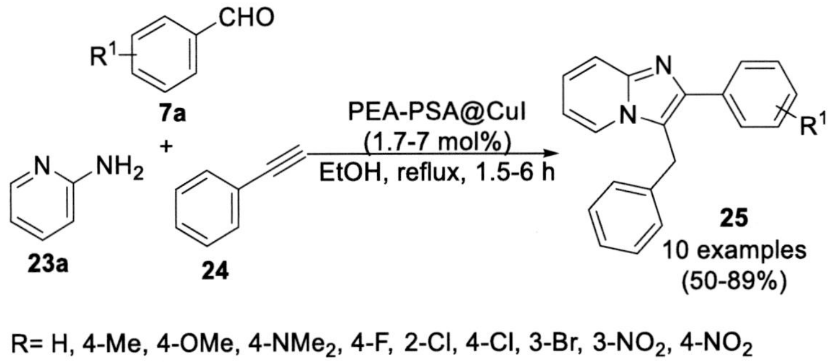

Scheme (6). PEA-PSA@CuI catalyzed synthesis of imidazo[1,2-*a*]pyridines (**25**) from 2-aminopyridine (**23a**), phenylacetylene (**24**), and benzaldehyde (**7a**).

FeNPs Catalyzed Synthesis of Heterocycles

Abbasabadi and Azarifar have reported a novel one-pot three-component synthesis of 1*H*-pyrazolo[1,2-*b*]phthalazine-5,10-dione (**27**) from 2,3-dihydrophthalazine-1,4-dione (**26**), malononitrile (**6b**) and aromatic aldehyde (**7b**); and 2,3-dihydroquinazolin-4(1*H*)-one derivative (**30**) from aryl aldehyde

(**7b**), arylamine (**28a**) and isatoic anhydride (**29**) in presence of magnetic β-alanine-functionalized-graphene oxide quantum dots Fe_3O_4@GOQDs-*N*-(--alanine) as a catalyst, and water as green solvent at rt (Scheme **7**) [34]. The respective steps for the synthesis of catalyst Fe_3O_4@GOQDs-*N*-(β-alanine) nanomagnetic sheets include the formation of GOQDs from freshly synthesized GO, and then the former was acyl chlorinated by treatment with thionyl chloride followed by nucleophilic reaction with β-alanine under reflux condition. Finally, the resulting GOQDs-*N*-(β-alanine) were nano-magnetized by co-precipitation of ferrous (Fe^{2+}) and ferric (Fe^{3+}) ions in the presence of the resulting aminated GOQDs. The characterization of nanomagnetic catalysts was done using FT-IR, XRD, FE-SEM, HR-TEM, TGA, EDX, and vibrating sample magnetometry (VSM). The catalysts have shown recycling up to six cycles with significant reusability.

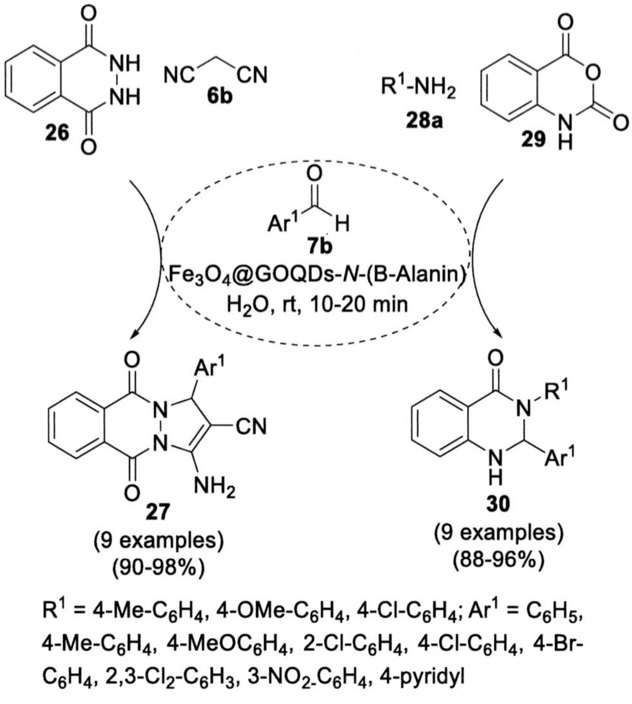

Scheme (7). One-pot three-component synthesis of 1*H*-pyrazolo[1,2-*b*] phthalazine-5,10-diones (**27**) and 2,3 dihydroquinazolin-4(1*H*)-one derivatives (**30**).

Rahmati *et al.* have performed the one-pot synthesis of tetrahydrobenzimidazo [2,1-*b*]quinazolin-1(2*H*)-ones (**32**) using aldehyde (**7c**), 2-aminobenzimidazole

(**31a**), and cyclic *β*-diketone (**9b**) catalyzed by silica-coated Fe_3O_4 NPs@silyl-propyl triethylammonium polyoxometalate in the presence of ethanol as a green solvent at 70-75 °C (Scheme **8**) [35]. The surface of Fe_3O_4 MNPs, prepared by aqueous co-precipitation using $FeCl_2 \cdot 4H_2O$ and $FeCl_3 \cdot 6H_2O$, have been treated with triethoxy silyl propyl chloride and triethylamine to construct nanomagnetic organic-inorganic hybrid catalysts (NM-OIHC), which have been characterized using atomic absorption spectroscopy (AAS), inductively coupled plasma-optical emission spectrometry (ICP-OES), TGA, FT-IR, SEM, EDX, and VSM. The catalysts demonstrated excellent reusability for up to ten runs without losing the inherent properties.

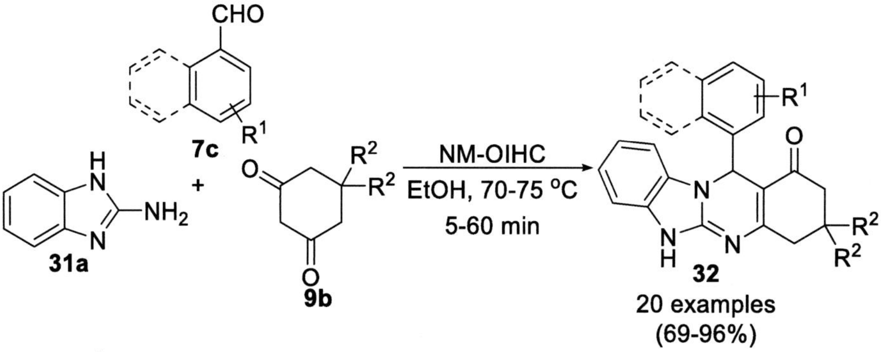

R¹ = H, 4-F, 2-Me, 4-*i*Pr, 4-Ph, 4-Cl, 2,4-(OMe)₂, 2,4-(Cl)₂, 4-Br, 2-Br, 3-F, 4-OH, 3-Br, 4-Me, 4-OMe, 4-NO₂, 1-naphthyl; R² = H, Me

Scheme (8). One-pot synthesis of tetrahydrobenzimidazo[2,1-*b*]quinazolin-1(2*H*)-ones (**32**).

Abbasabadi *et al.* performed one-pot three-component synthesis of dihydro pyrano[2,3-*c*]pyrazoles (**33**)/4*H*-chromenes (**10b**) from aldehyde (**7a**), malononitrile (**6b**) and 3-methyl-1-phenyl-1*H*-pyrazol-5(4*H*)-one (**13b**)/dimedone (**9a**) using MNPs loaded on functionalized graphene oxide quantum dots [Fe_3O_4@GOQD-*O*-(propane-1-sulfonic acid)] in deionised water at rt (Scheme **9**) [36]. The magnetic nanosheets were obtained by coprecipitation of ferrous (Fe^{2+}) and ferric (Fe^{3+}) ions in the presence of the GOQD-*O*-(propane-1-sulfonic acid) and have been characterized using FT-IR, XRD, EDX, TGA, FESEM, HR-TEM, and VSM. The NCs could be reused up to five times without considerable reduction of catalytic performance.

Arisawa *et al.* performed the liquid phase combinatorial synthesis of carbazole derivatives (**36**) from 1-azide-2-iodo(bromo) benzenes (**34**) and phenylboronic acids (**35**) catalyzed by sulfur-modified Au-supported Fe(0) [SAFe(0)] NPs in presence of 1,4-dioxane, potassium carbonate at 110 °C for 10 h and followed by removal of catalysts and addition of water under reflux for 12 h (Scheme **10**) [37]. MNPs have been prepared by the treatment of Au mesh with aqueous sulfuric acid and sodium thiosulphate (collectively called piranha solution) followed by immobilization of Fe(0)NPs using 1,4-bis(trimethylsilyl)-1,4-dihydropyrazine (Si-DHP) as a strong reducing agent. The characterization of MNPs has been achieved using SEM, TEM, X-ray absorption near edge structure (XANES), and XPS techniques. The MNPs have been found with excellent reusability characteristics for up to ten runs without losing the catalytic efficacy.

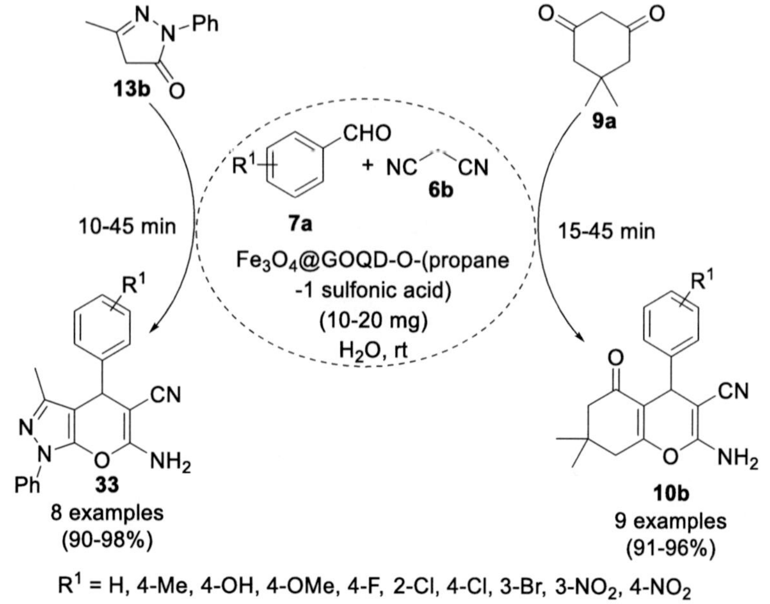

Scheme (9). Synthesis of dihydro pyrano[2,3-*c*]pyrazoles (**33**) and 4*H*-chromenes (**10b**) under the catalysis of Fe_3O_4@GOQD-*O*- (propane-1-sulfonic acid) NCs.

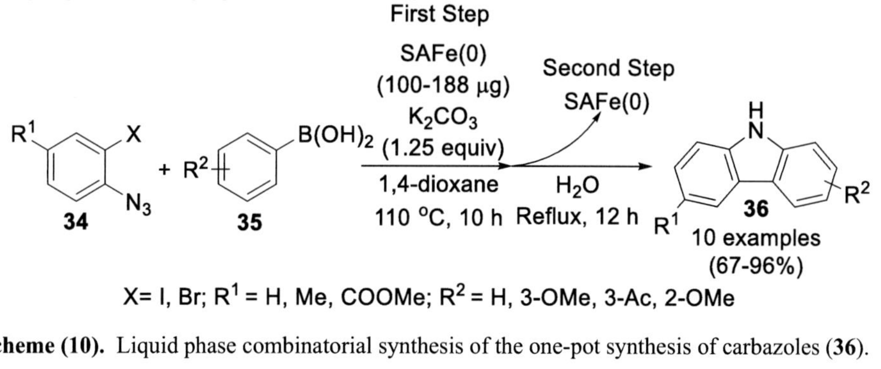

Scheme (10). Liquid phase combinatorial synthesis of the one-pot synthesis of carbazoles (**36**).

β-Cyclodextrin (β-CD) functionalized Fe_3O_4 MNPs [38] catalyzed pseudo-four component synthesis of dispiroheterocycles **(39)** from 6-amino (thio)uracil **(37a)**, *p*-toluidine **(28b)**, and 5-substituted isatins **(38a)** in an ethanol-water as co-solvent (Scheme 11) has been reported by Kumar *et al.* [39]. The nano character of these MNPs has been confirmed using FT-IR, SEM, TEM, EDX, and XRD. The recyclable and reusable MNPs have been recycled up to six cycles without considerable loss of their properties.

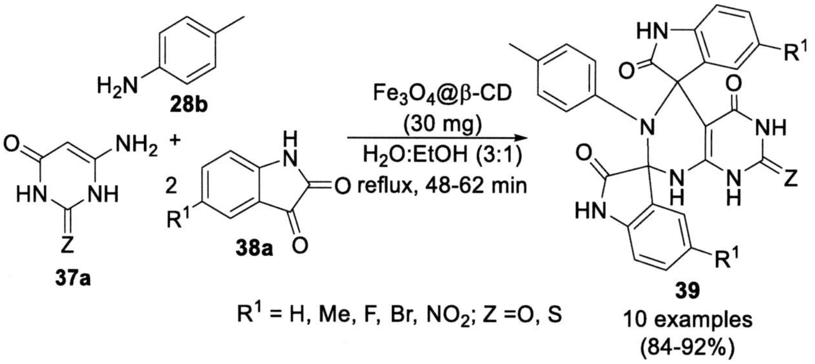

Scheme (11). Synthesis of dispiroheterocycles (**39**) using 6-amino (thio)uracil (**37a**), *p*-toluidine (**28b**) and 5-substituted isatins (**38a**) using β-CD functionalized Fe3O4 NPs.

Zolfigol *et al.* reported the synthesis of 2-aryl-quinoline-4-carboxylic acids **(41)** from α-naphthylamine **(28c)**, pyruvic acid **(40)**, and aryl carbaldehydes **(7d)** using $Fe_3O_4@SiO_2@(CH_2)_3$–urea–thiazole sulfonic acid chloride as the catalysts without using solvent for 12-30 minutes at 80 °C (**Scheme 12**) [40]. The MNPs have been prepared through the reaction of Fe_3O_4 nanoparticles with tetraethyl orthosilicate (TEOS), followed by the reaction of $Fe_3O_4@SiO_2$ with the urea-based ligand, and chlorosulfuric acid. The catalyst showed a good recovery and reusability rate up to six times.

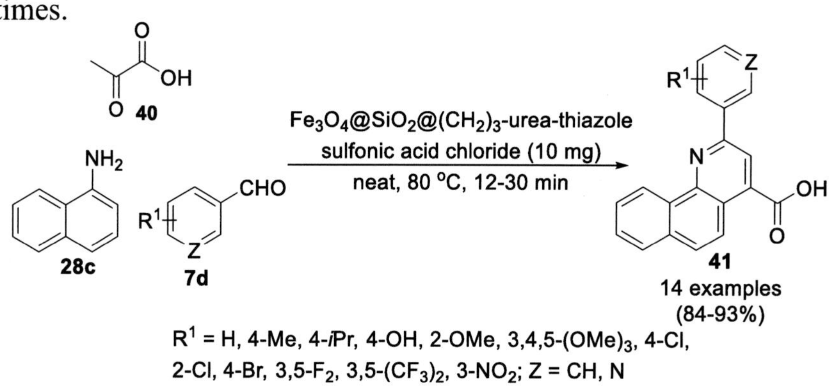

Scheme (12). Catalytic synthesis of 2-aryl-quinoline-4-carboxylic acids (**41**) catalyzed by Fe3O4@ SiO2@(CH2)3–urea–thiazole sulfonic acid chloride.

Mohammadpoor-Baltork *et al.* reported the one-pot synthesis of *mono-* (**45a/45c**) and *bis-4H*-pyrimido[2,1-*b*]benzothiazoles (**45b**) *via* sequential Knoevenagel-Michael-cyclocondensation of β-ketoester/1,3-diketone (**42a**), 2-amino benzo [*d*]thiazole (**43a**) and aryl aldehyde (**7e/45a/7g**) catalyzed by [NicTC] HSO4@MNPs in solvent-free condition at 25-50 °C (Scheme 13) [41]. The SiO_2@MNPs have been treated with 3-aminopropyltriethoxysilane (APTES), diisopropylethylamine (DIPEA), and trichlorotriazine/nicotine (TC) to final NCs [NicTC]HSO4@MNPs. The characterization of these NCs has been performed using different techniques, including FT-IR, VSM, TEM, SEM, EDX, TGA, XRD, and elemental analysis. The catalyst can be reused at least five times without any significant loss of its efficiency and activity.

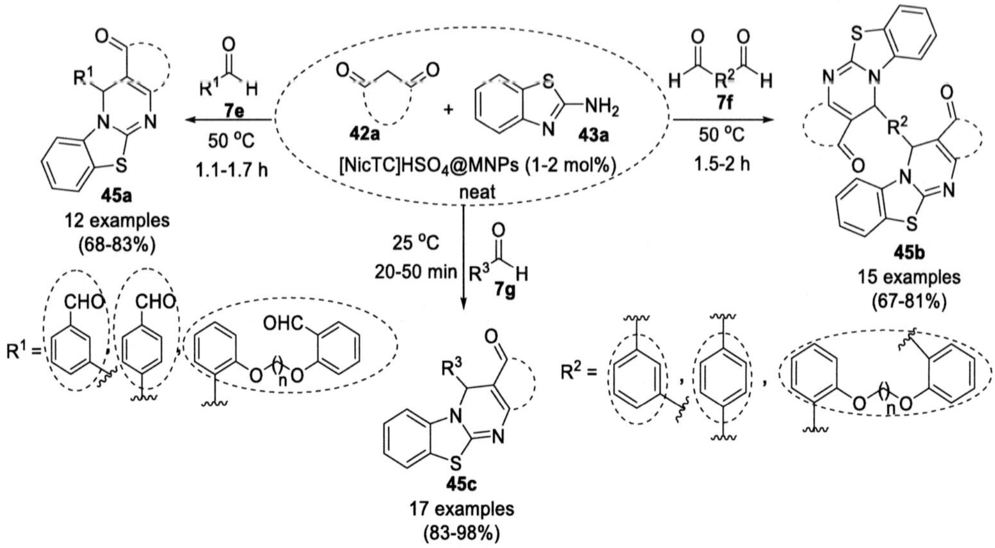

R^3 = 4-*i*Pr-C_6H_4, 4-*t*Bu-C_6H_4, 3-OMe-C_6H_4, 4-OBn-C_6H_4, 4-Ph-C_6H_4, 4-Br-C_6H_4, 2,4-$(Cl)_2$-C_6H_3, 3-NO_2-C_6H_4, 4-NO_2-C_6H_4, naphthalen-2-yl, cinnamyl, 5-methylfuran-2-yl, thiophen-2-yl, 6-chloro-4-oxo-4*H*-chromen-3-yl; n = 4, 6; **42a** = (m)ethyl acetoacetate, dimedone, cyclohexan-1,3-dione

Scheme (13). Synthesis of mono- and bis-4*H*-pyrimido[2,1-*b*]benzothiazoles (**45a/45b/45c**) catalyzed by [NicTC]HSO4@MNPs.

NiNPs Catalyzed Synthesis of Heterocycles

Zhou *et al.* have performed nano nickel oxides (NiO/Gr) catalyzed radical type dehydrogenation of *N*-heterocycles **(46a/48)** at 100-130 °C in dimethyl sulfoxide (DMSO, **Scheme 14**) [42]. The key catalysts have been prepared by atomic layer deposition using graphene nanoplates as a substrate, nickel acetoacetate precursor conjugated with *N, N, N', N'*-tetramethyl ethylenediamine [$Ni(acac)_2$(TMEDA)] at a sublimation temperature of 95 °C in the presence of ozone as an oxidizing agent. Further, they have been characterized by XRD, TEM, and XPS. The NiO/Gr catalysts gain several advantages such as feasible preparation, cost-efficiency, excellent functional tolerance, stability, and reusability up to five runs without losing the catalytic property. It ultimately served as a practical alternative to access nitrogen-containing aromatic compounds (**47a**/**49**).

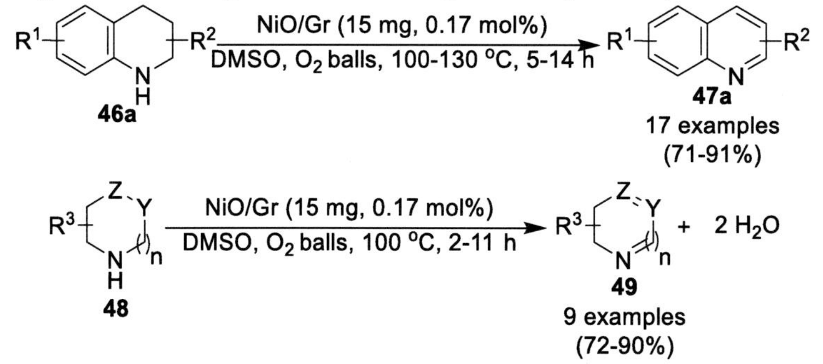

Z = C, N; Y = C, N; n = 0-1; R^1 = H, 5-Me, 6-Me, 6-OH, 6-OMe, 6-NH_2, 6-F, 6-Cl, 6-Br, 6-Cl, 6-COOH, 6-COOMe, 7-NO_2; R^2 = H, 2-Me, 3-Me, 2-Ph

Scheme (14). Dehydrogenation of *N*-heterocycles using nanonickel oxides (NiO/Gr) as catalysts.

Defilippi *et al.* have performed the conversion of levulinic acid (**50a**) into *N*-heterocycles *via* reductive amination using Ni-based catalyst (Ni^0, Ni^3N and Ni_3N+Ni^0) using continuous flow reactor at 100 °C in acetonitrile to synthesize 1-ethyl-5-methyl-1,3-dihydro-2*H*-pyrrol-2-one (**51**) and 1-ethyl-5-methylpyrroli-in-2-one (**52a**, Scheme **15**) [43]. The Ni-based catalysts have been prepared by the green sol-gel-based process involving ultrasonication of reaction mixture obtained from mixing of nickel acetate with ethanol, followed by the addition of urea as a nitrogenous source and its thermal treatment at 350-500 °C. Such Ni-based catalysts have shown promising results in such a reaction in the absence of other noble metals.

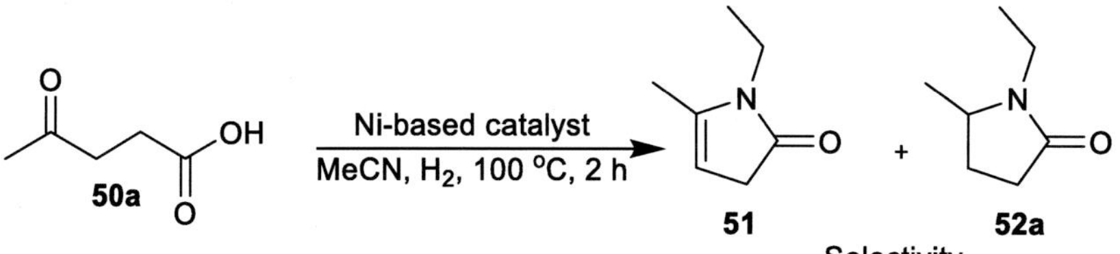

	Selectivity	
Catalysts	**51** (%)	**52a** (%)
Ni^0	19	81
Ni_3N	25	75
Ni_3N+Ni^0	12	88

Scheme (15). Conversion of levulinic acid (**50a**) towards *N*-heterocycles using Ni-based catalyst.

Lu *et al.* have performed the hydrogenation of ethyl levulinate (**53**) to synthesize γ-valerolactone (**54**) using Ni-based catalyst supported on activated carbon (Ni/AC) in *iso*-propanol at 250 °C for 2 h in 79% of yield (Scheme **16**) [44]. The catalyst has been prepared by metalorganic chemical vapor deposition method by mechanical mixing of nickel acetoacetate [Ni(acac)$_2$] with the activated carbon as support followed by thermal decomposition and reduction at 460 °C for 4 h in the flow of H_2/N_2 with a heating rate of 5 °C/min. The present catalysts prepared by the above method have been proved to be the best catalytic system for the hydrogenation of **53** to **54** in better yields as compared to the catalysts prepared by the impregnation method along with their recyclability up to four runs without losing the catalytic potential.

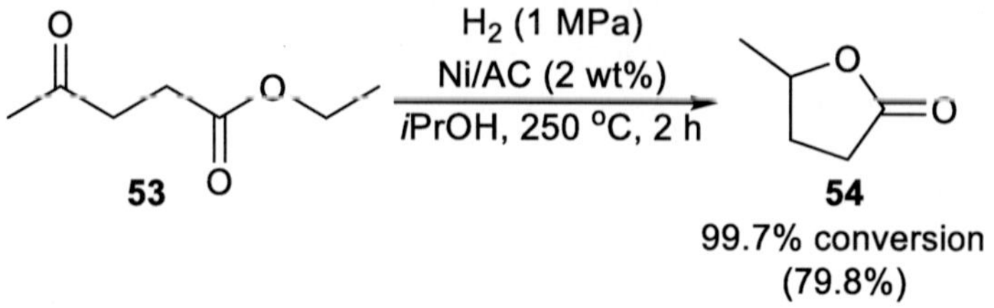

Scheme (16). Hydrogenation of ethyl levulinate (**53**) to γ-valerolactone (**54**) using Ni-based catalyst.

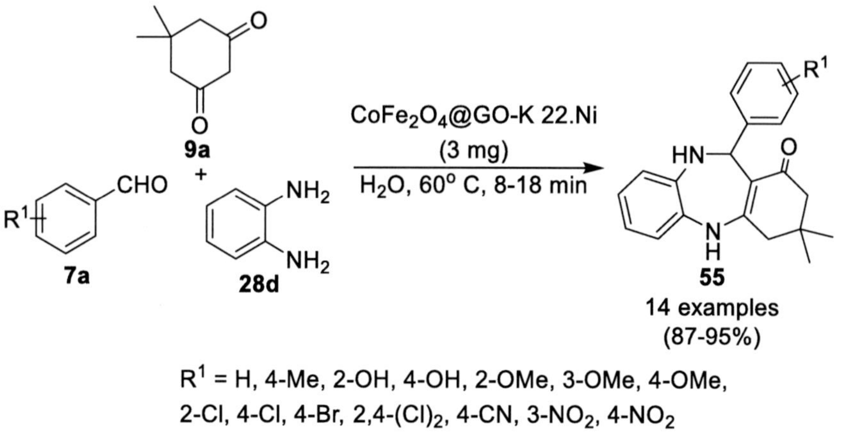

Scheme (17). Synthesis of dibenzo-1,4-diazepines (**55**) with $CoFe_2O_4$@GO–K 22.Ni NPs.

Ghadermazi *et al.* have synthesized various dibenzo-1,4-diazepine derivatives (**55**) from the reaction between aldehydes (**7a**), dimedone (**9a**), and *o*-phenylenediamine (**28d**) using $CoFe_2O_4$@GO-K 22.Ni as the catalysts and water as a solvent at 60 °C in 87-95% yields (Scheme **17**) [45]. NiNPs incorporated in

Kryptofix 22 conjugated magnetic nano-graphene oxide ($CoFe_2O_4$@GO–K 22. Ni) has been synthesized *via* the grafting of Kryptofix 22 moieties on the magnetic nanographene oxide surface, followed by complexation with nickel nitrate. These MNPs have been further characterized using FT-IR, FE-SEM, TGA, ICP, EDX, XRD, VSM, and BET. The catalysts have been successfully examined for six runs with a slight loss of catalytic activity.

PdNPs Catalyzed Synthesis of Heterocycles

Domine *et al.* have performed reductive amination of acetol (**56**) in presence of ethylenediamine (**28e**) using Pd supported catalyst (Pd/TiO_2-Al_2O_3) at 90 °C for 7h which yielded 2-methylpiperazine (**58a**, 68%, Scheme **18**) [46]. The metallic precursors have been dissolved in water along with mixed oxides and ammonia solution has been added to set the pH 9 followed by stirring, drying, and calcinated at 500 °C for 3 h under air with the rate of 2 °C/min to obtain the metallic mixed oxide support (TiO_2-Al_2O_3) and Pd has been incorporated on the support by incipient wetness impregnation method using an aqueous solution of tetraamine palladium(II) chloride hexahydrate [$Pd(NH_3)_4Cl_2 \cdot 6H_2O$] followed by drying and calcinated again under air at a rate of 3 °C/min to obtain final catalysts. Next, they have been characterized by ICP analysis, XRD, HR-TEM, ammonia thermo-programmed desorption (NH_3-TPD), and FT-IR. The catalysts have shown higher activity in shorter times with low catalytic loading, and also have been found reusable up to three runs. The regenerated catalysts, obtained at the end of the third cycle of reused catalysts by processing them under a hydrogenated environment at 400 °C for 2 h at the heating rate of 3 °C/min, have demonstrated almost similar catalytic potential as that of fresh ones.

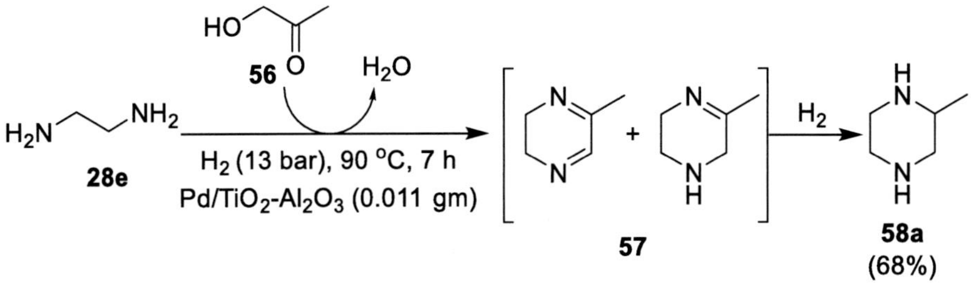

Scheme (18). Reductive amination of acetol (**56**) to 2-methylpiperazine (**58a**).

Li *et al.* have performed C-H activation of bromoarenes **(59a)** and thiazole **(44a)** catalyzed by diimine-palladium complex anchored on graphene oxide (Pd-DI@GO) at 120 °C in presence of pivalic acid (PivOH) for 12 h to get 5-arylated thiazoles in 20-99% yields (Scheme **19**) [47]. The catalyst has been prepared by reaction of diimine palladium complex with the suitable steric hindrance of isopropyl group followed by providing electron supply to produce palladium active centres, followed by the anchoring with graphene oxides (GO) to fabricate the final catalysts. The structural integrity of the final NPs has been characterized using FT-IR, XRD, SEM, and XPS. The catalyst could reproduce 90% of yield after the fourth run of catalysts recovered after recyclization. Further, they have also explored the same catalysts for Suzuki reaction of aryl boronic acids with bromobenzenes at 65 °C in the presence of K_2CO_3 and ethanol for 2 h to synthesize substituted biphenyls in 44-99% of yields.

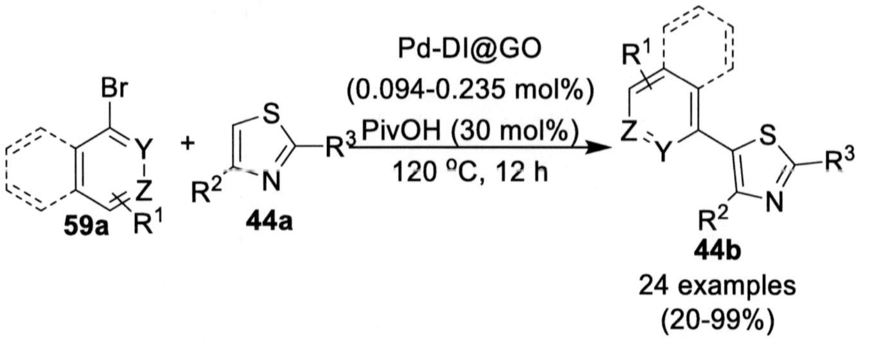

R¹ = H, 2-Me, 4-Me, 4-*t*Bu, 4-OMe, 4-F, 4-Cl, 4-Br, 5-Br, 6-Br, 3,5-Br_2, 4-CF_3, 4-CN, 4-COMe, 4-CHO, 4-COOMe; R^2/R^3 = H, Me; Y/Z= CH, N

Scheme (19). C-H direct arylation of brominated benzene (**59a**) and heterocyclic thiazole (**44a**) catalyzed by Pd-DI@GO.

Zhang *et al.* have performed epoxidation of styrene (**60**) using various PdNPs grafted on molecular organic frameworks such as MIL-101 and sulfonated GO (Pd@MIL-101@SGO-4 and Pd@MIL-101) using *tert*-butyl hydroperoxide (TBHP) as an oxidant in the presence of water at 343 K for 5 h (Scheme **20**) [48]. The key catalysts MIL-01@SGO have been prepared by reacting MIL-101, made up of chromium ions and terephthalic acids, with previously sulfonated and oxidized SGO (graphite powder) by hydrothermal method, followed by encapsulation of 1 wt% palladium on it to obtain the final catalysts. The catalysts have been characterized using XRD, SEM, TEM, EDS, FT-IR, TGA, nitrogen physisorption measurements, PXRD, and ICP-AES. Pd@MIL-101@SGO has also been explored as an effective and heterogeneous catalyst for the synthesis of β-alkoxy alcohol.

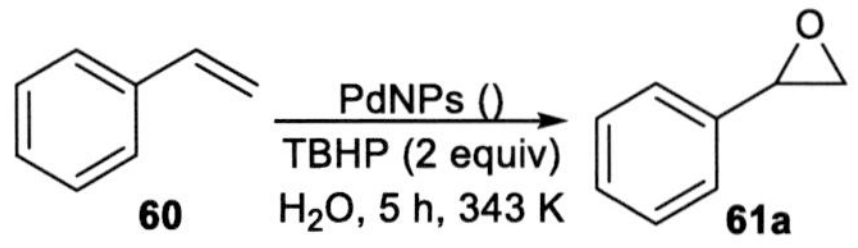

Catalysts	Yields (%)
Pd@MIL-101@SGO-4	38.05
Pd@MIL-101	29.04

Scheme (20). PdNPs catalyzed epoxidation of styrene (**60**).

PtNPs Catalyzed Synthesis of Heterocycles

Liu *et al.* have achieved platinum nanoparticles (PtNPs) loaded on cellulose-derived carbon supports (Pt/c-C) catalyzed chemoselective reduction of 2-nitroacylbenzenes (**7h**) at 30 °C for 3 h under 1 atm pressure of H_2 gas in methanol to synthesize 2,1-benzisoxazole (**62**) in 83-97% yields (Scheme **21**) [49].

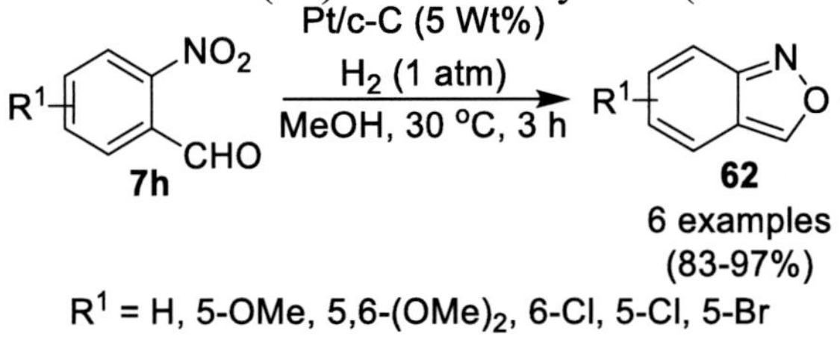

Scheme (21). Reduction of 2-nitrobenzaldehydes from the reaction with H_2 using Pt/c-C catalyst.

The scope of the present catalysts has been also extended for the reductive amination of levulinic acid (**50b**) to pyrrolidones (**52b**) using amines (**28a**) under similar conditions in 46-100% yields within 3-10 h (Scheme **22**). The Pt/c-C catalysts have been prepared using the impregnation reduction method wherein the obtained PtNPs (2 nm) have been uniformly loaded on the surface of the c-C support. These NPs have been characterized by methods such as FT-IR, TEM, XPS, ICP-MS, NH_3-TPD, and CO_2-TPD. Pt/c-C NPs have reproduced the catalytic effect up to five runs and been found to lose their catalytic power after that.

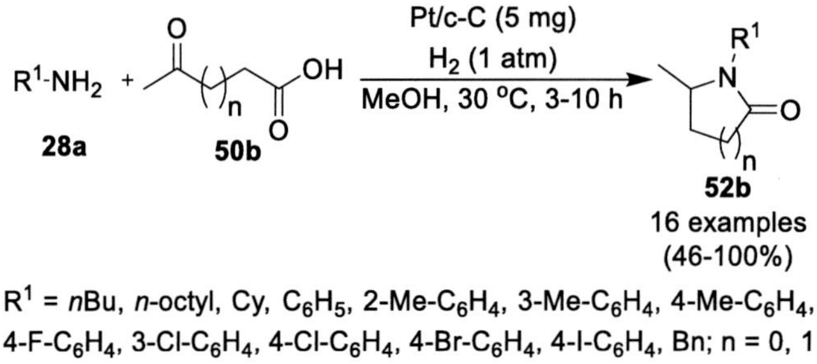

Scheme (22). Pt/c-C assisted reductive amination of levulinic acid (**50b**) to pyrrolidones (**52b**).

ZnNPs Catalyzed Synthesis of Heterocycles

ZnO NPs catalyzed synthesis of tetrazole derivatives (**66**) have been reported by Mohamed and co-workers from ethyl esters of amino acids (**63**) with triethyl orthoformate (**64**) and sodium azide (**65**) in acetonitrile/water (1:1) as co-solvent under sonicated conditions in good to excellent yields (Scheme **23**) [50]. ZnO NPs have been obtained by chemical reduction of zinc precursor using ascorbic acid, trisodium citrate under the basic environment of caustic soda. The prepared suspension of NPs has been centrifuged at 9000 rpm for 15 min followed by 2-3 times wash with Milli-Q water. Further, these NPs have been characterized by FT-IR, NMR, HRMS, liquid chromatography/mass spectrometry (LC/MS), XRD, TEM, SEM, and EDX. The yields of **66** have been improved along with the reduction in time due to co-operative effects of NPs and ultrasonication. Further, ZnO NPs have been recycled up to five runs without notable loss in their catalytic performance.

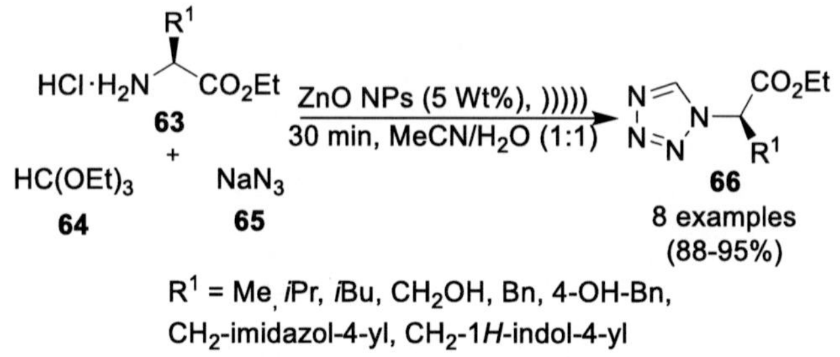

Scheme (23). Synthesis of tetrazole derivatives (**66**) from ethyl esters of amino acids (**63**) under the catalytic influence of ZnO NPs.

Other NPs Catalyzed Synthesis of Heterocycles

The three-component sequential reflux reaction of phenylhydrazine (**67**), *o*-aminoacetophenone (**28f**), and cyclic ketones (**38a/69/70/71**) catalyzed by terbium doped TiO_2 NPs has been reported by Kumar and co-workers for the synthesis of spiroannulated indolo[3,2-*c*]quinolines-mimetic scaffolds of isocryptolepine (**68a/68b/68c/68d**, Scheme **24**) [51]. The Tb doped TiO_2 NPs have been prepared from the reaction of titanium *isopropoxide* and terbium chloride as Tb precursors in presence of thioglycerol followed by heating at 80 °C for 2.5 h and drying at 100 °C in the oven. The nanostructured catalyst has been characterized by XRD, TEM, EDX, and FT-IR. TiO_2 NPs have higher catalytic efficiency and facilitate the reaction to provide the products in excellent yields with special features like operational simplicity, atom-economy, mild reaction conditions, environmental sustainability, and high synthetic efficiency along with recyclability and reusability. The catalytic effects of Tb doped TiO_2 NPs are conserved up to five runs of recyclization.

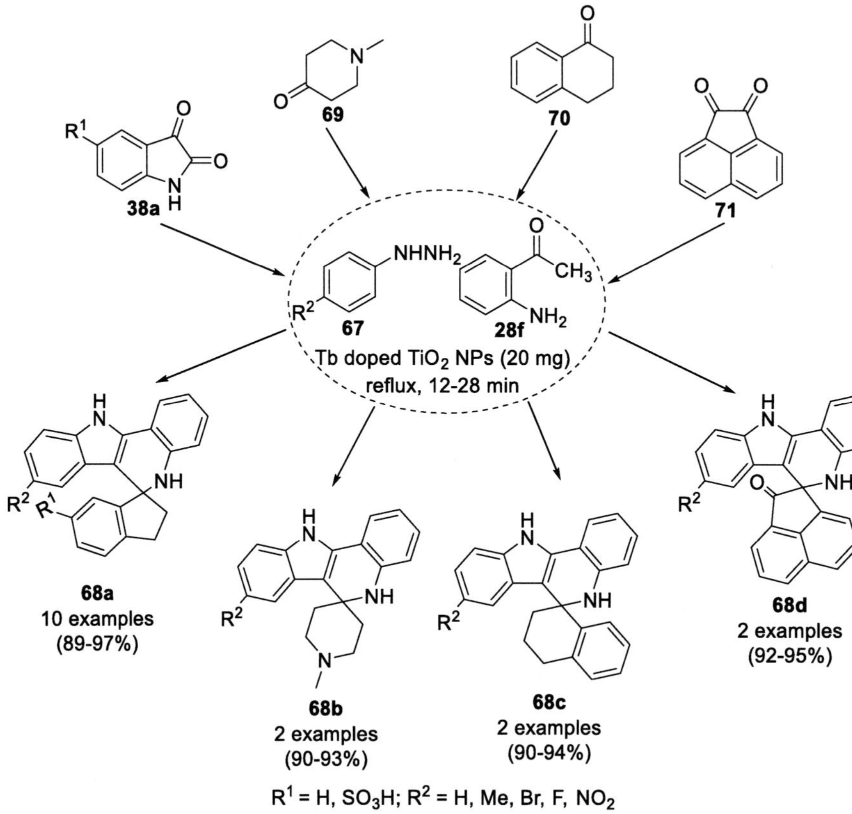

Scheme (24). Synthesis of spiroannulated indolo[3,2-*c*]quinolines-mimetics of isocryptolepine (**68a/68b/68c/68d**) assisted by Tb doped TiO_2 NPs.

Dong *et al.* have performed oxidative dehydrogenation reaction of 1,2,3,4-tetrahydroquinoline (**46b**), 1,2,3,4-tetrahydroisoquinolines (**72a**), and 1*H*-indolines (**73a**) in methanol with potassium carbonate as base and O_2 balloon at 80 °C for 12h using catalyst Cobalt nano species finely supported on dendritic fibrous silica nanospheres (KCC-1@Co-N-C-800) (Scheme **25**) [52]. They have also performed the catalytic transfer hydrogenation of quinolines (**47c**), isoquinolines (**74b**), pthalazines (**76**) by reaction with formic acid and toluene in the sealed tube using the same catalyst at 130 °C for 6-12 h (Scheme **26**). The catalysts have been prepared in two steps; synthesis of KCC-1 followed by synthesis of KCC-1@Co-N-C-800. KCC-1 has been prepared by reaction of the mixture containing urea and 1-hexadecylpyridinium bromide monohydrate dissolved in water.

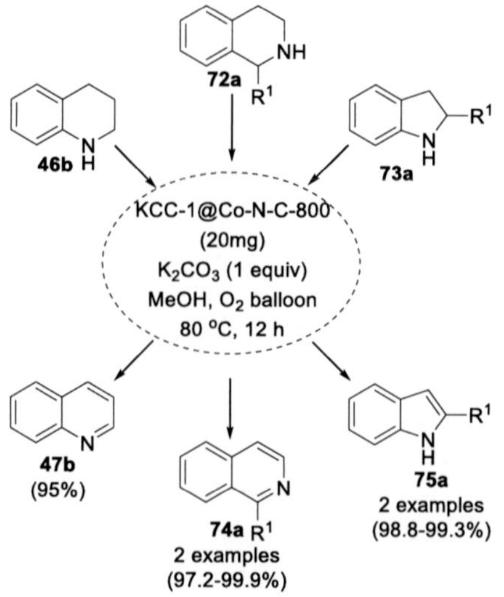

R^1 = H, CH_3

Scheme (25). Oxidative dehydrogenation of 1,2,3,4-tetrahydro(iso)quinoline (**46b/72a**) and indolines (**73a**). The conversions from starting materials have been provided in parenthesis obtained from GC-MS.

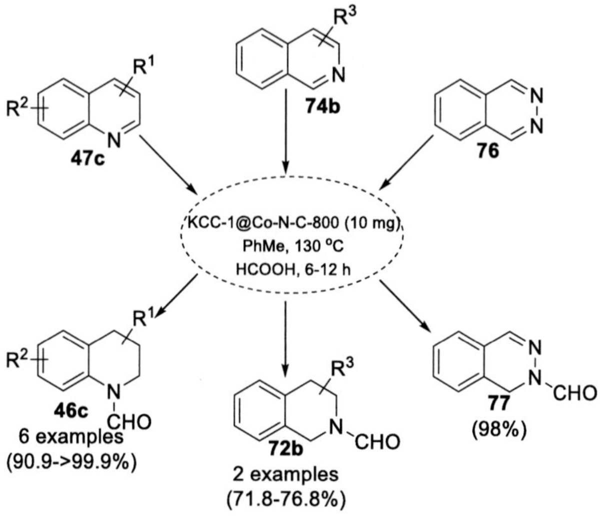

R^1 = H, 2-Me, 4-Me; R^2 = 6-OMe, 8-OH; R^3 = H, 4-Br

Scheme (26). Hydrogenation of *N*-heteroaromatic compounds (**47c/74b/76**). The values in parenthesis represent the % conversion (GC-MS).

A mixture containing tetramethyl orthosilicate, cyclohexane, and 1-pentanol and treated in the teflon-sealed hydrothermal reactor at 120 °C for 5 h followed by Jonnalagadda *et al.* have synthesized novel [1, 3, 4]thiadiazolo[3,2-*a*]pyrimidines (**80a/80b/80c**, Scheme **27**) and benzo [4, 5]thiazolo[3,2-*a*]pyrimidines (**45d**, Scheme **28**) *via* the multicomponent reaction between 1,3,4-thiadiazole-2-amines

(**79**) or 2-amino-benzothiazole (**43a**), aldehydes (**7a**) and active methylene compounds (**6c/78/42b**) in ethanol at rt using vanadia on fluorapatite (V_2O_5/FAp) as catalysts [53]. The catalysts have been prepared by reaction between trisodium phosphate dodecahydrate, sodium fluoride, and calcium nitrate tetrahydrate followed by drying and calcination and their characterization has been achieved using TEM, BET, XRD, FT-IR, SEM, and EDX. In comparison with other reported conditions such as catalyst-free condition [54], sodium hydroxide [55], polyethylene glycol-400 [56], phosphorus pentoxide [57], *p*-toluene sulfonic acid [58], Fe_3O_4@nano-cellulose/Cu(II) [59], nano-kaolin/Ti^{4+}/Fe_3O_4 [60], nano-cellulose/BF_3/Fe_3O_4 [61], and nano-Fe_3O_4@SiO_2–$TiCl_3$ [62], the V_2O_5/FAp gives high yields of products at rt in shorter reaction time and also it has been found easy to synthesize, recyclable, eco-friendly catalysts with no need of column chromatography. The catalyst can give its catalytic effect up to five runs of recyclization.

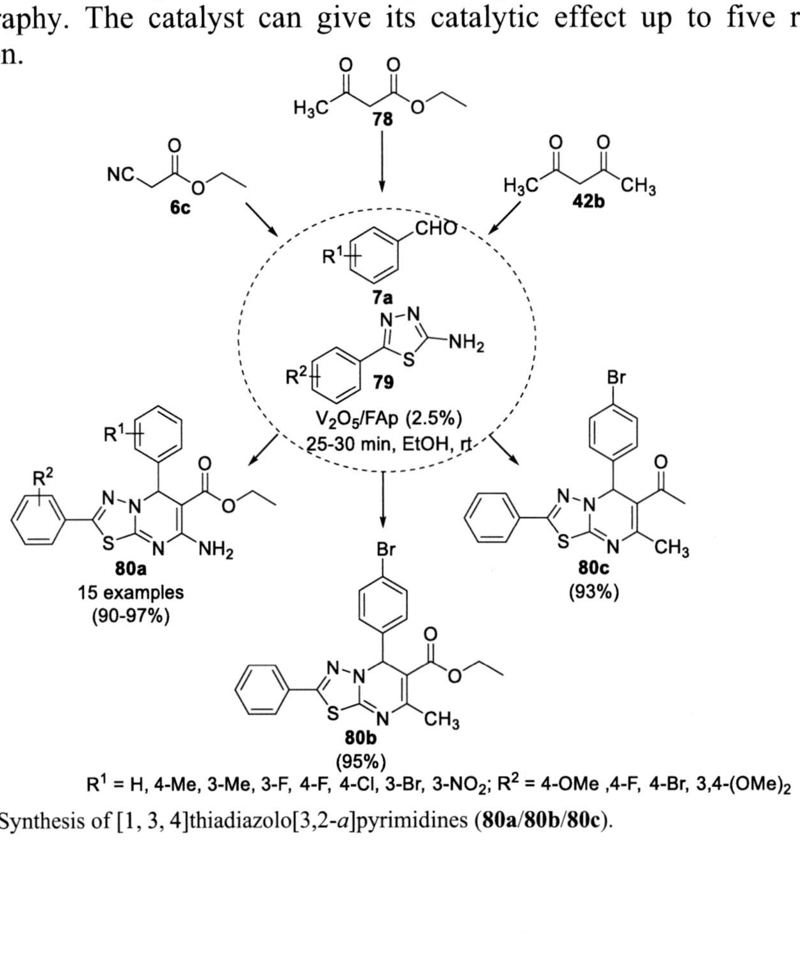

Scheme (27). Synthesis of [1, 3, 4]thiadiazolo[3,2-*a*]pyrimidines (**80a/80b/80c**).

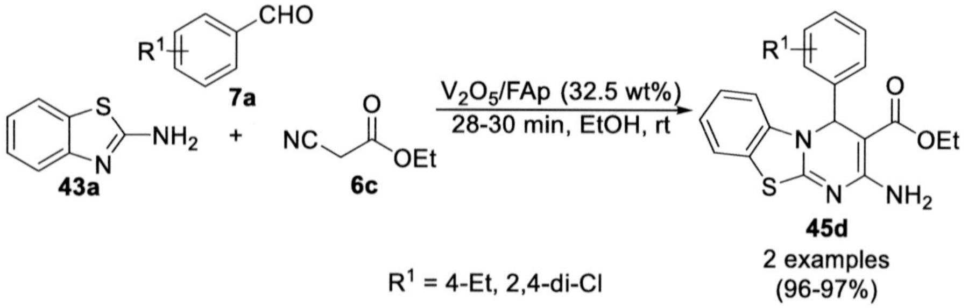

Scheme (28). Synthesis of benzo [4, 5]thiazolo[3,2-*a*]pyrimidines (**45d**).

Bimetallic NPs Catalyzed Synthesis of Heterocycles

Ghorbani-Vaghei *et al.* have used the nano-magnetic pyridinium-tribromide (TBP-SCMNPs) for the synthesis of pyrimidines (**83**) from substituted acetophenone (**81a**), triethyl orthoformate (**64**), and ammonium acetate (**82**) at 90 °C in deep eutectic solvent (Des) in 60-94% yields (Scheme **29**) [63]. The synthesis of NPs has been achieved by the treatment of silica-coated Fe_3O_4 NPs with 3-chloropropyltrimethoxysilane, followed by reaction with 4-aminopyridine, and salt formation using Br_2 in CCl_4 and they have been characterized using XRD, SEM, TEM, FT-IR, TGA, VSM, and EDX. NPs played the key role as Lewis acid to catalyze the cyclocondensation through the activation of **81a**, **64** and formed intermediates during the catalytic cycle. After the completion of the reaction, NPs have been recycled up to seven runs without losing the catalytic activity.

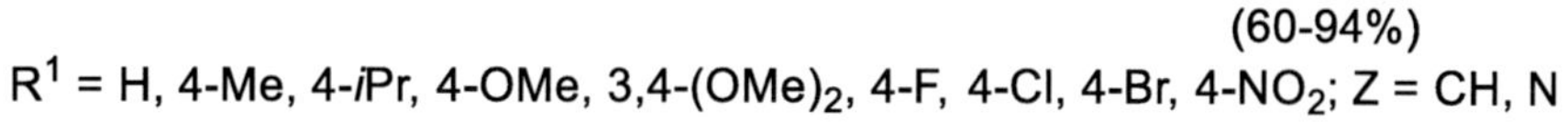

Scheme (29). NPs catalyzed the synthesis of pyrimidines (**83**).

Khalaj has used MgO-$MgAl_2O_4$ composite nano-powder for the synthesis of benzo [4, 5]thiazolo[3,2-*a*]chromeno[4,3-*d*]pyrimidin-6-ones (**45e**) from benzaldehyde (**7a**), 4-hydroxycoumarin (**11b**), and 2-aminobenzothiazole (**43b**) without using solvent by three different conditions such as thermal heating, ultrasonic irradiation and high-speed ball milling (HSBM) technique (Scheme **30**) [64]. The MgO-$MgAl_2O_4$ composites nano-powder was prepared from the co-precipitation of $Mg(NO_3)_2$ and $Al(NO_3)_3$ salts followed by its characterization using XRD, TEM, EDS, FTIR. The catalysts promoted the reaction between **7a**

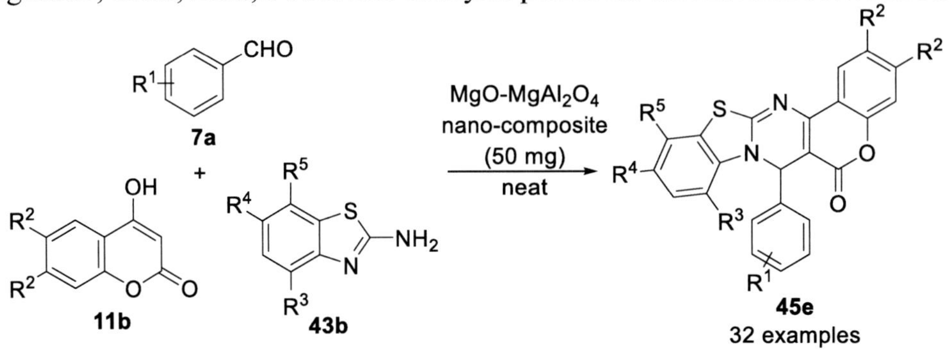

Method	Temp. (°C)	Time (h)	Yields (%)
conventional	100	1.2-3	70-95
))))	80	1-2.5	69-97
HSBM	70	0.7-2.5	67-97

R^1 = H, Cy, 2-Me, 4-Me, 4-OH, 2-OMe, 3-OMe, 4-OMe, 3,4-$(OMe)_2$, 4,5-$(OMe)_2$, 3,4,5-$(OMe)_3$, 4-NMe_2, 2-Cl, 3-Cl, 4-Cl, 2-Br, 3-Br, 4-Br, 3,4-$(Cl)_2$, 3-NO_2, 4-NO_2; R^2 = H, OMe, Cl; R^3 = H, Me; R^4 = H, Me, Cl, Br, NO_2; R^5 = H, Cl

Scheme (30). Preparation of benzo [4, 5]thiazolo[3,2-*a*]chromeno[4,3-*d*]pyrimidin-6-ones (**45e**) using MgO-$MgAl_2O_4$ nanocomposites.

and **11b** by the virtue of its dual acidic and basic character to form enedione intermediate, which underwent condensation with **43c** to afford the desired product. After the completion of the reaction, nano-powder composites have been recycled up to five times without appreciable loss of catalytic activity. Ciotonea *et al.* have achieved the catalytic hydrogenation of *N*-heteroarenes using phyllosilicate derived nickel-cobalt bimetallic NPs at 80°C in 12-99% yields (Scheme **31**) [65]. The final catalysts have been prepared by deposition-precipitation initiated by urea using purely silicic acid (SBA-15) as support and different Ni/Co weight ratios (3:1, 2:2, 1:3) from nickel and cobalt chloride as catalytic precursors. These NPs have been characterized by powder X-ray diffraction (PXRD), N_2 physisorption, TEM, and a high-resolution scanning transmission electron microscope (STEM-EDX). The catalytic performance of NPs has been found to decrease upon recycling after the fifth run by 51% as compared to the first catalytic run, suggesting the necessity of a different mode of reactivation between each cycle. The authors have explored the same catalytic protocol for the hydrogenation of acyclic imines and oximes.

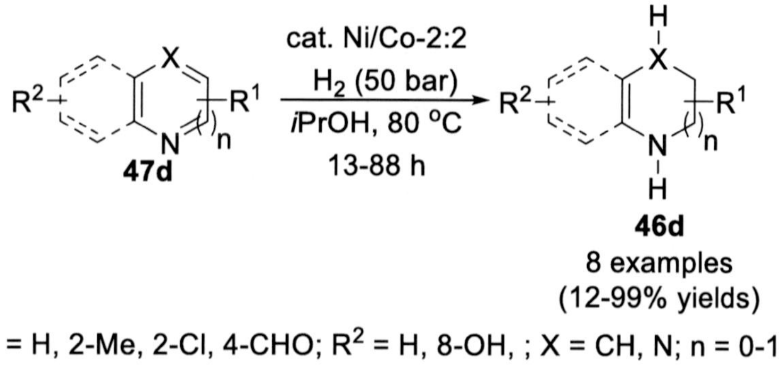

Scheme (31). Hydrogenation of *N*-heterocycles (**47d**) by phyllosilicate derived nickel-cobalt bimetallic NPs.

The multicomponent reaction of sulfur, 4-substituted acetophenones (**81b**), and 5-substituted indoles (**75b**) catalyzed by magnetic nanoparticle-supported $[Urea]_4[ZnCl_2]$ deep eutectic solvent (DES, DES@MNP) has been reported by Nguyen and Tran for the synthesis of thieno[2,3-*b*]indoles (**84**) at 140 °C for 12 h in 54-89% yields (Scheme **32**) [66]. The final catalysts have been synthesized by treatment of silica-supported Fe_3O_4 NPs with (3-chloropropyl)triethoxysilane followed by its loading on DES prepared from urea and zinc chloride. The nano-particulate structure of the final bimetallic NPs has been characterized by various techniques including FT-IR, Raman, TGA, SEM, TEM, XRD, VSM, EDX, and ICP-MS. The DES@MNP played the key role as powerful catalysts to promote the reaction proceeds *via* the initial formation of 1-(dimethyl-amino)-2-phenyl-1*H*-their-1-ium from **81b**, DMF, and S_8 followed by subsequent ring-opening addition by **75b,** and cyclocondensation to afford **84**. The MNPs could be recovered using an external magnet and reused up to five consecutive runs without a considerable loss of catalytic performance.

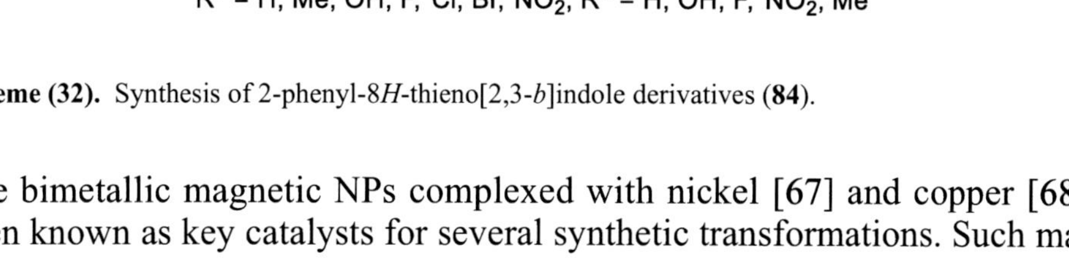

Scheme (32). Synthesis of 2-phenyl-8*H*-thieno[2,3-*b*]indole derivatives (**84**).

The bimetallic magnetic NPs complexed with nickel [67] and copper [68] have been known as key catalysts for several synthetic transformations. Such magnetic carbon nanotube (Fe_3O_4-CNT-In) have been used in the synthesis of pyrazole-fused isocoumarins (**87**) from the mixture of ninhydrin (**85**) and arylhydrazones (**86**) at 100 °C for 15 min without using a solvent in good to moderate yields (88-94%, Scheme **33**) [69]. Magnetic carbon nanotube has been prepared by reacting ferric oxide carbon nanotube and indium chloride. The structural and chemical nature of prepared magnetic bio-nano composites has been confirmed by FT-IR, TGA, SEM, VSM, EDX, and XRD. The NCs have accelerated the condensation of ninhydrin and aryl hydrazones through their coordinating capacity with starting materials and formed intermediates as Lewis acid. The catalyst has been claimed sustainable after the recycling experiment involving five consecutive runs without any notable reduction in its efficiency.

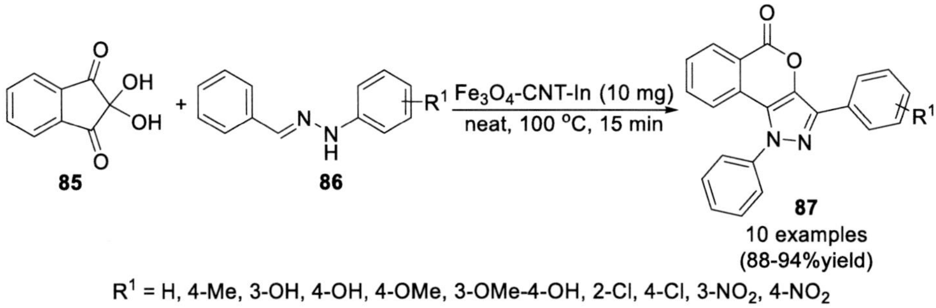

Scheme (33). Synthesis of pyrazole fused isocoumarins (**87**) catalyzed by Fe_3O_4-CNT-In.

Sheikhhosseini *et al.* have achieved the synthesis of 2-oxospiro[indoline-3,4′- [1, 3]di-thine]-5′-carbonitrile derivatives (**89**) from barbituric acids (**5b**), carbon disulfide (**88**), and 2-(2-oxoindolin-3-ylidene)malononitrile (**73b**) at rt in the presence of spiro[indoline-3,4′- [1, 3]dithiine]@$Cu(NO_3)_2$ supported on Fe_3O_4@gly@CE MNPs [Fe_3O_4@gly@thiophen@$Cu(NO_3)_2$, (Scheme **34**) [70]. The preparation of catalysts has been carried out by loading Fe_3O_4 NPs synthesized from $FeCl_2{\cdot}4H_2O$ loaded on glycine, followed by the subsequential complexation with 6′-amino-2′-(tetrahydropyrimidin-5(2*H*)-ylidene)-2-oxospiro [indoline-3,4′- [1, 3]dithiine]-5′-carbonitrile with $Cu(NO_3)_2{\cdot}3H_2O$. The structural and chemical nature of prepared magnetic nanoparticles has been confirmed by EDX, XRD, SEM, VSM, TGA, FT-IR, and ICP-OES. The present MNPs catalyzed protocol has been found more productive in terms of reaction time and yields of **89** as compared to triethylamine as base-catalyzed synthesis. The MNPs have promoted the synthesis of **89** by activation of **5b** and **88** to form thia adduct followed by reaction with **73b** to undergo sequential MNPs assisted.

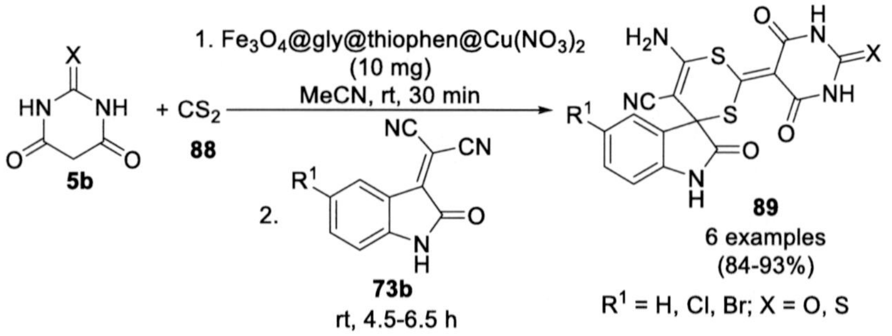

Scheme (34). Multicomponent synthesis of 6′-amino-2′-(tetrahydropyrimidin-5(2*H*)-ylidene)-2-oxospiro[indoline-3,4′- [1, 3]dithiine]-5′-carbonitrile derivatives (**89**).

Michael addition, proton shift, ring closure, and tautomerism. NPs have been reused up to six times to claim the reusability of the catalysts. The assessment of synthetic targets (**89**) has been further performed in search of potential anti-bacterial, anti-fungal, and antioxidant agents. Further, they have also reported the synthesis of 3,4-dihydro-2*H*-pyran derivatives (**90**) by spiro[indoline-3,4′- [1, 3]dithiine]@$Ni(NO_3)_2$ supported on the Fe_3O_4@SiO_2@CPS from *N*-cyclohexyl isocyanide (**6d**), oxirane (**61b**), and malononitrile (**6b**) at reflux in ethanol in 61-96% yields (Scheme **35**) [71]. For the preparation of catalyst, 6′-amino-2′-(-,4-dimethyl-2,6-dioxocyclohexylidene)-2-oxospiro[indoline-3,4′- [1, 3]dithiine]-5′-carbonitrile has been treated with spiro[indoline-3,4′- [1, 3]dithiine] supported on the Fe_3O_4@SiO_2@CPS followed by complexation with $Ni(NO_3)_2{\cdot}3H_2O$. The structural characterization of NPs has been confirmed using EDX, SEM, TEM, XRD, TGA, FT-IR, ICP-OES, VSM, BET analysis. The coordination of MNPs with the oxygen of **61b** promoted nucleophilic attack by **6d** to undergo ring-opening, followed by MNPs assisted nucleophilic attack of **6b** and ring closure to yield **90**. NPs can be reused up to five times without any decrease in efficiency.

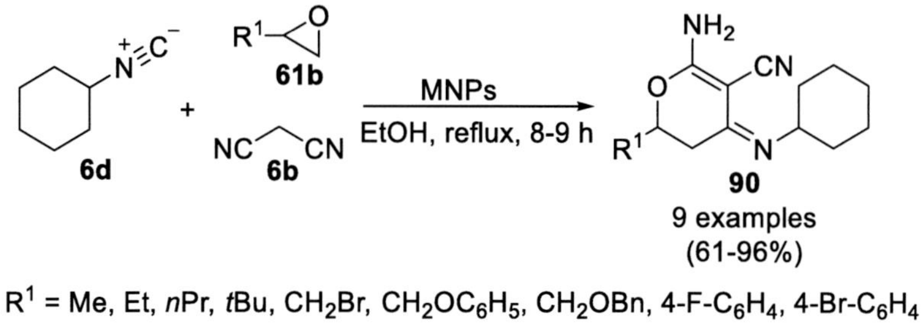

Scheme (35). Synthesis of 3,4-dihydro-2*H*-pyran derivatives (**90**) by spiro [indoline-3,4′- [1, 3]dithiine]@$Ni(NO_3)_2$ supported on the Fe_3O_4@SiO_2@CPS.

Hajizadeh *et al.* have used novel and green $NiFe_2O_4$/geopolymer NCs based on bentonite for the synthesis of imidazole heterocycles from aldehyde (**7a**), benzil (**91**), and ammonium acetate using ultrasonic irradiation at 80 °C (Scheme **36**) [72]. The $NiFe_2O_4$ NPs, prepared by coprecipitation method from nickel nitrate hexahydrate ($Ni(NO_3)_2{\cdot}6H_2O$), iron (III) nitrate nonahydrate ($Fe(NO_3)_3{\cdot}9H_2O$), have been intercalated on calcined bentonite clay-based geopolymer and have been characterized using FT-IR, EDX, FE-SEM, TEM, BET, XRD, TGA, and VSM analysis. The synthesis of **92** has been potentiated under the influence of NCs as Lewis acid *via* activation of aldehyde and formed intermediates during the catalytic cycle. The catalyst recovered through an external magnet has been reused up to eight times with the conserved catalytic potential.

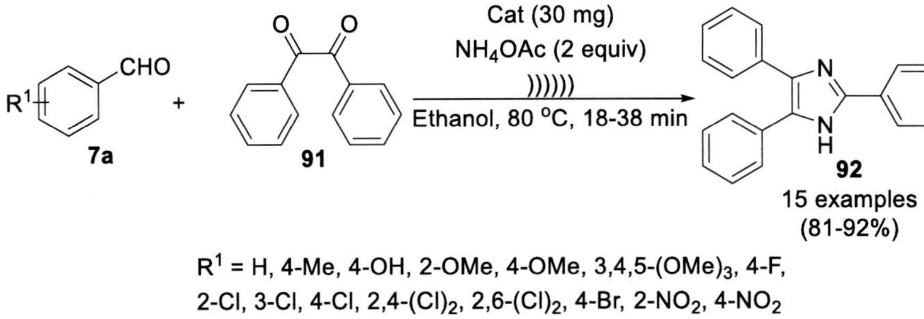

Scheme (36). Synthesis of 2,4,5-trisubstituted imidazoles (**92**) from substituted benzaldehyde (**7a**), benzil (**91**), and ammonium acetate.

Tannic acid-grafted silica-coated Fe_3O_4 NPs (Fe_3O_4@SiO_2@Tannic acid) catalyzed synthesis of 5-pyrazolin-1,2,4-triazolidine-3-ones or thiones (**95**) has been reported by Nikpass and co-workers from pyrazole-4-carbaldehydes (**93**) and semicarbazide (**94**) at rt in water in good to excellent yields (Scheme **37**) [73]. The synthesis of these NPs has been achieved by ultrasonic treatment of Fe_3O_4@SiO_2 NPs with 3-chloropropyl-trimethoxysilane (CPTES) to form Fe_3O_4@SiO_2-Cl MNPs followed by its functionalization using tannic acid in an alkaline medium. These NPs have been characterized by using XRD, EDX, TEM, FE-SEM, VSM, and FT-IR. The NCs promoted the nucleophilic attack of the free amine of **94** on the carbonyl carbon of **93** by coordination with the carbonyl oxygen of **93** followed by dehydration and cyclization.

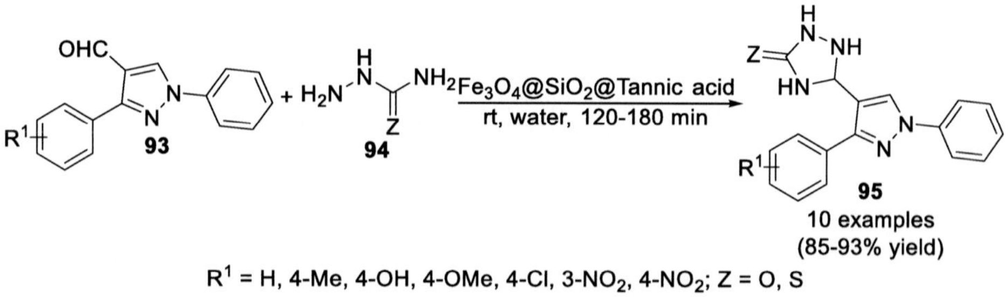

Scheme (37). Fe_3O_4@SiO_2@Tannic acid promoted the synthesis of 5-pyrazolin-1,2,4-triazolidine-3-ones or thiones (**95**).

Zolfigol *et al.* have used silica-coated MNPs loaded on 2-aminoethyl dihydrogen phosphate [Fe_3O_4@SiO_2@$(CH_2)_3NH(CH_2)_2O_2P(OH)_2$] in the synthesis of benzo[*h*]quinoline-4-carboxylic acids from 1-naphthylamine (**28c**), pyruvic acid (**40**), and aryl aldehyde (**7b**) at 80 °C in 80-89% yields (Scheme **38**) [74]. The NCs have been prepared by functionalization of Fe_3O_4@SiO_2 using 2-aminoethyl dihydrogen phosphate and have been characterized by TGA, DTA, TEM, FE-SEM, EDS, EMA, FT-IR, and VSM. The phosphoric acids of NCs have coordinated with the reactants and intermediates to assist the formation of imine from **28c** and **7b**, followed by its Mannich reaction with **40** to undergo ring closure reaction and cooperative anomeric based oxidation to yield **96**. These NCs have been recovered magnetically and recycled up to six runs successfully. Hazeri *et al.* have used Fe_3O_4@gly@thiophen@$Cu(NO_3)_2$ MNPs for the synthesis of pyrimidines (**45f**) from substituted aryl carbaldehydes (**7a**), 4-hydroxycoumarin (**11a**), and 2-aminobenzothiazole (**43a**) at 50 °C for 30-120 min (Scheme **39**) [75]. The synthesis of these MNPs has been achieved by the reaction of nano-Fe_3O_4 coated with glycine and thiophene-2-carboxaldehyde to form Fe_3O_4@gly@thiophen, followed by its reaction with $Cu(NO_3)_2$. The final MNPs have been characterized using SEM, XRD, EDX, FT-IR, TGA, and VSM. These recyclable MNPs have also exhibited anti-bacterial and anti-fungal activity against the selected bacterial and fungal strains.

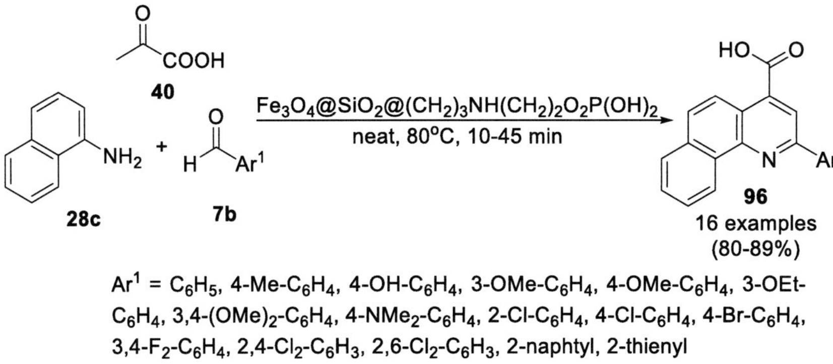

Scheme (38). NCs catalyzed the preparation of benzo[*h*]quinoline-4-carboxylic acids.

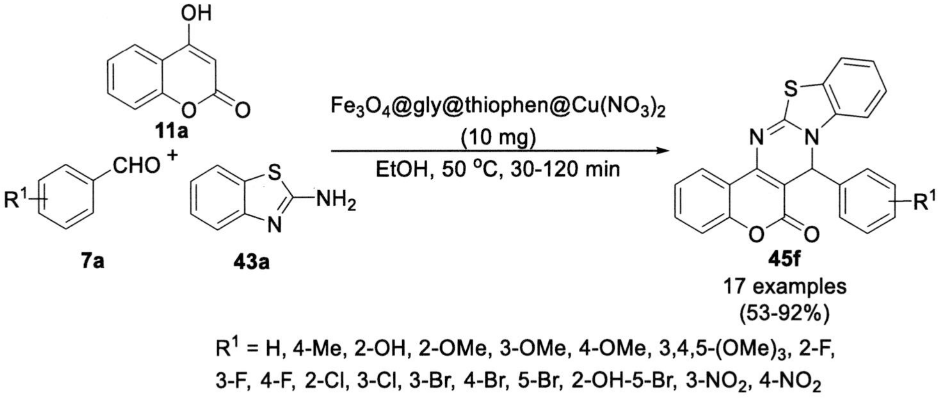

Scheme (39). Synthesis of pyrimidine derivatives (**45f**) from aromatic aldehyde (**7a**), 4-hydroxycoumarin (**11a**), and 2-aminobenzothiazole (**43a**) in the presence of Fe_3O_4@gly@thiophen@$Cu(NO_3)_2$.

Ebrahimiasl *et al.* have used Fe_3O_4@Co^{II} (macrocyclic Schiff base ligand) as a catalyst for the synthesis of 2-amino-4-aryl-6-(phenylsulfanyl)pyridine-3, 5-dicarbonitriles **(23b)** *via* Knoevenagel-thia-Michael type addition using the multicomponent reaction of thiophenol **(97)**, malononitrile **(6b)** and substituted benzaldehydes **(7a)** under the solvent-free condition at 100 °C in excellent yields (Scheme **40**) [76]. The same catalysts have been also found effective for the synthesis of 2-amino-5,10-dihydro-4*H*-benzo[*g*]chromene-3-carbonitriles **(98)** using 2-hydroxy-1,4-naphthoquinone **(20)** in place of **97** (Scheme **41**) [76]. First, the Fe_3O_4 NPs and macrocyclic Schiff base [previously prepared from 2,2'-(1,--diazepane-1,4-diyl)dianiline] ligand were treated in acetonitrile to obtain Fe_3O_4@macrocyclic Schiff base ligand followed by complexation with cobalt chloride hexahydrate ($CoCl_2{\cdot}6H_2O$) for the preparation of the catalyst. These Co-based MNPs have been characterized using FT-IR, SEM, TEM, XRD, EDX, TGA, VSM, DRS, ICP, and elemental analysis. The catalysts demonstrated recovery and reusability up to five runs without any greater reduction in catalytic power.

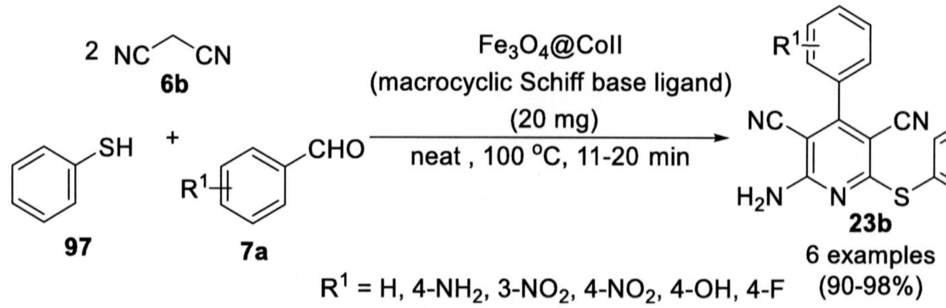

Scheme (40). Synthesis of 2-amino-4-aryl-6-(phenylsulfanyl)pyridine-3,5-dicarbonitriles (**23b**) in the presence of Fe_3O_4@Co^{II}(macrocyclic Schiff base ligand) from thiophenol (**97**), malononitrile (**6b**), and benzaldehyde (**7a**).

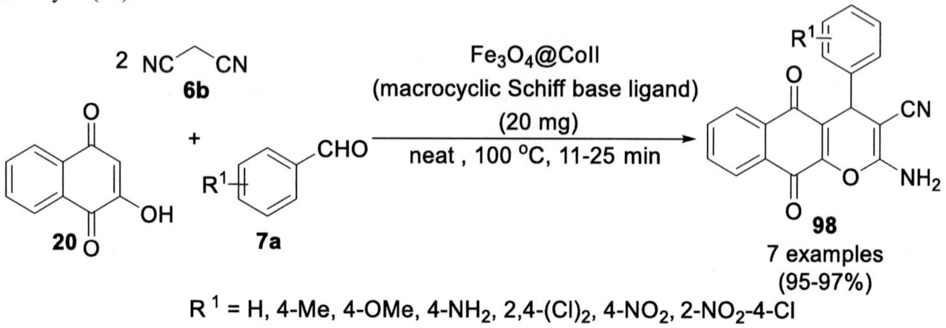

Scheme (41). Synthesis of 2-amino-5,10-dihydro-4*H*-benzo[g]chromene-3-carbonitriles (**98**).

The Ullmann-type C-N coupling through *N*-arylation of halobenzenes (**59b**) with alicyclic amines (**58b**) such as piperidines and morpholines has been reported by Hemmati *et al.* using Cu(I)-anchored polyvinyl alcohol coated MNPs (Fe_3O_4@PVA/CuCl) as heterogenous NCs at 100 °C under a nitrogen atmosphere (Scheme **42**) [77]. The present protocol has been also explored successfully for the synthesis of *N*-arylated indoles and imidazoles. The prepared magnetic Fe_3O_4 NPs using the chemical co-precipitation technique have been coated with PVA to obtain Fe_3O_4@PVA magnetic nanocomposite. Subsequently, it has been complexed with CuCl in water to obtain final NCs. The characterization of catalysts has been achieved using FT-IR, FESEM, EDX, TEM, XRD, and VSM. The catalysts recovered under the external magnetic influence have been reused up to seven consecutive runs with only a 5% drop in yields as compared to fresh catalytic runs.

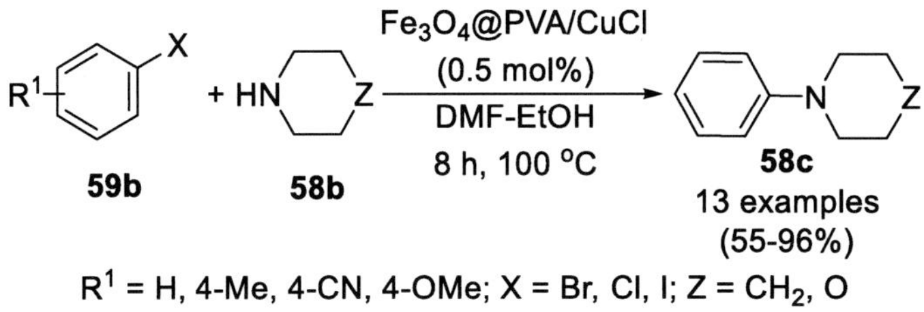

Scheme (42). Amination of halobenzenes (**59b**) with alicyclic amines (**58b**) using Fe_3O_4@PVA/CuCl.

Jafari and co-workers have demonstrated catalytic applications of the guanidine complex of Cu supported boehmite (BO) NPs (Cu-Guanidine@BO-NPs) for the synthesis of tetrazoles (**100**) from benzonitrile/2-phenylacetonitrile (**99**) and sodium azide (**65**) at 120 °C in PEG as solvent (Scheme **43**) [78]. The catalysts have been successfully applied for the sulfoxidation of sulfide using hydrogen peroxide for the synthetic construction of sulfoxides under the solvent free condition at rt. The previously prepared 3-chloropropyltrimethoxysilane MNPs (nPr-Cl@BO-NPs) have been complexed with guanidine nitrate followed by immobilization of copper using copper nitrate hexahydrate [$Cu(NO_3)_2.9H_2O$] to get the final NCs. The nano-characteristics of final MNPs have been confirmed using several spectroscopic tools such as AAS, SEM, WDX, EDS, TGA, and BET. The heterogeneous NCs have been recovered and recycled with 92% of average yields during six catalytic runs confirming the good reusability of this catalyst.

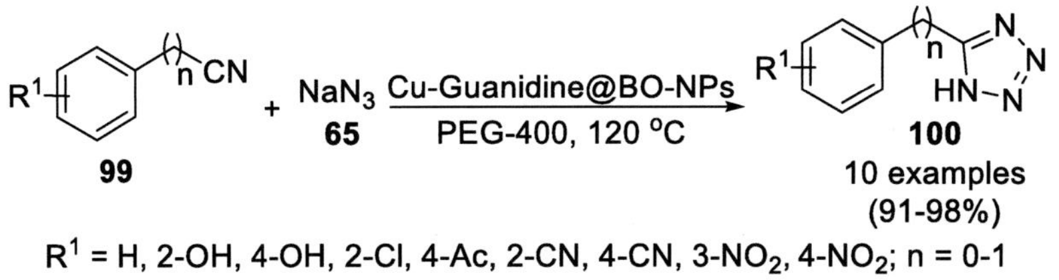

Scheme (43). Synthesis of 1,2,3,4-tetrazoles (**100**) from benzonitrile or 2-phenyl acetonitrile (**99**) and sodium azide (**65**) in presence of Cu-Guanidine@BO-NPs.

Bagherzade *et al.* have used novel biocompatible core/shell Fe_3O_4 NPs coated with nanofiber cellulose and complexed with cobalt (II) [Fe_3O_4@NFC@Co(II)] as the new catalysts for one-pot synthesis of 4*H*-pyran derivatives (**102**) and pyranopyrazoles (**14b**) from benzaldehyde (**7a**), malononitrile (**6b**) and ethyl acetoacetate (**101**) under aqueous medium in good to excellent yields (Scheme **44**) [79]. The preparation of Fe_3O_4@NFC@Co(II) MNPs has been carried by reacting ferric oxide with nanofiber cellulose (NFC) solution to form Fe_3O_4@NFC precipitates followed by complexation with cobalt(II)acetate. The prepared catalysts have been further characterized by FT-IR, XRD, VSM, FESEM, EDAX, TEM, ICP, TGA. The catalyst can be reused at least five times in the present Knoevenagel-Michael-cyclocondensation without a noticeable loss of activity.

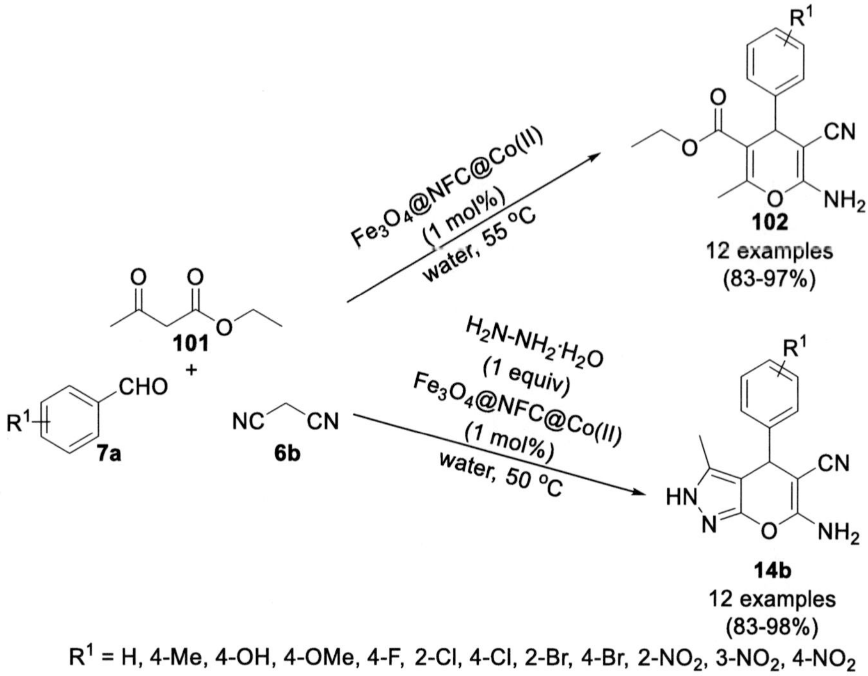

Scheme (44). One-pot synthesis of 4*H*-pyrans (**102**) and pyranopyrazoles (**14b**) from benzaldehyde (**7a**), malononitrile (**6b**), and ethyl acetoacetate (**101**).

Kazemnejadi *et al.* have applied the catalytic use of Fe_3O_4@SiO_2 MNPs immobilized on chlorophyll *b* anchored with magnesium (Fe_3O_4@SiO_2@Chl-Mg) for the synthesis of benzimidazole (**31b**) by reacting *o*-phenylenediamines (**28g**) with benzaldehyde (**7i**) in ethanol at 60 °C in excellent yields (Scheme **45**) [80]. These MNPs have been also found to catalyze the successful synthesis of spirooxindole (**103a/b**) from dimedone (**9b**) or 4-hydroxycoumarin (**11a**), malononitrile (**6b**), and isatins (**38b**, Scheme **46**). Fe_3O_4@SiO_2 NPs have been prepared by co-precipitation of iron salts followed by coating with silica shell and immobilization on chlorophyll b, extracted from leaves of *Heliotropium europaeum* plant. The structure of MNPs has been characterized by FT-IR, UV-vis, XRD, EDX, TGA, VSM, FE-SEM, dynamic light scattering spectroscopy

(DLS). These catalysts could be recycled for at least eight consecutive cycles without any significant reactivity loss and the least amount of leaching of magnesium content from chlorophyll.

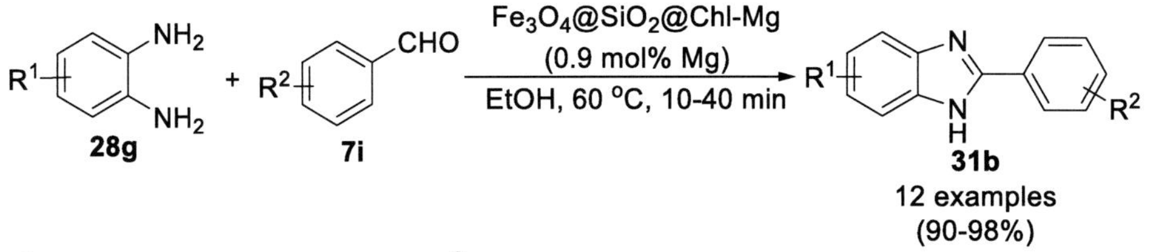

R^1 = H, 3-Me, 4-Me, 4-OMe, 3-Cl; R^2 = H, 4-OH, 4-OMe, 4-NMe$_2$, 2-Cl, 4-Cl, 3-NO$_2$, 4-NO$_2$

Scheme (45). Fe_3O_4@SiO_2@Chl-Mg catalyzed synthesis of 2-phenyl benzimidazole (**31b**) from *o*-phenylenediamines (**28g**) and benzaldehyde (**7i**).

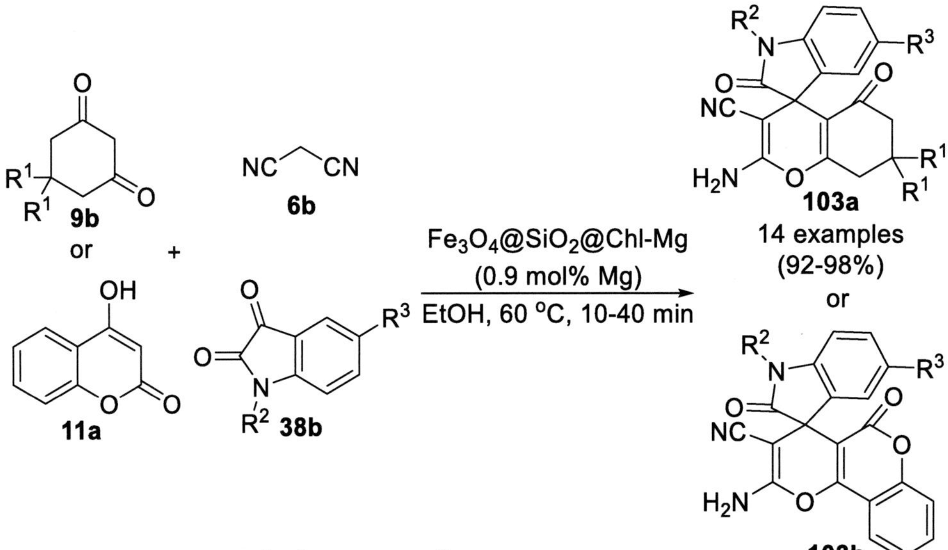

R^1/R^2 = H, Me; R^3 = H, Me, OMe, Cl, NO$_2$

Scheme (46). Preparation of spirooxindoles (**103a/b**) from dimedone (**9b**) and 4-hydroxycoumarin (**11a**), malononitrile (**6b**), and isatins (**38b**) catalyzed by Fe_3O_4@SiO_2@Chl-Mg.

Mokhtary *et al.* have achieved the synthesis of indenopyrido[2,3-*d*]pyrimidines (**105a**) or pyrimido[4,5-*b*]quinolines (**105b**) by the three-component reaction of aldehydes, 6-amino-1,3-dimethyluracil (**37b**) and 1,3-dicarbonyl compounds such as 1*H*-indene-1,3(2*H*)-dione (**104**) or dimedone/cyclohexanone (**9b**) in the presence of glycolic acid-supported cobalt ferrite $CoFe_2O_4$@SiO_2@Si

$(CH_2)_3NHCOOCH_2COOH$] as MNPs in ethanol under reflux (Scheme **47**) [81]. The NCs have been prepared by the treatment of 4,4-diethoxy-9-oxo-3,10-dioxa-8-aza-4-siladodecan-12-oic acid with silica-coated cobalt ferrite NPs ($CoFe_2O_4$@SiO_2) and characterized by FT-IR, XRD, EDS, TEM, SEM, VSM, and TGA. The recovered catalysts have been reused for five consecutive cycles without any considerable loss to yield Knoevenagel-Michael-cyclized products.

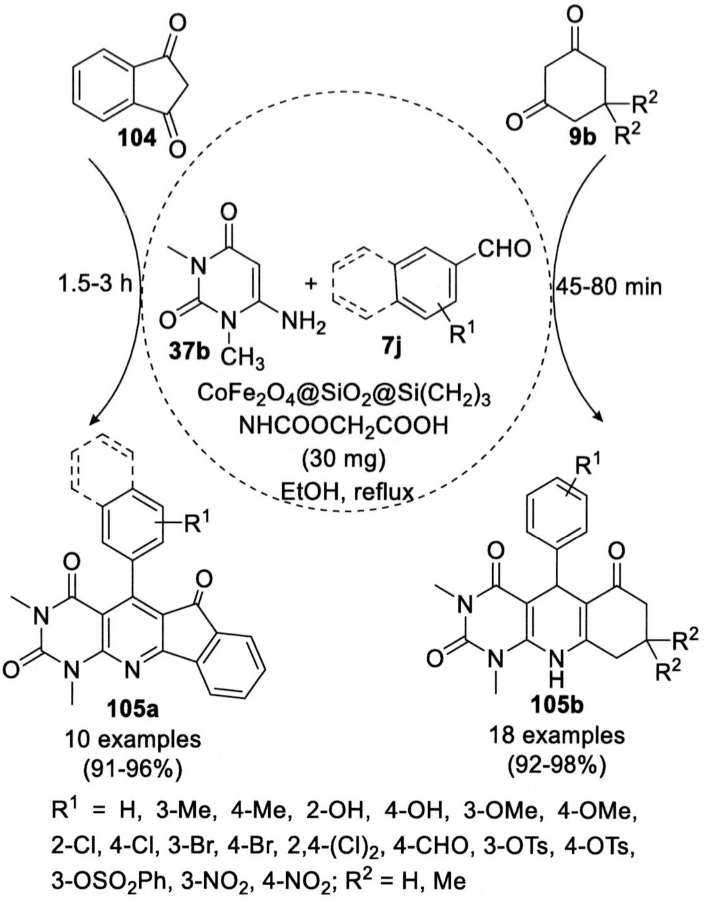

Scheme (47). Synthesis of pyrimido[4,5-*b*]quinolines (**105b**) and indenopyrido[2,3-*d*]pyrimidine (**105a**) in the presence of glycolic acid—supported cobalt ferrite $CoFe_2O_4@SiO_2@Si(CH_2)_3NHCOOCH_2COOH$.

Heydari *et al.* have achieved the Fe_2O_3@[proline]–CuMgAl–Layered Double) Hydroxide) (Fe_2O_3@[proline]-Cu-LDH) catalyzed synthesis of quinolines (**47e/47f/47g**, Scheme **48**) or 2-phenyl-2*H*-indazole (**106**, Scheme **49**) from 2-

bromo benzaldehyde (**7k**) with arylalkyl ketones (**81c**)/ pentane-2,4-dione (**42c**)/cyclic 1,3-diones (**9b**) or aryl amines (**28h**) [82]. To synthesize these bimetallic NCs, Fe_3O_4 NPs, prepared using the co-precipitation method, has been treated with LDH formed from magnesium nitrate [$Mg(NO_3)_2{\cdot}6H_2O$], aluminium nitrate [$Al(NO_3)_3{\cdot}9H_2O$] and copper nitrate [$Cu(NO_3)_2{\cdot}3H_2O$] followed by calcination of the synthesized Fe_2O_3@CuMgAl-LDH and subsequential intercalation with L-proline. These NPs have been characterized by FT-IR, XRD, VSM, SEM, EDS, ICP, and TGA. The recyclable catalysts have been reused and recycled up to five times without loss of catalytic power.

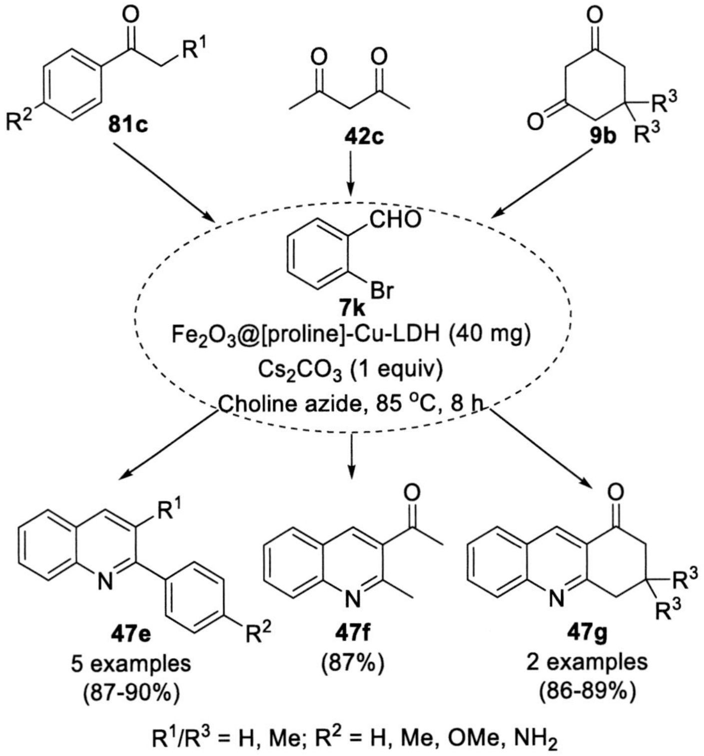

Scheme (48). Synthesis of quinolines (**47e/47f/47g**) catalyzed by Fe_2O_3@[proline]-Cu-LDH.

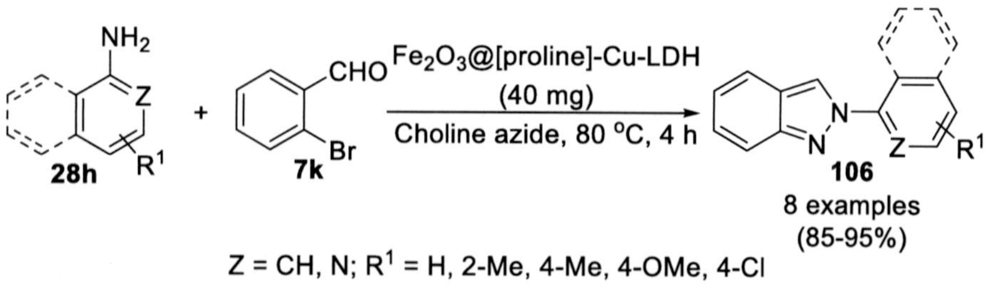

Scheme (49). Synthesis of 2*H*-indazoles (**106**) by condensation of amines (**28h**) with 2-bromobenzaldehyde (**7k**).

Ghasemzadeh and Ghaffarian have synthesized xanthenes (**111**)/quinazolines (**110**)/acridines (**108**) *via* Knoevenagel-Michael-cyclocondensation of aldehydes (**7a**), dimedone (**9a**) and β-naphthol (**107**)/urea (**109**)/substituted aniline (**28i**) in the presence of reusable (six times) cobalt ferrite NPs supported on ($CoFe_2O_4$@OCMC/Cu) under ultrasonic irradiation in good to excellent yields (Scheme **50**) [83]. For the preparation of the catalysts, cobalt ferrite NPs immobilized on carboxymethyl chitosan ($CoFe_2O_4$@OCMC) have been treated with polyvinylpyrrolidone (PVP) followed by the inclusion of a mixture of copper (II) nitrate and terephthalic acid dissolved in DMF.

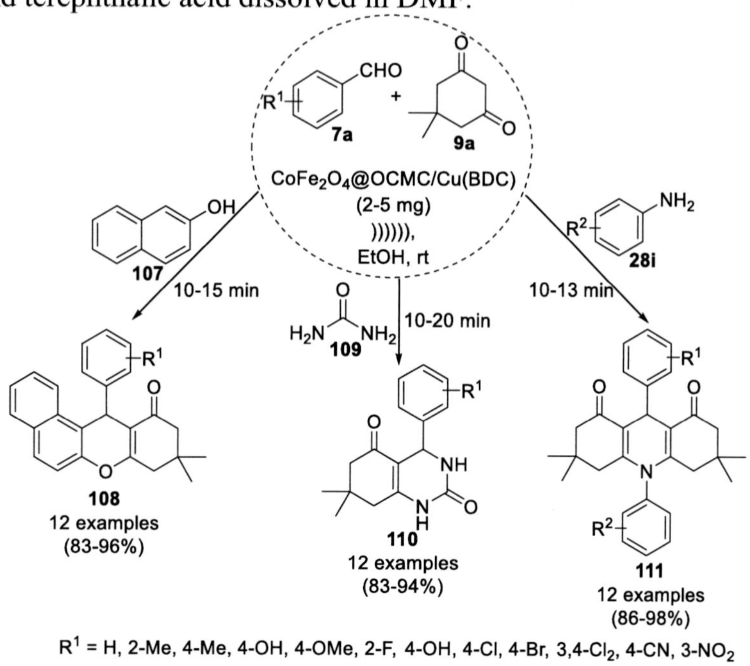

Scheme (50). Synthesis of acridines (**108**), quinazolines (**110**), and xanthenes (**111**) in the presence of $CoFe_2O_4$@OCMC/Cu (BDC) under ultrasonic irradiation.

CONCLUDING REMARKS

The selection of the suitable catalyst for the desired organic transformation has been always full of several challenges, wherein synthetic constructions of desired scaffolds have been always attempted through trial and error. Due to the advancement in nano-technology and the techniques for characterization of MNPs, several research groups have embarked on the MNPs catalyzed synthesis of aza- and oxa-heterocycles through several organic transformations. The predominant MNPs such as AgNPs, CuNPs, FeNPs, NiNPs, PdNPs, PtNPs, ZnNPs, other NPs, or bimetallic MNPs catalyzed synthesis of aza- and oxa-heterocycles have been compiled in the present work to address the selection of suitable MNPs to achieve the synthesis of desired heterocycles. Mostly, these MNPs acted as Lewis or Lowry-Brønsted acid to promote the reaction by either bringing the reactive termini of reactants near each other or forming hydrogen bonds, ionic interactions, coordination covalent bonds, *etc*. These MNPs have been also found suitable tools for sustainable green chemistry through their excellent reusability and recyclability due to their heterogeneous character, which has been recognized as the evergreen field in the recent era of organic synthesis.

CONSENT FOR PUBLICATION

Not applicable.

CONFLICT OF INTEREST

The author declares no conflict of interest, financial or otherwise.

ACKNOWLEDGEMENTS

The motivation and invaluable support provided by Prof. Asit K. Chakraborti, National Institute of Pharmaceutical Education and Research (NIPER), S.A.S. Nagar, Punjab (India) has been deeply acknowledged.

REFERENCES

[1] Dhameliya, T.M.; Bhakhar, K.A.; Gajjar, N.D.; Patel, K.A.; Devani, A.A.; Hirani, R.V. Recent Advancements and Developments in Search of Anti-tuberculosis Agents: A Quinquennial Update and Future Directions. *J. Mol. Struct.,* **2022**, *1248*, 131473.
[http://dx.doi.org/10.1016/j.molstruc.2021.131473]

[2] Pancholia, S.; Dhameliya, T.M.; Shah, P.; Jadhavar, P.S.; Sridevi, J.P.; Yogeshwari, P.; Sriram, D.; Chakraborti, A.K. Benzo[*d*]thiazol-2-yl(piperazin-1-yl)methanones as new anti-mycobacterial chemotypes: Design, synthesis, biological evaluation and 3D-QSAR studies. *Eur. J. Med. Chem.,* **2016**, *116*, 187-199.
[http://dx.doi.org/10.1016/j.ejmech.2016.03.060] [PMID: 27061982]

[3] Dhameliya, T.M.; Tiwari, R.; Banerjee, A.; Pancholia, S.; Sriram, D.; Panda, D.; Chakraborti, A.K. Benzo[*d*]thiazole-2-carbanilides as new anti-TB chemotypes: Design, synthesis, biological evaluation,

and structure-activity relationship. *Eur. J. Med. Chem.,* **2018**, *155*, 364-380. [http://dx.doi.org/10.1016/j.ejmech.2018.05.049] [PMID: 29902722]

[4] Dhameliya, T.M.; Patel, K.I.; Tiwari, R.; Vagolu, S.K.; Panda, D.; Sriram, D.; Chakraborti, A.K. Design, synthesis, and biological evaluation of benzo[*d*]imidazole-2-carboxamides as new anti-TB agents. *Bioorg. Chem.,* **2021**, *107*, 104538. [http://dx.doi.org/10.1016/j.bioorg.2020.104538] [PMID: 33349456]

[5] Jadhavar, P.S.; Dhameliya, T.M.; Vaja, M.D.; Kumar, D.; Sridevi, J.P.; Yogeeswari, P.; Sriram, D.; Chakraborti, A.K. Synthesis, biological evaluation and structure-activity relationship of 2-styrylquinazolones as anti-tubercular agents. *Bioorg. Med. Chem. Lett.,* **2016**, *26*(11), 2663-2669. [http://dx.doi.org/10.1016/j.bmcl.2016.04.012] [PMID: 27095514]

[6] Jadhavar, P.S.; Patel, K.I.; Dhameliya, T.M.; Saha, N.; Vaja, M.D.; Krishna, V.S.; Sriram, D.; Chakraborti, A.K. Benzimidazoquinazolines as new potent anti-TB chemotypes: Design, synthesis, and biological evaluation. *Bioorg. Chem.,* **2020**, *99*, 103774. [http://dx.doi.org/10.1016/j.bioorg.2020.103774] [PMID: 32224336]

[7] Jadhavar, P.S.; Vaja, M.D.; Dhameliya, T.M.; Chakraborti, A.K. Oxazolidinones as Anti-tubercular Agents: Discovery, Development and Future Perspectives. *Curr. Med. Chem.,* **2015**, *22*(38), 4379-4397. [http://dx.doi.org/10.2174/0929867323666151106125759] [PMID: 26549430]

[8] Sureja, D.K.; Vadalia, K.R. $POCl_3$ Catalyzed, One-Step, Solvent-Free Synthesis of Some Novel Thieno[2,3-*d*]Pyrimidin-4(3*H*)-One Derivatives as Antimicrobial Agent. *J. Saudi Chem. Soc.,* **2018**, *22*(2), 248-253. [http://dx.doi.org/10.1016/j.jscs.2016.07.004]

[9] Chhabria, M.; Rathod, I.; Vala, K.; Patel, P. Identification of Important Structural Features of Thiazolo[4,5-*d*]pyrimidines Required for Potent Antifungal Activity. *Med. Chem. Res.,* **2011**, *20*(9), 1450-1454. [http://dx.doi.org/10.1007/s00044-010-9378-2]

[10] De Clercq, E. Strategies in the design of antiviral drugs. *Nat. Rev. Drug Discov.,* **2002**, *1*(1), 13-25. [http://dx.doi.org/10.1038/nrd703] [PMID: 12119605]

[11] Vitaku, E.; Smith, D.T.; Njardarson, J.T. Analysis of the structural diversity, substitution patterns, and frequency of nitrogen heterocycles among U.S. FDA approved pharmaceuticals. *J. Med. Chem.,* **2014**, *57*(24), 10257-10274. [http://dx.doi.org/10.1021/jm501100b] [PMID: 25255204]

[12] Delost, M.D.; Smith, D.T.; Anderson, B.J.; Njardarson, J.T. From Oxiranes to Oligomers: Architectures of U.S. FDA Approved Pharmaceuticals Containing Oxygen Heterocycles. *J. Med. Chem.,* **2018**, *61*(24), 10996-11020. [http://dx.doi.org/10.1021/acs.jmedchem.8b00876] [PMID: 30024747]

[13] Blakemore, D.C.; Castro, L.; Churcher, I.; Rees, D.C.; Thomas, A.W.; Wilson, D.M.; Wood, A. Organic synthesis provides opportunities to transform drug discovery. *Nat. Chem.,* **2018**, *10*(4), 383-394. [http://dx.doi.org/10.1038/s41557-018-0021-z] [PMID: 29568051]

[14] Roughley, S.D.; Jordan, A.M. The medicinal chemist's toolbox: an analysis of reactions used in the pursuit of drug candidates. *J. Med. Chem.,* **2011**, *54*(10), 3451-3479. [http://dx.doi.org/10.1021/jm200187y] [PMID: 21504168]

[15] Dhameliya, T.M.; Chourasiya, S.S.; Mishra, E.; Jadhavar, P.S.; Bharatam, P.V.; Chakraborti, A.K. Rationalization of Benzazole-2-carboxylate *versus* Benzazine-3-one/Benzazine-2,3-dione Selectivity Switch during Cyclocondensation of 2-Aminothiophenols/Phenols/Anilines with 1,2-Biselectrophiles in Aqueous Medium. *J. Org. Chem.,* **2017**, *82*(19), 10077-10091. [http://dx.doi.org/10.1021/acs.joc.7b01548] [PMID: 28846411]

[16] Chakraborti, A.K.; Roy, S.R.; Kumar, D.; Chopra, P. Catalytic Application of Room Temperature

Ionic Liquids: [Bmim][MeSO4] as a Recyclable Catalyst for Synthesis of Bis(indolyl)methanes. Ion-Fishing by MALDI-TOF-TOF MS and MS/MS Studies to Probe the Proposed Mechanistic Model of Catalysis. *Green Chem.,* **2008**, *10*(10), 1111-1118.
[http://dx.doi.org/10.1039/b807572g]

[17] Chakraborti, A.K.; Roy, S.R. On catalysis by ionic liquids. *J. Am. Chem. Soc.,* **2009**, *131*(20), 6902-6903.
[http://dx.doi.org/10.1021/ja900076a] [PMID: 19413313]

[18] Roy, S.R.; Chakraborti, A.K. Supramolecular assemblies in ionic liquid catalysis for aza-Michael reaction. *Org. Lett.,* **2010**, *12*(17), 3866-3869.
[http://dx.doi.org/10.1021/ol101557t] [PMID: 20690631]

[19] Seth, K.; Purohit, P.; Chakraborti, A.K. Cooperative catalysis by palladium-nickel binary nanocluster for Suzuki-Miyaura reaction of ortho-heterocycle-tethered sterically hindered aryl bromides. *Org. Lett.,* **2014**, *16*(9), 2334-2337.
[http://dx.doi.org/10.1021/ol500587m] [PMID: 24720556]

[20] Purohit, P.; Seth, K.; Kumar, A.; Chakraborti, A.K. C-O Bond Activation by Nickel-Palladium Hetero-Bimetallic Nano-Particles for Suzuki-Miyaura Reaction of Bioactive Heterocycle-Tethered Sterically Hindered Aryl Carbonates. *ACS Catal.,* **2017**, *7*(4), 2452-2457.
[http://dx.doi.org/10.1021/acscatal.6b02912]

[21] Dandia, A.; Parewa, V.; Sharma, A. An Approach Towards Green Switch Through Nanocatalysis for the Synthesis of Biodynamic Heterocycles. In: *Green Chemistry: Synthesis of Bioactive Heterocycles*; Ameta, K.L.; Dandia, A., Eds.; Springer India: New Delhi, **2014**; pp. 129-161.

[22] Astruc, D.; Lu, F.; Aranzaes, J.R. Nanoparticles as recyclable catalysts: the frontier between homogeneous and heterogeneous catalysis. *Angew. Chem. Int. Ed.,* **2005**, *44*(48), 7852-7872.
[http://dx.doi.org/10.1002/anie.200500766] [PMID: 16304662]

[23] Polshettiwar, V.; Varma, R.S. Green Chemistry by Nano-Catalysis. *Green Chem.,* **2010**, *12*(5), 743-754.
[http://dx.doi.org/10.1039/b921171c]

[24] De Jong, W.H.; Borm, P.J.A. Drug delivery and nanoparticles:applications and hazards. *Int. J. Nanomedicine,* **2008**, *3*(2), 133-149.
[http://dx.doi.org/10.2147/IJN.S596] [PMID: 18686775]

[25] Sarfraz, N.; Khan, I. Plasmonic Gold Nanoparticles (AuNPs): Properties, Synthesis and their Advanced Energy, Environmental and Biomedical Applications. *Chem. Asian J.,* **2021**, *16*(7), 720-742.
[http://dx.doi.org/10.1002/asia.202001202] [PMID: 33440045]

[26] Shah, P.; Dhameliya, T.M.; Bansal, R.; Nautiyal, M.; Kommi, D.N.; Jadhavar, P.S.; Sridevi, J.P.; Yogeeswari, P.; Sriram, D.; Chakraborti, A.K. *N*-Arylalkylbenzo[*d*]Thiazole-2-Carboxamides as Anti-mycobacterial Agents: Design, New Methods of Synthesis and Biological Evaluation. *MedChemComm,* **2014**, *5*(10), 1489-1495.
[http://dx.doi.org/10.1039/C4MD00224E]

[27] Khodashenas, B.; Ghorbani, H.R. Synthesis of Silver Nanoparticles with Different Shapes. *Arab. J. Chem.,* **2019**, *12*(8), 1823-1838.
[http://dx.doi.org/10.1016/j.arabjc.2014.12.014]

[28] Zhang, X.; Chen, K.H.; Zhou, Z.H.; He, L.N. Reduced Graphene Oxide Supported Ag Nanoparticles: An Efficient Catalyst for CO_2 Conversion at Ambient Conditions. *ChemCatChem,* **2020**, *12*(19), 4825-4830.
[http://dx.doi.org/10.1002/cctc.202000738]

[29] Ghosh, S.; Khan, T.S.; Ghosh, A.; Chowdhury, A.H.; Haider, M.A.; Khan, A.; Islam, S.M. Utility of Silver Nanoparticles Embedded Covalent Organic Frameworks as Recyclable Catalysts for the Sustainable Synthesis of Cyclic Carbamates and 2-Oxazolidinones *via* Atmospheric Cyclizative CO_2

Capture. *ACS Sustain. Chem.& Eng.,* **2020**, *8*(14), 5495-5513.
[http://dx.doi.org/10.1021/acssuschemeng.9b06704]

[30] Saneinezhad, S.; Mohammadi, L.; Zadsirjan, V.; Bamoharram, F.F.; Heravi, M.M. Silver nanoparticles-decorated Preyssler functionalized cellulose biocomposite as a novel and efficient catalyst for the synthesis of 2-amino-4H-pyrans and spirochromenes. *Sci. Rep.,* **2020**, *10*(1), 14540.
[http://dx.doi.org/10.1038/s41598-020-70738-z] [PMID: 32884026]

[31] Abiraman, T.; Rajavelu, K.; Rajakumar, P.; Balasubramanian, S. Sub 1 nm Poly(acrylic acid)-Capped Copper Nanoparticles for the Synthesis of 1,2,3-Triazole Compounds. *ACS Omega,* **2020**, *5*(14), 7815-7822.
[http://dx.doi.org/10.1021/acsomega.9b03995] [PMID: 32309691]

[32] Mollabagher, H.; Taheri, S.; Mojtahedi, M.M.; Seyedmousavi, S. Cu-Metal Organic Frameworks (Cu-MOF) as an Environment-Friendly and Economical Catalyst for One Pot Synthesis of Tacrine Derivatives. *RSC Advances,* **2020**, *10*(4), 1995-2003.
[http://dx.doi.org/10.1039/C9RA10111J]

[33] Hamidi Dastjerdi, F.; Ghorbani-Vaghei, R.; Alavinia, S. Copper Iodide Nanoparticles Immobolized Porous Polysulfonamide: An Effective Nanocatalyst for Synthesis of Imidazo. *Catal. Lett.,* **2020**, *150*(12), 3514-3522. [1,2-a].
[http://dx.doi.org/10.1007/s10562-020-03265-1]

[34] Khaleghi Abbasabadi, M.; Azarifar, D. β-Alanine-Functionalized Magnetic Graphene Oxide Quantum Dots: An Efficient and Recyclable Heterogeneous Basic Catalyst for the Synthesis of 1*H*-Pyrazolo[1,2-*b*]phthalazine-5,10-dione and 2,3-Dihydroquinazolin-4(1*H*)-one Derivatives. *Appl. Organomet. Chem.,* **2020**, *34*(10), e5872.
[http://dx.doi.org/10.1002/aoc.5872]

[35] Moussa, A.; Rahmati, A. Synthesis and Characterization of Silica-Coated Fe_3O_4 Nanoparticle@silylpropyl Triethylammonium Polyoxometalate as an Organic–Inorganic Hybrid Heterogeneous Catalyst for the One-Pot Synthesis of Tetrahydrobenzimidazo[2,1-*b*] quinazolin-1(2*H*)-ones. *Appl. Organomet. Chem.,* **2020**, *34*(11), e5894.
[http://dx.doi.org/10.1002/aoc.5894]

[36] Khaleghi Abbasabadi, M.; Azarifar, D.; Esmaili Zand, H.R. Sulfonic Acid-Functionalized Fe_3O_4-Supported Magnetized Graphene Oxide Quantum Dots: A Novel Organic-Inorganic Nanocomposite as an Efficient and Recyclable Nanocatalyst for the Synthesis of Dihydropyrano[2,3-*c*]Pyrazole and 4*H*-Chromene Derivatives. *Appl. Organomet. Chem.,* **2020**, *34*(12), e6004.
[http://dx.doi.org/10.1002/aoc.6004]

[37] Akiyama, T.; Wada, Y.; Yamada, M.; Shio, Y.; Honma, T.; Shimoda, S.; Tsuruta, K.; Tamenori, Y.; Haneoka, H.; Suzuki, T.; Harada, K.; Tsurugi, H.; Mashima, K.; Hasegawa, J.Y.; Sato, Y.; Arisawa, M. Self-Assembled Multilayer Iron(0) Nanoparticle Catalyst for Ligand-Free Carbon-Carbon/Carbo--Nitrogen Bond-Forming Reactions. *Org. Lett.,* **2020**, *22*(18), 7244-7249.
[http://dx.doi.org/10.1021/acs.orglett.0c02574] [PMID: 32903001]

[38] Doustkhah, E.; Nanoporous Silica-Supported Organocatalyst, R.S. A Heterogeneous and Green Hybrid Catalyst for Organic Transformations. *J. Magn. Magn. Mater.,* **2015**, *386*, 111-116.

[39] Tailor, Y.K.; Khandelwal, S.; Verma, K.; Gopal, R.; Kumar, M. Multicomponent synthesis of dispiroheterocycles using a magnetically separable and reusable heterogeneous catalyst. *RSC Advances,* **2020**, *10*(60), 36713-36722.
[http://dx.doi.org/10.1039/D0RA06676A]

[40] Ghasemi, P.; Yarie, M.; Zolfigol, M.A.; Taherpour, A.A.; Torabi, M. Ionically tagged magnetic nanoparticles with urea linkers: application for preparation of 2-aryl-quinoline-4-carboxylic Acids *via* an anomeric-based oxidation mechanism. *ACS Omega,* **2020**, *5*(7), 3207-3217.
[http://dx.doi.org/10.1021/acsomega.9b03277] [PMID: 32118136]

[41] Alishahi, N.; Nasr-Esfahani, M.; Mohammadpoor-Baltork, I.; Tangestaninejad, S.; Mirkhani, V.;

Moghadam, M. Nicotine-based ionic liquid supported on magnetic nanoparticles: an efficient and recyclable catalyst for selective one-pot synthesis of mono- and Bis-4*H*-Pyrimido. *Appl. Organomet. Chem.,* **2020**, *34*(8), e5681. [2,1-b].
[http://dx.doi.org/10.1002/aoc.5681]

[42] Du, L.; Shi, L.; Liu, Y.; Ling, Y.; Zhang, Y.; Zhou, C.; Xiong, B. Nanonickel oxides prepared by atomic layer deposition as efficient catalyst for the dehydrogenation of *N*-Heterocycles. *ChemistrySelect,* **2020**, *5*(38), 11811-11816.
[http://dx.doi.org/10.1002/slct.202003410]

[43] Defilippi, C.; Rodríguez-Padrón, D.; Luque, R.; Giordano, C. Simplifying levulinic acid conversion towards a sustainable biomass valorisation. *Green Chem.,* **2020**, *22*(9), 2929-2934.
[http://dx.doi.org/10.1039/C9GC04306C]

[44] Ji, N.; Liu, Z.; Diao, X.; Bao, J.; Yu, Z.; Song, C.; Liu, Q.; Ma, D.; Lu, X. A novel Ni/AC catalyst prepared by MOCVD method for hydrogenation of ethyl levulinate to γ-valerolactone. *Mol. Catal.,* **2020**, *495*, 111155.
[http://dx.doi.org/10.1016/j.mcat.2020.111155]

[45] Mozafari, R.; Ghadermazi, M. A nickel nanoparticle engineered cofe2o4@go-kryptofix 22 composite: a green and retrievable catalytic system for the synthesis of 1,4-benzodiazepines in water. *RSC Advances,* **2020**, *10*(26), 15052-15064.
[http://dx.doi.org/10.1039/D0RA01671C]

[46] Mazarío, J.; Raad, Z.; Concepción, P.; Cerdá-Moreno, C.; Domine, M.E. Pd supported on mixed metal oxide as an efficient catalyst for the reductive amination of bio-derived acetol to 2-methylpiperazine. *Catal. Sci. Technol.,* **2020**, *10*(23), 8049-8063.
[http://dx.doi.org/10.1039/D0CY01423K]

[47] Sun, Y.; Li, T. Fabrication and Application of Graphene Supported Diimine-Palladium Complex Catalyst for Organic Synthesis. *ChemistrySelect,* **2020**, *5*(4), 1431-1438.
[http://dx.doi.org/10.1002/slct.201904488]

[48] Tang, H.; Zhou, M.L.; Li, X.; Zhang, Y.Y.; Han, Z.B. A Catalyst of Pd@MIL-101@SGO Catalyzes Epoxidation and Hydroxymethoxylation Tandem Reactions of Styrene. *ChemistrySelect,* **2020**, *5*(12), 3724-3729.
[http://dx.doi.org/10.1002/slct.202000111]

[49] Wu, Y.; Wu, Y.; Zhao, Y.; Zhao, Y.; Wang, H.; Wang, H.; Zhang, F.; Zhang, F.; Li, R.; Li, R.; Xiang, J.; Wang, Z.; Han, B.; Han, B.; Liu, Z.; Liu, Z.; Liu, Z. Ambient reductive synthesis of *N*-heterocyclic compounds over cellulose-derived carbon supported pt nanocatalyst under H_2 atmosphere. *Green Chem.,* **2020**, *22*(12), 3820-3826.
[http://dx.doi.org/10.1039/D0GC01177K]

[50] Mohamed, Y.M.A.; Attia, Y.A. The Influence of Ultrasonic Irradiation on Catalytic Performance of ZnO Nanoparticles toward the Synthesis of Chiral 1-Substituted-1H-Tetrazolederivatives from α-Amino Acid Ethyl Esters. *Appl. Organomet. Chem.,* **2020**, *34*(9), e5758.
[http://dx.doi.org/10.1002/aoc.5758]

[51] Verma, K.; Tailor, Y.K.; Khandelwal, S.; Agarwal, M.; Rushell, E.; Pathak, S.; Kumari, Y.; Awasthi, K.; Kumar, M. Synthesis and Characterization of Terbium Doped TiO_2 Nanoparticles and Their Use as Recyclable and Reusable Heterogeneous Catalyst for Efficient and Environmentally Sustainable Synthesis of Spiroannulated Indolo[3,2-*c*]quinolines- Mimetic Scaffolds of Isoc. *Appl. Organomet. Chem.,* **2020**, *34*(10), e5836.
[http://dx.doi.org/10.1002/aoc.5836]

[52] Xu, D.; Zhao, H.; Dong, Z.; Ma, J. Catalytically Active Co−Nx Species Stabilized on Nitrogen-Doped Porous Carbon for Efficient Hydrogenation and Dehydrogenation of *N*-Heteroarenes. *ChemCatChem,* **2020**, *12*(17), 4406-4415.
[http://dx.doi.org/10.1002/cctc.202000826]

[53] Kerru, N.; Gummidi, L.; Maddila, S.N.; Bhaskaruni, S.V.H.S.; Maddila, S.; Jonnalagadda, S.B. Green Synthesis and Characterisation of Novel [1,3,4]Thiadiazolo/Benzo[4,5]Thiazolo[3,2-*a*] Pyrimidines via multi component reaction using vanadium oxide loaded on fluorapatite as a robust and sustainable catalyst. *RSC Advances,* **2020**, *10*(34), 19803-19810. [http://dx.doi.org/10.1039/D0RA02298E]

[54] Zhao, B.; Xu, Y.; Deng, Q.G.; Liu, Z.; Wang, L.Y.; Gao, Y. One-Pot, Three Component Synthesis of Novel 5*H*-[1,3,4]Thiadiazolo[3,2-*a*]pyrimidine-6-carboxylate Derivatives by Microwave Irradiation. *Tetrahedron Lett.,* **2014**, *55*(33), 4521-4524. [http://dx.doi.org/10.1016/j.tetlet.2014.06.073]

[55] Tiwari, S.V.; Seijas, J.A.; Vazquez-Tato, M.P.; Sarkate, A.P.; Lokwani, D.K.; Nikalje, A.P.G. Ultrasound Mediated One-Pot, Three Component Synthesis, Docking and ADME Prediction of Novel 5-Amino-2-(4-chlorophenyl)-7-substituted phenyl-8,8a-dihydro-7*H*-(1,3,4)thiadiazolo(3,2-*α*)pyrimidine-6-carbonitrile Derivatives as Anticancer Agents. *Molecules,* **2016**, *21*(8), 894. [http://dx.doi.org/10.3390/molecules21080894]

[56] Kasaboina, S.; Bollu, R.; Gomedhika, P.M.; Ramineni, V.; Nagarapu, L.; Dumala, N.; Grover, P.; Nanubolu, J.B. A Green Protocol for One-Pot Synthesis of Benzosuberone Tethered Thiadiazolopyrimidine-6-carboxylates Using PEG-400 as Potent Anti-proliferative Agents. *Tetrahedron Lett.,* **2018**, *59*(31), 3015-3019. [http://dx.doi.org/10.1016/j.tetlet.2018.06.068]

[57] Dong, H.R.; Gao, Z.L.; Li, R.S.; Hu, Y.M.; Dong, H.S.; Xie, Z.X. One-Pot Synthesis of 5*H*-1,3,--Thiadiazolo[3,2-*a*]pyrimidin-5-one Derivatives. *RSC Advances,* **2014**, *4*(99), 55827-55831. [http://dx.doi.org/10.1039/C4RA02714K]

[58] Wadhwa, P.; Kaur, T.; Singh, N.; Singh, U.P.; Sharma, A. *p*-Toluenesulfonic Acid-Mediated Three-Component Reaction "On-Water" Protocol for the Synthesis of Novel Thiadiazolo[2,3-*b*]quinazolin-6(7*H*)-ones. *Asian J. Org. Chem.,* **2016**, *5*(1), 120-126. [http://dx.doi.org/10.1002/ajoc.201500397]

[59] Safajoo, N.; Mirjalili, B.B.F.; Bamoniri, A. Fe_3O_4@nano-Cellulose/Cu(Ii): A Bio-Based and Magnetically Recoverable Nano-Catalyst for the Synthesis of 4*H*-Pyrimido[2,1-*b*]Benzothiazole Derivatives. *RSC Advances,* **2019**, *9*(3), 1278-1283. [http://dx.doi.org/10.1039/C8RA09203F]

[60] Mirjalili, B.B.F.; Soltani, R. Nano-Kaolin/Ti_4^+/Fe_3O_4: A Magnetic Reusable Nano-Catalyst for the Synthesis of Pyrimido. *RSC Advances,* **2019**, *9*(33), 18720-18727. [2,1-b]. [http://dx.doi.org/10.1039/C9RA01767D]

[61] Mirjalili, B.B.F.; Aref, F. Nano-Cellulose/BF_3/Fe_3O_4: A Magnetic Bio-Based Nano-Catalyst for the Synthesis of Pyrimido[2,1-*b*]benzothiazoles under Solvent-Free Conditions. *Res. Chem. Intermed.,* **2018**, *44*(7), 4519-4531. [http://dx.doi.org/10.1007/s11164-018-3401-5]

[62] Fazeli-Attar, S.A.; Mirjalili, B.B.F. Nano-Fe_3O_4@SiO_2–$TiCl_3$ as a Novel Nano-Magnetic Catalyst for the Synthesis of 4*H*-Pyrimido. *Res. Chem. Intermed.,* **2018**, *44*(10), 6419-6430. [2,1-b]. [http://dx.doi.org/10.1007/s11164-018-3498-6]

[63] Kharazmi, A.; Ghorbani-vaghei, R.; Alavinia, S. Synthesis of Pyrimidine Derivatives Catalyzed by Nanomagnetic Pyridinium-Tribromide Ionic Liquid. *ChemistrySelect,* **2020**, *98*(4), 1424-1430. [http://dx.doi.org/10.1002/slct.201904697]

[64] Khalaj, M. Preparation of Benzo[4,5]thiazolo[3,2-*a*]chromeno[4,3-*d*]pyrimidin-6-One Derivatives Using MgO-$MgAl_2O_4$ Composite Nano-Powder. *Arab. J. Chem.,* **2020**, *13*(8), 6403-6411. [http://dx.doi.org/10.1016/j.arabjc.2020.05.041]

[65] Ciotonea, C.; Hammi, N.; Dhainaut, J.; Marinova, M.; Ungureanu, A.; El Kadib, A.; Michon, C.; Royer, S. Phyllosilicate-Derived Nickel-Cobalt Bimetallic Nanoparticles for the Catalytic

Hydrogenation of Imines, Oximes and *N*-Heteroarenes. *ChemCatChem,* **2020**, *12*(18), 4652-4663. [http://dx.doi.org/10.1002/cctc.202000704]

[66] Nguyen, T.T.; Tran, P.H. One-Pot multicomponent synthesis of thieno[2,3-*b*] indoles catalyzed by a magnetic nanoparticle-supported [Urea]$_4$[$ZnCl_2$] deep eutectic solvent. *RSC Advances,* **2020**, *10*(16), 9663-9671.
[http://dx.doi.org/10.1039/D0RA00773K]

[67] Aqeel Ashraf, M.; Liu, Z.; Li, C.; Zhang, D. Synthesis of heterocycles using nanomagnetic nickel catalysts. *Synth. Commun.,* **2020**, *50*(19), 2906-2923.
[http://dx.doi.org/10.1080/00397911.2020.1789168]

[68] Aqeel Ashraf, M.; Liu, Z.; Yang, Y.; Zhang, D. Magnetic Nanoparticles Supported Copper Catalysts: Synthesis of Heterocyclic Scaffolds. *Synth. Commun.,* **2020**, *50*(19), 2885-2905.
[http://dx.doi.org/10.1080/00397911.2020.1789167]

[69] Akbarzadeh, P.; Koukabi, N. Magnetic Carbon Nanotube as a Highly Stable Support for the Heterogenization of $InCl_3$ and Its Application in the Synthesis of Isochromeno[4,3-*c*]Pyrazole-5(1*H*)-One Derivatives. *Appl. Organomet. Chem.,* **2020**, *34*(9), e5746.
[http://dx.doi.org/10.1002/aoc.5746]

[70] Moghaddam-manesh, M.; Sheikhhosseini, E.; Ghazanfari, D. Bioorganic Chemistry on Fe_3O_4@gly@CE MNPs as Efficient Catalyst and Evaluation of Biological Activity. *Bioorg. Chem.,* **2020**, *98*, 103751.
[http://dx.doi.org/10.1016/j.bioorg.2020.103751] [PMID: 32182517]

[71] Moghaddam-Manesh, M.; Ghazanfari, D.; Sheikhhosseini, E.; Akhgar, M. Synthesis of Bioactive Magnetic Nanoparticles Spiro[Indoline-3,4'-[1,3]Dithiine]@Ni $(NO_3)_2$ Supported on Fe_3O_4@SiO_2@CPS as Reusable Nanocatalyst for the Synthesis of Functionalized 3,4-Dihydro--H-Pyran. *Appl. Organomet. Chem.,* **2020**, *34*(4), e5543.
[http://dx.doi.org/10.1002/aoc.5543]

[72] Hajizadeh, Z.; Radinekiyan, F.; Eivazzadeh-Keihan, R.; Maleki, A. Development of novel and green $NiFe_2O_4$/geopolymer nanocatalyst based on bentonite for synthesis of imidazole heterocycles by ultrasonic irradiations. *Sci. Rep.,* **2020**, *10*(1), 11671.
[http://dx.doi.org/10.1038/s41598-020-68426-z] [PMID: 32669578]

[73] Pourkarim, Z.; Nikpassand, M. Synthesis, Characterization and Application of Fe_3O_4@SiO_2@Tannic Acid Nanoparticles: A Novel and Magnetically Recyclable Catalyst for One-Pot Synthesis of Novel 5-Pyrazolin-1,2,4-Triazolidine-3-Ones (Thiones). *J. Mol. Struct.,* **2020**, *1217*, 128433.
[http://dx.doi.org/10.1016/j.molstruc.2020.128433]

[74] Karimi, F.; Yarie, M.; Zolfigol, M.A. Synthesis and characterization of Fe_3O_4@SiO_2@$(CH_2)_3NH(CH_2)_2O_2P(OH)_2$ and its catalytic application in the synthesis of Benzo-[*h*]quinoline-4-carboxylic Acids*via* a cooperative anomeric based oxidation mechanism. *Mol. Catal.,* **2020**, *489*, 110924.
[http://dx.doi.org/10.1016/j.mcat.2020.110924]

[75] Hosseinzadegan, S.; Hazeri, N.; Maghsoodlou, M.T. Synthesis of Novel Thiazolo[3,2-*a*] chromeno[4,3-*d*]pyrimidine-6(7*H*)-ones by Bioactive Fe_3O_4@gly@thiophen@$Cu(NO_3)_2$ as Reusable Magnetic Nanocatalyst. *Appl. Organomet. Chem.,* **2020**, *34*(9), e5797.
[http://dx.doi.org/10.1002/aoc.5797]

[76] Ebrahimiasl, H.; Azarifar, D.; Rakhtshah, J.; Keypour, H.; Mahmoudabadi, M. Application of Novel and Reusable Fe_3O_4@CoII(Macrocyclic Schiff Base Ligand) for Multicomponent Reactions of Highly Substituted Thiopyridine and 4*H*-Chromene Derivatives. *Appl. Organomet. Chem.,* **2020**, *34*(9), 5769.
[http://dx.doi.org/10.1002/aoc.5769]

[77] Hemmati, S.; Ahany Kamangar, S.; Yousefi, M.; Hashemi Salehi, M.; Hekmati, M. Cu(I)-anchored polyvinyl alcohol coated-magnetic nanoparticles as heterogeneous nanocatalyst in ullmann-type C–N coupling reactions. *Appl. Organomet. Chem.,* **2020**, *34*(5), 5611.

[http://dx.doi.org/10.1002/aoc.5611]

[78] Jafari, F.; Ghorbani-Choghamarani, A.; Hasanzadeh, N. Guanidine complex of copper supported on boehmite nanoparticles as practical, recyclable, chemo and homoselective organic–inorganic hybrid nanocatalyst for organic reactions. *Appl. Organomet. Chem.,* **2020**, *34*(11), 5901. [http://dx.doi.org/10.1002/aoc.5901]

[79] Kargar, P.G.; Bagherzade, G.; Eshghi, H. Novel biocompatible core/shell Fe_3O_4@NFC@Co(II) as a new catalyst in a multicomponent reaction: an efficient and sustainable methodology and novel reusable material for one-pot synthesis of 4*H*-pyran and pyranopyrazole in aqueous media. *RSC Advances,* **2020**, *10*(61), 37086-37097. [http://dx.doi.org/10.1039/D0RA04698A]

[80] Shamaei, A.; Mahmoudi, B.; Kazemnejadi, M.; Nasseri, M.A. Mg-catalyzed one-pot preparation of benzimidazoles and spirooxindoles by an immobilized chlorophyll b on magnetic nanoparticles. *Appl. Organomet. Chem.,* **2020**, *34*(12), e5997. [http://dx.doi.org/10.1002/aoc.5997]

[81] Gholami, A.; Mokhtary, M.; Nikpassand, M. Glycolic acid-supported cobalt ferrite-catalyzed one-pot synthesis of pyrimido[4,5-*b*]quinoline and Indenopyrido[2,3-*d*]pyrimidine Derivatives. *Appl. Organomet. Chem.,* **2020**, *34*(12), e6007. [http://dx.doi.org/10.1002/aoc.6007]

[82] Esfandiary, N.; Heydari, A. Fe_2O_3@[Proline]–CuMgAl–LDH: a magnetic bifunctional copper and organocatalyst system for one-pot synthesis of quinolines and 2*H*-Indazoles in green media. *Appl. Organomet. Chem.,* **2020**, *34*(9), e5760. [http://dx.doi.org/10.1002/aoc.5760]

[83] Ghasemzadeh, M.A.; Ghaffarian, F. Preparation of core/shell/shell $CoFe_2O_4$/OCMC/Cu (BDC) nanostructure as a magnetically heterogeneous catalyst for the synthesis of substituted xanthenes, quinazolines and acridines under ultrasonic irradiation. *Appl. Organomet. Chem.,* **2020**, *34*(5), e5580. [http://dx.doi.org/10.1002/aoc.5580]

CHAPTER 8

Nanomaterials-Based Photocatalytic Reactions

Soniya Agarwal[1] and **Kalyanjyoti Deori**[1,*]

[1] *Department of Chemistry, Dibrugarh University, Dibrugarh-786004, Assam, India*

Abstract: Mounting energy demand and alternatives to its non-renewable sources have led to an interest in the arena of photocatalysis. This field has emerged as a promising greener solution to environmental glitches. Nanomaterials are known to be potential semiconductors because they exhibit unique chemical and physical properties. Henceforth, the search and development of nanomaterial photocatalyst is in full swing and also need of the hour. Nanomaterials such as metal oxides, sulfides, and nitrides have cropped up as efficient photocatalysts to meet up energy demands by catalysing a variety of reactions. In this article, we have discussed different types of nanomaterials and their might to display various photocatalytic reactions.

Keywords: Band gap, Green, Nanomaterials, Photocatalyst, Semiconductor, Sustainable.

INTRODUCTION

The major universal concern at present is environmental pollution and global warming leading to a life-threatening crisis. The scientific community has the huge responsibility of finding and giving greener and sustainable solutions for the factors responsible for this catastrophe. One such remedy to this foreseen tragedy is a technique called photocatalysis which uses light as the clean energy source for catalysing various reactions with the help of a catalyst known as photocatalyst [1]. These photocatalysts are substances that can absorb light and modify the rate of photoreaction without being consumed. Among various substances, semiconducting nanomaterials are known to be good and efficient photocatalysts compared to their bulk counterparts. This change of property of a material in the nanoscale is because of the easy tunability of the bandgap due to the quantum confinement effect [2]. A wide range of nanomaterials has been developed for photocatalysis which includes various metal oxides [3], chalcogenides [4], nitrides, while a few are carbon and metal-organic framework based [5]. Also, several types of bimetallic nanoparticles, both bare and supported [6, 7], are employed as photo-

* **Corresponding author Kalyanjyoti Deori:** Department of Chemistry, Dibrugarh University, Dibrugarh, Assam; E-mail: kalchemdu@gmail.com

Vijai K. Rai, Manorama Singh and Ankita Rai

catalysts for organic conversions. However, photocatalysis is not just limited to organic conversions, but many processes like dye degradation [8], Cr(VI) reduction [9] for waste water treatment, water splitting reaction for hydrogen and oxygen evolution [10], CO_2 reduction [11], *etc.* are also demonstrating good efficacy and getting tremendous attention in photocatalytic research.

In this chapter, we will discuss the fundamental theories and mechanism of photocatalysis, properties making nanomaterial reliable and excellent photo-catalyst, and also an overview of the nano photocatalysts that have been designed and developed so far for this purpose (Fig. **1**).

Fig. (1). Schematic representation showing various nanomaterials with different bandgaps used for a variety of photocatalytic reactions.

NANO-PHOTOCATALYST

Photocatalysts are the materials that act as a catalyst for various chemical conversions when light is irradiated on them. As mentioned, this interesting property of photocatalyst is due to its semiconducting property. This is where the role of nanomaterials comes to play. When a bulk substance is reduced to the nanoscale, the energy levels become discrete as a result of which the bandgap of the material increases. All this is feasible because of the confinement of electrons in the nanomaterial which helps in electron transfer between the valence band and conduction band. This fascinating property leads to the conversion of an insulator to a semiconductor. Another property that is crucial for an effective photocatalyst is the possession of a large surface area. More the surface area available, more will be the light absorbed for enhanced photocatalysis. The motive is to inhibit recombination of charge carriers to increase photocatalytic performance [2].

The utmost important entity to focus on while synthesizing a semiconductor photocatalyst is the bandgap. A high bandgap results in absorbing the ultra-violet (UV) light rather than visible light energy and also as observed the majority of the reported photocatalyst ensure the same. UV light comprises only 4% of the solar spectrum compared to visible light with 45%. Moreover, the inherent health hazards of UV exposure are also a matter of concern, and expensive quartz reactors are needed to deal with it.

Nano-photocatalyst can be of varied types concerning their composition, size, bandgap, morphology, and electronic structure which can be tuned according to the need [12]. A plethora of nanomaterials has been synthesized to date to serve as a photocatalyst. Transition metal-based oxides and sulfides like TiO_2, ZnO [13], CuS, ZnS [14], CdS [12]; carbonaceous materials like graphite, reduced graphene oxide (rGO), carbon nanotubes [15]; oxyhalides like BiOX (X = Cl, Br and I) [16]; bimetallic nanoparticles such as AuAg, AuPd, PtAu; and a composite or a hybrid of noble metals [17] with the mentioned compounds serves as potential photocatalyst.

Fundamentals and Mechanism of Photocatalysis

Photocatalysis, in simpler words, is the harvesting of light energy for various chemical conversions. It is a type of advanced oxidation process where an eco-friendly technique based on the formation of reactive oxygen species for pollutant degradation is responsible. Photocatalyst on exposure to light of the desired wavelength generates electron-hole pair. The light energy *i.e.*, photons absorbed by the semiconductor photocatalyst promotes an electron from the valence band (VB) to the conduction band (CB) leaving behind a hole in the former (Fig. **2**). These generated electrons reduce the acceptor molecule and the respective holes oxidise the donor molecule which further produces free radicals. Reactive species with one unpaired electron are very short-lived and therefore readily and effortlessly reacts with many chemical species to initiate the desired secondary reactions which are otherwise arduous to carry out. Parameters like charge recombination rate, visible light absorption, and unsuitable band gap can seldom act as limitations to the photocatalytic process [2, 12, 18].

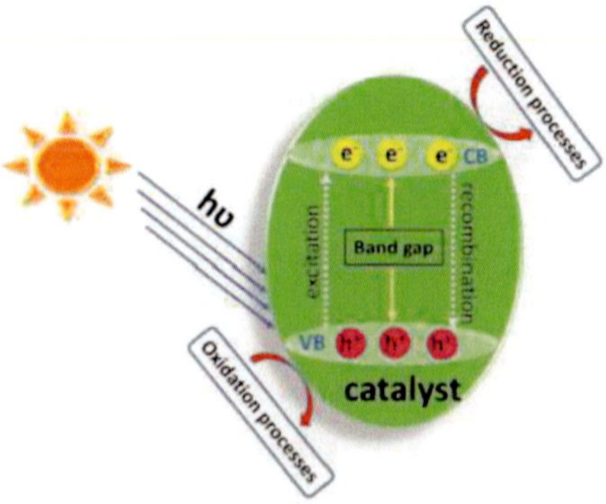

Fig. (2). Mechanism of photocatalysis.

Oxide-based Nanophotocatalyst

The field of photocatalysis has been dominated by various metal oxides (pure or decorated). The earliest report of photocatalysis is found with the use of TiO_2 (titania or titanium dioxide) and ZnO for isopropanol oxidation [19]. The past two decades have seen a surge in the research and advancement of TiO_2 as a photocatalyst. Titania exists in three polymorphs: anatase, rutile, and brookite, out of which anatase is the most active ensuring a bandgap of 3.2 eV. Strategies like tuning the morphology, crystal phase/crystallinity, crystal size, defects, crystal facets, surface modification, and catalyst support are employed to increase the activity and efficiency [20]. In one of the cases, three types of TiO_2 were synthesized with different exposed crystal facets: {001}, {101} and co-exposed {001}-{101} for CO_2 reduction. On enquiry co-exposed {001}-{101} TiO_2 enhanced CO_2 reduction performance because of the charge transfer at {001}-{101} junction boosted the charge separation (Fig. **3**) [21]. This co-exposed titania with an optimal ratio 58/42 for {001}/{101} exhibited a higher rate for CH_4 production than the commercial TiO_2 [22].

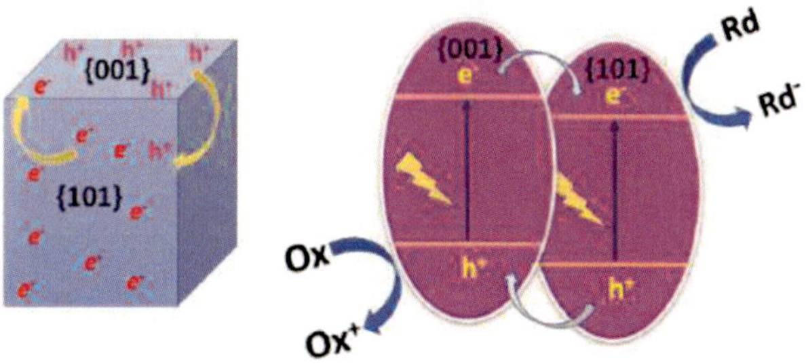

Fig. (3). Charge transfer between crystal facets of TiO_2.

ZnO has transpired as the successor to TiO_2 for promising efficient photocatalysis. The band gap of ZnO (3.37 eV) is almost similar to that of TiO_2 (3.2 eV) with exciton binding energy at room temperature of 60 meV. ZnO appears as a white hexagonal crystal crystallizing in wurtzite structure. The optical, electrical, and magnetic properties of ZnO can be altered by precisely controlling its nanoscale features. Moreover, it is cheaper than titania so could be used in large-scale photocatalytic waste water treatment. ZnO being compatible with living beings and the ecosystem as a whole furnish the elasticity to test its photocatalytic activity for a broad range of applications. ZnO has the potential to absorb in a wide range of solar spectrum compared to other metal oxide semiconducting devices. Nevertheless, it has its drawbacks like wide bandgap and photocorrosion restricting its appliance and limiting its relevance. Here the concept of doping or hybridisation is employed to overcome the aforementioned shortcomings. In a praiseworthy example, nano-heterostructure Bi_2O_3-ZnO was synthesized with an improved band gap of 2.98 eV having a mixture of nanochain and nanofiber (Fig. **4a**) structures displaying high photocatalytic activity for degradation of Acid Black 1 than pure ZnO, Bi_2O_3 and commercial titania [23]. Following similar strategy nanomaterials such as Cu doped TiO_2/ZnO (2.2 eV) [24], nanocomposite Ag/Ag_2O/ZnO (1.75 eV) [25] (Fig. **4b**), heterostructure SnO_2/ZnO (3.65 eV) [26] (Fig. **4c**), Bi_2O_3-CeO_2-ZnO (2.62 eV) [27] (Fig. **4d**) and Cr decorated ZnO nanoparticles (3.09 eV) [28] were formulated for photocatalysis.

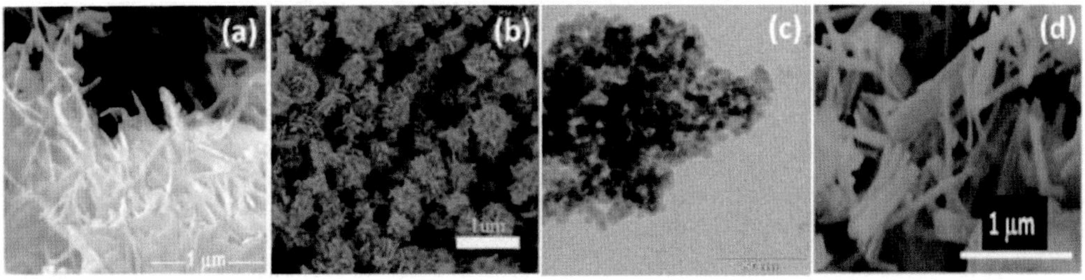

Fig. (4). (a) HR-SEM image of Bi_2O_3-ZnO. Reprinted with permission from ref [23]. Copyright 2012 American Chemical Society. **(b)** SEM image of Ag/Ag_2O/ZnO nanocomposite. Reprinted with permission from ref [25]. Copyright 2019 Elsevier. **(c)** TEM image of SnO_2/ZnO nanocomposite. Reprinted with permission from ref [26]. Copyright 2020 The Royal Society of Chemistry. **(d)** FE-SEM image of Bi_2O_3-CeO_2-ZnO. Reprinted with permission from ref [27]. Copyright 2017 Elsevier.

Chalcogenide Based Nanophotocatalyst

Metal chalcogenide nanoparticles are tremendously investigated as photocatalysts due to their unique physical and chemical properties. They belong to a class of semiconductor photocatalysts with narrower band gaps compared to their oxide counterparts to exercise visible light directly. One popular sulfide photocatalyst of this category is CdS possessing a bandgap of 2.4 eV. CdS nanopowder, solid solutions, and quantum dots have been used for the reaction involving H_2 production and water decontamination [29]. Moreover, CdS quantum dots (radius

= 1.8 nm) catalysed benzyl alcohol oxidation to benzaldehyde with 99% selectivity *via* visible light absorption [30]. When the synthesized CdS does not fulfil the eligibility criteria for the desired reaction, a heterojunction can be introduced by blending with noble metals, other oxides and sulfides, and decorating or coupling with cocatalysts. The formation of heterojunction prevents carrier recombination and photocorrosion that resulting enhanced photocatalytic efficiency. Nano-heterostructures like ZnO-CdS, CdS@g-C_3N_4 (core@shell), MoS_2-CdS [31] (Fig. **6a**), NiN_3-CdS [12], Au-CdS [32], CdSe/CdS/Au [33] (Fig. **6b**), CuO-CdS [34], Cu_2WS_4 [35] (Fig. **6c**) and many more have been therefore introduced in recent years for this purpose. In one paradigm, two sulfide nanomaterials viz., Co_3S_4 nanosheets and CdS nanoparticles were coupled, for visible light-driven H_2O splitting reaction (Fig. **5**) [36].

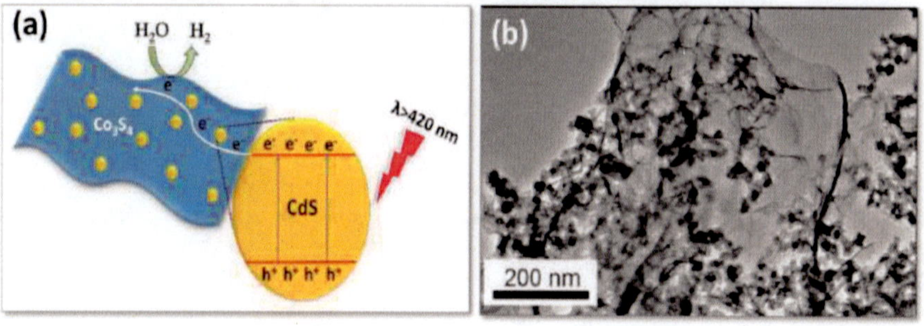

Fig. (5). (a) Forwarded mechanism for H_2 production *via* CdS/Co_3S_4 nanophotocatalyst, **(b)** TEM image of CdS/Co_3S_4. Reprinted with permission from ref [36]. Copyright 2018 Elsevier.

The current area of research on metal sulfide photocatalyst mainly involves the address to two concerning issues: cadmium and lead-free sulfide photocatalyst and hunt for noble metal-free cocatalysts without compromising the efficacy. In this regard, CuS and ZnS have gained tremendous popularity. Many photocatalytic reports are available on ZnS for organic pollutant degradation, CO_2 photoreduction, dehalogenation, and water splitting reactions [37]. Similarly, CuS and its composite with other carbonaceous materials like g-C_3N_4 have been proven as efficient visible light propelled photocatalysts for degradation of various stubborn dyes [38].

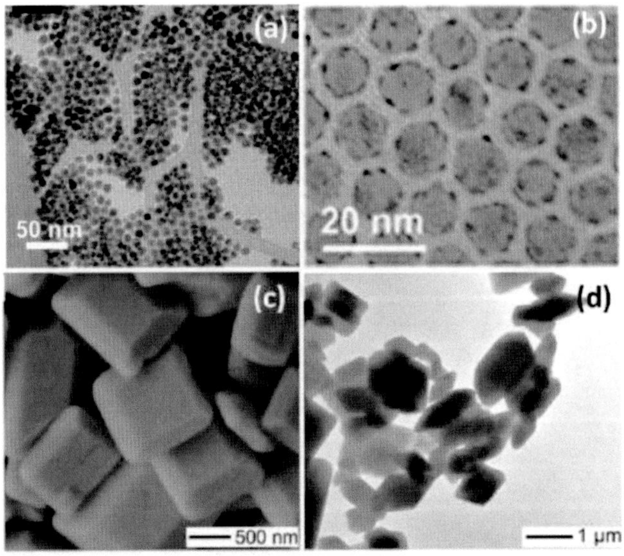

Fig. (6). (a)TEM image of MoS_2-CdS nanohybrid. Reprinted with permission from ref [31]. Copyright 2015 Wiley-VCH. **(b** TEM image of CdSe/CdS/Au. Reprinted with permission from ref [33]. Copyright 2019 American Chemical Society. **(c)** SEM image of Cu_2WS_4, **(d)** TEM image of Cu_2WS_4. Both **(c)** and **(d)** reprinted with permission from ref [35]. Copyright 2014 The Royal Society of Chemistry.

Keeping the popularity of sulfide based photocatalyst aside, other chalcogenide based nanomaterials like CdSe [39], Ba_2ZnSe_3 [40], ZnSe, $Cd_xZn_{(1-x)}S_ySe_{(1-y)}$ [41], are also being explored for photocatalytic activity.

Carbonaceous Nanophotocatalyst

Outstanding attractive features of carbonaceous materials have surfaced its nomination for promising photocatalyst. Carbon dots, activated carbon, carbon nanotube, carbon nanofiber, fullerene, graphene, and 3-D carbons have now been developed with various compositions and structures for this purpose. A facile metal-free and inexpensive route of photocatalysis with earth-abundant Carbon has been put forward by this class of nanomaterials [42]. The discussion is incomplete if we do not talk about the most exciting carbonaceous material *i.e.*, graphene. Apart from its excellent properties like high surface area, conductivity, and mechanical strength, it is known to behave as a macromolecular photosensitizer and narrow the bandgap of photocatalyst semiconductors. Despite ZnS being a wide-gap semiconductor, its composite with graphene makes it active under visible light [43]. Many nanocomposites of graphene to date have been synthesized but one noteworthy case is of Co_3O_4/graphene nanocomposites. Nanocrystals of Co_3O_4 grown on rGO are demonstrated as an efficient photocatalyst for both oxygen evolution reaction and oxygen reduction reaction.

This activity is further enhanced by doping nitrogen on graphene. This surprising high unusual activity is attributed to synergistic chemical coupling between Co_3O_4 and rGO [44]. In another example, rGO acts as an electron transport mediator in photocatalytic systems with Z-scheme composed of Ru-$SrTiO_3$:Rh, $BiVO_4$, and rGO (Fig. **7**). This photoreduced graphene oxide can ferry electrons from the oxygen-evolving $BiVO_4$ to hydrogen evolving Ru-$SrTiO_3$:Rh for visible light-mediated water-splitting reaction [45].

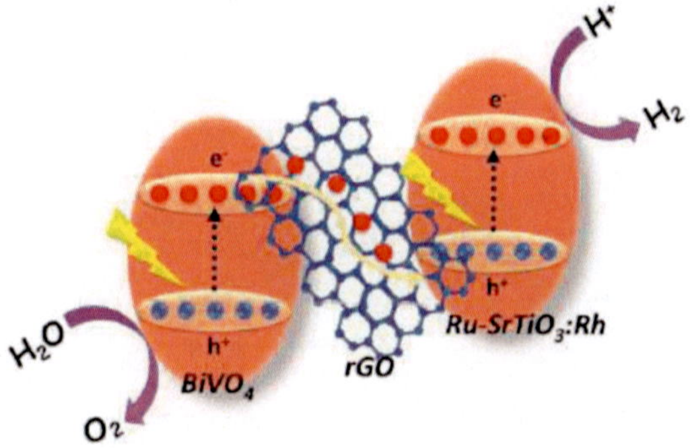

Fig. (7). Schematic diagram of rGO as a mediator for electron transport between $BiVO_4$ and Ru-$SrTiO_3$:Rh.

The other allotrope of carbon, carbon nanotube (CNT), offers an affordable catalytic route for hydrogen evolution reaction (HER). The pristine CNTs with open ends form different structures of 5 membered rings which offer high active sites for HER than traditional 6 membered structures [46].

Bimetallic Nanophotocatalyst

Bimetallic nanoparticles show unique optical, electronic, and catalytic properties in comparison to their monometallic counterparts. Besides displaying the properties of individual metal atoms an additional novel property results out of the synergy between the metals. This synergic effect enhances the photocatalytic efficiency of the nanomaterial [47]. Reported photocatalysis through bimetallic nanoparticles is mostly composed of noble metals or a combination of noble-non noble metals. To identify and comprehend the reason behind it we need to discuss the fascinating optical property exhibited by noble metals: localized surface plasmon resonance. Surface plasmon resonance is the collective oscillation of the conduction electrons and these oscillating electrons are in resonance with the electromagnetic field of the incident light. These interactions generate plasmonic electrons *via* nonradiative excitation.

In most of the cases the bimetallic nanoparticles are decorated on oxides like TiO_2 [48], $BiPO_4$ [49] (Fig. **8a** and **8b**) and ZnO [50] (Fig. **8c** and **8d**), graphene [51] and graphitic carbon nitride [52], and others to provide increased surface area for substrate-catalyst interaction and high charge separation. In a recent report, a hybrid of bimetallic PtAu alloy nanoparticles and g-C_3N_4 was responsible for efficient photocatalytic conversion of H_2O to H_2. The bandgap of the hybrid material decreases to 2.12 eV from 2.6 eV of g-C_3N_4. The bimetallic nanoparticle here acts as a co-catalyst for g-C_3N_4 sheets uniformly distributed on it [53] (Fig. **8e** and **8f**). ZnO decorated Au/Pd nanocomposite was synthesized for degradation of methylene blue on irradiation with visible light and increased photocatalytic activity of the nanocomposite was observed compared to bare ZnO [54].

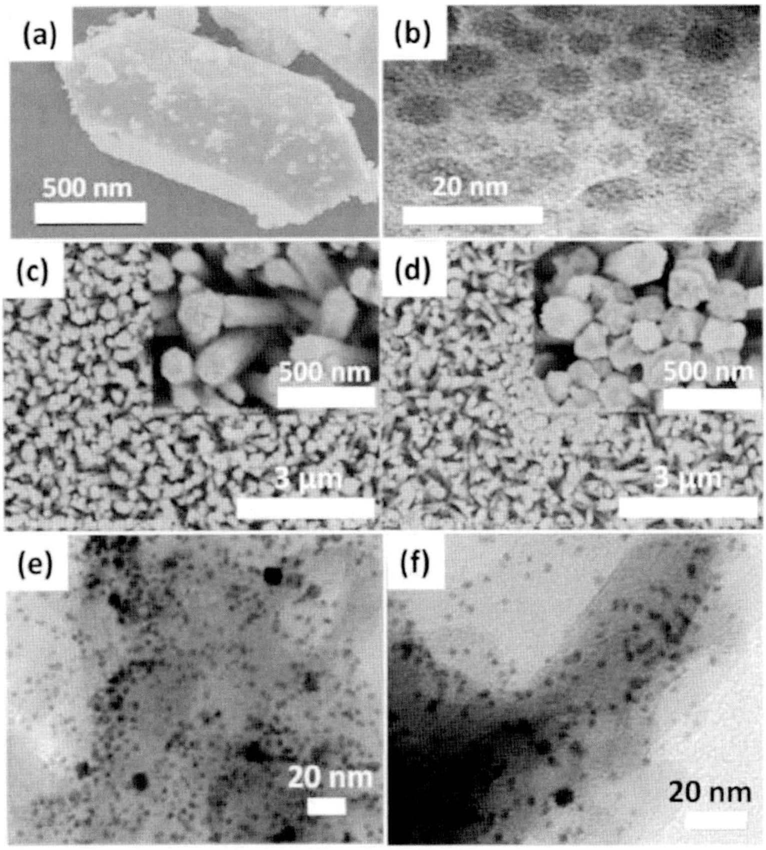

Fig. (8). (a) SEM **(b)** HR-TEM image of $BiPO_4$/Au-Pd nanorods. Both **(a)** and **(b)** reprinted with permission from ref [49]. Copyright 2017 Elsevier. SEM image of ZnO NRs/AuAg **(c)** nanoflowers, **(d)** mesopopcorns. Both **(c)** and **(d)** reprinted with permission from ref [50]. Copyright 2021 Elsevier. TEM images of PtAu/g-C_3N_4 (Pt:Au) **(e)** 63:37 **(f)** 35:65. Both **(e)** and **(f)** reprinted with permission from ref [53]. Copyright 2018 American Chemical Society.

CONCLUSION

The last decade has seen a hike in the research arena of photocatalysis which also is the need of the hour. An overview of semiconductor photocatalysis for environmental remediation and energy production has been presented. As discussed in the chapter, bandgap, charge recombination rate, crystal phase, exposed crystal plane, and morphology are the intrinsic material parameters upon which the apparent photocatalytic performance is dependent. Molding the semiconductor nanomaterial with a co-catalyst is observed to increase the charge separation. Till now, water and air treatment through semiconductor photo-catalysis has shown promising results. The use of UV light, affordable industrial-scale synthesis and application are few of the pitfalls of the process. Despite recent development in this field, the influence and study of composition and morphology on the photochemical activity still needs to be uncovered for the further design and development of novel efficient photocatalyst.

CONSENT FOR PUBLICATION

Not applicable.

CONFLICT OF INTEREST

The author declares no conflict of interest, financial or otherwise.

ACKNOWLEDGEMENTS

Declared none.

REFERENCES

[1] Lee, J.S.; Jang, J. Hetero-structured semiconductor nanomaterials for photocatalytic applications. *J. Ind. Eng. Chem.*, **2014**, *20*(2), 363-371. [http://dx.doi.org/10.1016/j.jiec.2013.11.050]

[2] Som, I.; Roy, M.; Saha, R. Advances in nanomaterial-based water treatment approaches for photocatalytic degradation of water pollutants. *ChemCatChem,* **2020**, *12*(13), 3409-3433. [http://dx.doi.org/10.1002/cctc.201902081]

[3] Jing, L.; Zhou, W.; Tian, G.; Fu, H. Surface tuning for oxide-based nanomaterials as efficient photocatalysts. *Chem. Soc. Rev.*, **2013**, *42*(24), 9509-9549. [http://dx.doi.org/10.1039/c3cs60176e] [PMID: 24048255]

[4] Lee, G.J.; Hou, Y.H.; Huang, H.T.; Wang, W.; Lyu, C.; Wu, J.J. Microwave-Assisted Solvothermal Synthesis of Chalcogenide Composite Photocatalyst and Its Photocatalytic CO_2 Reduction Activity under Simulated Solar Light. *Catalysts,* **2020**, *10*(7), 789. [http://dx.doi.org/10.3390/catal10070789]

[5] Zhu, L.; Meng, L.; Shi, J.; Li, J.; Zhang, X.; Feng, M. Metal-organic frameworks/carbon-based materials for environmental remediation: A state-of-the-art mini-review. *J. Environ. Manage.*, **2019**, *232*, 964-977. [http://dx.doi.org/10.1016/j.jenvman.2018.12.004] [PMID: 33395765]

[6] Hu, B.; Yuan, J.Y.; Tian, J.Y.; Wang, M.; Wang, X.; He, L.; Zhang, Z.; Wang, Z.W.; Liu, C.S. Co/Fe-bimetallic organic framework-derived carbon-incorporated cobalt-ferric mixed metal phosphide as a highly efficient photocatalyst under visible light. *J. Colloid Interface Sci.,* **2018**, *531*, 148-159. [http://dx.doi.org/10.1016/j.jcis.2018.07.037] [PMID: 30031257]

[7] Gallo, A.; Marelli, M.; Psaro, R.; Gombac, V.; Montini, T.; Fornasiero, P.; Dal Santo, V. Bimetallic Au–Pt/TiO_2 photocatalysts active under UV-A and simulated sunlight for H 2 production from ethanol. *Green Chem.,* **2012**, *14*(2), 330-333. [http://dx.doi.org/10.1039/C2GC16112E]

[8] Sankar, R.; Manikandan, P.; Malarvizhi, V.; Fathima, T.; Shivashangari, K.S.; Ravikumar, V. Green synthesis of colloidal copper oxide nanoparticles using Carica papaya and its application in photocatalytic dye degradation. *Spectrochim. Acta A Mol. Biomol. Spectrosc.,* **2014**, *121*, 746-750. [http://dx.doi.org/10.1016/j.saa.2013.12.020] [PMID: 24388701]

[9] Velegraki, G.; Miao, J.; Drivas, C.; Liu, B.; Kennou, S.; Armatas, G.S. Fabrication of 3D mesoporous networks of assembled CoO nanoparticles for efficient photocatalytic reduction of aqueous Cr (VI). *Appl. Catal. B,* **2018**, *221*, 635-644. [http://dx.doi.org/10.1016/j.apcatb.2017.09.064]

[10] Chen, W.; Huang, G.B.; Song, H.; Zhang, J. Efficient and stable charge transfer channels for photocatalytic water splitting activity of CdS without sacrificial agents. *J. Mater. Chem. A Mater. Energy Sustain.,* **2020**, *8*(40), 20963-20969. [http://dx.doi.org/10.1039/D0TA06177H]

[11] Bie, C.; Zhu, B.; Xu, F.; Zhang, L.; Yu, J. *In situ* grown monolayer N-doped graphene on CdS hollow spheres with seamless contact for photocatalytic CO_2 reduction. *Adv. Mater.,* **2019**, *31*(42), e1902868. [http://dx.doi.org/10.1002/adma.201902868] [PMID: 31490585]

[12] Yuan, Y.J.; Chen, D.; Yu, Z.T.; Zou, Z.G. Cadmium sulfide-based nanomaterials for photocatalytic hydrogen production. *J. Mater. Chem. A Mater. Energy Sustain.,* **2018**, *6*(25), 11606-11630. [http://dx.doi.org/10.1039/C8TA00671G]

[13] Reddy, C.V.; Reddy, K.R.; Shetti, N.P.; Shim, J.; Aminabhavi, T.M.; Dionysiou, D.D. Hetero-nanostructured metal oxide-based hybrid photocatalysts for enhanced photoelectrochemical water splitting–A review. *Int. J. Hydrogen Energy,* **2020**, *45*(36), 18331-18347. [http://dx.doi.org/10.1016/j.ijhydene.2019.02.109]

[14] Hao, H.; Lang, X. Metal Sulfide Photocatalysis: Visible-Light-Induced Organic Transformations. *ChemCatChem,* **2019**, *11*(5), 1378-1393. [http://dx.doi.org/10.1002/cctc.201801773]

[15] Kandy, M.M. Carbon-based photocatalysts for enhanced photocatalytic reduction of CO_2 to solar fuels. *Sustain. Energy Fuels,* **2020**, *4*(2), 469-484. [http://dx.doi.org/10.1039/C9SE00827F]

[16] Li, J.; Yu, Y.; Zhang, L. Bismuth oxyhalide nanomaterials: layered structures meet photocatalysis. *Nanoscale,* **2014**, *6*(15), 8473-8488. [http://dx.doi.org/10.1039/C4NR02553A] [PMID: 24975748]

[17] Wax, T.J.; Dey, S.; Chen, S.; Luo, Y.; Zou, S.; Zhao, J. Excitation wavelength-dependent photoluminescence decay of hybrid gold/quantum dot nanostructures. *ACS Omega,* **2018**, *3*(10), 14151-14156. [http://dx.doi.org/10.1021/acsomega.8b01959] [PMID: 31458107]

[18] Raizada, P.; Sudhaik, A.; Singh, P. Photocatalytic water decontamination using graphene and ZnO coupled photocatalysts: A review. *Mater. Sci. Energy Technol.,* **2019**, *2*(3), 509-525. [http://dx.doi.org/10.1016/j.mset.2019.04.007]

[19] Filimonov, V.N. Photocatalytic oxidation of gaseous isopropanol on ZnO and TiO_2. *Dokl. Akad. Nauk,* **1964**, *154*, 922-925.

[20] Liu, G.; Yu, J.C.; Lu, G.Q.M.; Cheng, H.M. Crystal facet engineering of semiconductor photocatalysts: motivations, advances and unique properties. *Chem. Commun. (Camb.),* **2011**, *47*(24), 6763-6783.
[http://dx.doi.org/10.1039/c1cc10665a] [PMID: 21448488]

[21] Yu, J.; Low, J.; Xiao, W.; Zhou, P.; Jaroniec, M. Enhanced photocatalytic CO_2-reduction activity of anatase TiO_2 by coexposed 001 and 101 facets. *J. Am. Chem. Soc.,* **2014**, *136*(25), 8839-8842.
[http://dx.doi.org/10.1021/ja5044787] [PMID: 24918628]

[22] He, Z.; Wen, L.; Wang, D.; Xue, Y.; Lu, Q.; Wu, C.; Song, S. Photocatalytic reduction of CO_2 in aqueous solution on surface-fluorinated anatase TiO_2 nanosheets with exposed {001} facets. *Energy Fuels,* **2014**, *28*(6), 3982-3993.
[http://dx.doi.org/10.1021/ef500648k]

[23] Balachandran, S.; Swaminathan, M. Facile fabrication of heterostructured Bi_2O_3–ZnO photocatalyst and its enhanced photocatalytic activity. *J. Phys. Chem. C,* **2012**, *116*(50), 26306-26312.
[http://dx.doi.org/10.1021/jp306874z]

[24] Khaki, M.R.D.; Shafeeyan, M.S.; Raman, A.A.A.; Daud, W.M.A.W. Evaluating the efficiency of nano-sized Cu doped TiO2/ZnO photocatalyst under visible light irradiation. *J. Mol. Liq.,* **2018**, *258*, 354-365.
[http://dx.doi.org/10.1016/j.molliq.2017.11.030]

[25] Ding, W.; Zhao, L.; Yan, H.; Wang, X.; Liu, X.; Zhang, X.; Tang, B. Bovine serum albumin assisted synthesis of $Ag/Ag_2O/ZnO$ photocatalyst with enhanced photocatalytic activity under visible light. *Colloids Surf. A Physicochem. Eng. Asp.,* **2019**, *568*, 131-140.
[http://dx.doi.org/10.1016/j.colsurfa.2019.02.015]

[26] Uddin, M.T.; Hoque, M.E.; Bhoumick, M.C. Facile one-pot synthesis of heterostructure SnO_2/ZnO photocatalyst for enhanced photocatalytic degradation of organic dye. *RSC Advances,* **2020**, *10*(40), 23554-23565.
[http://dx.doi.org/10.1039/D0RA03233F]

[27] Hezam, A.; Namratha, K.; Drmosh, Q.A.; Yamani, Z.H.; Byrappa, K. Synthesis of heterostructured Bi_2O_3–CeO_2–ZnO photocatalyst with enhanced sunlight photocatalytic activity. *Ceram. Int.,* **2017**, *43*(6), 5292-5301.
[http://dx.doi.org/10.1016/j.ceramint.2017.01.059]

[28] Elamin, N.; Modwi, A.; Aissa, M.A.B.; Taha, K.K.; Al-Duaij, O.K.; Yousef, T.A. Fabrication of Cr–ZnO photocatalyst by starch-assisted sol–gel method for photodegradation of congo red under visible light. *J. Mater. Sci. Mater. Electron.,* **2021**, *32*(2), 2234-2248.
[http://dx.doi.org/10.1007/s10854-020-04988-y]

[29] Sharma, S.; Dutta, V.; Raizada, P.; Hosseini-Bandegharaei, A.; Singh, P.; Nguyen, V.H. Tailoring cadmium sulfide-based photocatalytic nanomaterials for water decontamination: a review. *Environ. Chem. Lett.,* **2020**.
[http://dx.doi.org/10.1007/s10311-020-01066-x]

[30] McClelland, K.P.; Weiss, E.A. Selective photocatalytic oxidation of benzyl alcohol to benzaldehyde or C–C coupled products by visible-light-absorbing quantum dots. *ACS Appl. Energy Mater.,* **2018**, *2*(1), 92-96.
[http://dx.doi.org/10.1021/acsaem.8b01652]

[31] Chen, J.; Wu, X.J.; Yin, L.; Li, B.; Hong, X.; Fan, Z.; Chen, B.; Xue, C.; Zhang, H. Li. B.; Hong, X.; Fan, Z.; Zhang, H. One-pot synthesis of CdS nanocrystals hybridized with single-layer transition-metal dichalcogenide nanosheets for efficient photocatalytic hydrogen evolution. *Angew. Chem.,* **2015**, *127*(4), 1226-1230.
[http://dx.doi.org/10.1002/ange.201410172]

[32] Ibrahim, I.; Lim, H.N.; Abou-Zied, O.K.; Huang, N.M.; Estrela, P.; Pandikumar, A. Cadmium sulfide nanoparticles decorated with au quantum dots as ultrasensitive photoelectrochemical sensor for

selective detection of copper (II) ions. *J. Phys. Chem. C,* **2016**, *120*(39), 22202-22214. [http://dx.doi.org/10.1021/acs.jpcc.6b06929]

[33] Liu, E.; Zhu, H.; Yi, J.; Kobbekaduwa, K.; Adhikari, P.; Liu, J.; Shi, Y.; Zhang, J.; Li, H.; Oprisan, A.; Rao, A.M.; Sanabria, H.; Chen, O.; Gao, J. Manipulating charge transfer from core to shell in CdSe/CdS/Au heterojunction quantum dots. *ACS Appl. Mater. Interfaces,* **2019**, *11*(51), 48551-48555. [http://dx.doi.org/10.1021/acsami.9b17339] [PMID: 31782302]

[34] Senobari, S.; Nezamzadeh-Ejhieh, A. A comprehensive study on the photocatalytic activity of coupled copper oxide-cadmium sulfide nanoparticles. *Spectrochim. Acta A Mol. Biomol. Spectrosc.,* **2018**, *196*, 334-343. [http://dx.doi.org/10.1016/j.saa.2018.02.043] [PMID: 29475182]

[35] Li, N.; Liu, M.; Zhou, Z.; Zhou, J.; Sun, Y.; Guo, L. Charge separation in facet-engineered chalcogenide photocatalyst: a selective photocorrosion approach. *Nanoscale,* **2014**, *6*(16), 9695-9702. [http://dx.doi.org/10.1039/C4NR02068E] [PMID: 24993804]

[36] Zhang, F.; Zhuang, H.Q.; Song, J.; Men, Y.L.; Pan, Y.X.; Yu, S.H. Coupling cobalt sulfide nanosheets with cadmium sulfide nanoparticles for highly efficient visible-light-driven photocatalysis. *Appl. Catal. B,* **2018**, *226*, 103-110. [http://dx.doi.org/10.1016/j.apcatb.2017.12.046]

[37] Lee, G.J.; Wu, J.J. Recent developments in ZnS photocatalysts from synthesis to photocatalytic applications—A review. *Powder Technol.,* **2017**, *318*, 8-22. [http://dx.doi.org/10.1016/j.powtec.2017.05.022]

[38] Chen, X.; Li, H.; Wu, Y.; Wu, H.; Wu, L.; Tan, P.; Pan, J.; Xiong, X. Facile fabrication of novel porous graphitic carbon nitride/copper sulfide nanocomposites with enhanced visible light driven photocatalytic performance. *J. Colloid Interface Sci.,* **2016**, *476*, 132-143. [http://dx.doi.org/10.1016/j.jcis.2016.05.024] [PMID: 27209398]

[39] Patidar, D.; Yadav, A.; Paul, D.R.; Sharma, A.; Nehra, S.P. Nanohybrids cadmium selenide-reduced graphene oxide for improving photo-degradation of methylene blue. *Physica E,* **2019**, *114*, 113560. [http://dx.doi.org/10.1016/j.physe.2019.113560]

[40] Reshak, A.H. Active photocatalytic water splitting solar-to-hydrogen energy conversion: Chalcogenide photocatalyst Ba_2ZnSe_3 under visible irradiation. *Appl. Catal. B,* **2018**, *221*, 17-26. [http://dx.doi.org/10.1016/j.apcatb.2017.09.018]

[41] Lopes, P.A.; Maia, D.L.D.S.; Silva, L.A. Chalcogenide nanoparticles like $Cd_xZn_{(1-x)}S_ySe_{(1-y)}$ applied to photocatalytic hydrogen production from natural seawater under visible light irradiation. *Mater. Today Commun.,* **2020**, *25*, 101503. [http://dx.doi.org/10.1016/j.mtcomm.2020.101503]

[42] Ge, J.; Zhang, Y.; Park, S.J. Recent advances in carbonaceous photocatalysts with enhanced photocatalytic performances: A mini review. *Materials (Basel),* **2019**, *12*(12), 1916. [http://dx.doi.org/10.3390/ma12121916] [PMID: 31200594]

[43] Zhang, Y.; Zhang, N.; Tang, Z.R.; Xu, Y.J. Graphene transforms wide band gap ZnS to a visible light photocatalyst. The new role of graphene as a macromolecular photosensitizer. *ACS Nano,* **2012**, *6*(11), 9777-9789. [http://dx.doi.org/10.1021/nn304154s] [PMID: 23106763]

[44] Liang, Y.; Li, Y.; Wang, H.; Zhou, J.; Wang, J.; Regier, T.; Dai, H. Co_3O_4 nanocrystals on graphene as a synergistic catalyst for oxygen reduction reaction. *Nat. Mater.,* **2011**, *10*(10), 780-786. [http://dx.doi.org/10.1038/nmat3087] [PMID: 21822263]

[45] Iwase, A.; Ng, Y.H.; Ishiguro, Y.; Kudo, A.; Amal, R. Reduced graphene oxide as a solid-state electron mediator in Z-scheme photocatalytic water splitting under visible light. *J. Am. Chem. Soc.,* **2011**, *133*(29), 11054-11057. [http://dx.doi.org/10.1021/ja203296z] [PMID: 21711031]

[46] Holmberg, N.; Laasonen, K. Theoretical insight into the hydrogen evolution activity of open-ended carbon nanotubes. *J. Phys. Chem. Lett.*, **2015**, *6*(19), 3956-3960. [http://dx.doi.org/10.1021/acs.jpclett.5b01846] [PMID: 26722898]

[47] Zaleska-Medynska, A.; Marchelek, M.; Diak, M.; Grabowska, E. Noble metal-based bimetallic nanoparticles: the effect of the structure on the optical, catalytic and photocatalytic properties. *Adv. Colloid Interface Sci.*, **2016**, *229*, 80-107. [http://dx.doi.org/10.1016/j.cis.2015.12.008] [PMID: 26805520]

[48] Kavitha, R.; Kumar, S.G. Review on bimetallic-deposited TiO_2: preparation methods, charge carrier transfer pathways and photocatalytic applications. *Chem. Pap.*, **2020**, *74*(3), 717-756. [http://dx.doi.org/10.1007/s11696-019-00995-4]

[49] Zhang, Y.; Park, S.J. Au–pd bimetallic alloy nanoparticle-decorated $BiPO_4$ nanorods for enhanced photocatalytic oxidation of trichloroethylene. *J. Catal.*, **2017**, *355*, 1-10. [http://dx.doi.org/10.1016/j.jcat.2017.08.007]

[50] Putri, A.E.; Roza, L.; Budi, S.; Umar, A.A.; Fauzia, V. Tuning the photocatalytic activity of nanocomposite ZnO nanorods by shape-controlling the bimetallic AuAg nanoparticles. *Appl. Surf. Sci.*, **2021**, *536*, 147847. [http://dx.doi.org/10.1016/j.apsusc.2020.147847]

[51] Darabdhara, G.; Das, M.R. Bimetallic Au-Pd nanoparticles on 2D supported graphitic carbon nitride and reduced graphene oxide sheets: A comparative photocatalytic degradation study of organic pollutants in water. *Chemosphere*, **2018**, *197*, 817-829. [http://dx.doi.org/10.1016/j.chemosphere.2018.01.073] [PMID: 29407845]

[52] Zhu, Y.; Marianov, A.; Xu, H.; Lang, C.; Jiang, Y. Bimetallic Ag–Cu supported on graphitic carbon nitride nanotubes for improved visible-light photocatalytic hydrogen production. *ACS Appl. Mater. Interfaces*, **2018**, *10*(11), 9468-9477. [http://dx.doi.org/10.1021/acsami.8b00393] [PMID: 29465987]

[53] Bhunia, K.; Chandra, M.; Khilari, S.; Pradhan, D. Bimetallic PtAu alloy nanoparticles-integrated g-C_3N_4 hybrid as an efficient photocatalyst for water-to-hydrogen conversion. *ACS Appl. Mater. Interfaces*, **2019**, *11*(1), 478-488. [http://dx.doi.org/10.1021/acsami.8b12183] [PMID: 30525406]

[54] Lee, S.J.; Jung, H.J.; Koutavarapu, R.; Lee, S.H.; Arumugam, M.; Kim, J.H.; Choi, M.Y. ZnO supported Au/Pd bimetallic nanocomposites for plasmon improved photocatalytic activity for methylene blue degradation under visible light irradiation. *Appl. Surf. Sci.*, **2019**, *496*, 143665. [http://dx.doi.org/10.1016/j.apsusc.2019.143665]

CHAPTER 9

Multicomponent Reactions Using Nanocatalysts

Savita Meena[1], Krishan Kumar[1], Surendra Saini[1], Anshu Dandia[1,*] and **Vijay Parewa[1,*]**

[1] *Centre of Advanced Studies, Department of Chemistry, University of Rajasthan, Jaipur-302004, India*

Abstract: This chapter intends to deliver a wide variety of multicomponent reactions (MCRs) using nanomaterials as a catalyst to produce value-added chemicals *via* sustainable and environmentally benign protocols. MCRs provide the inherent creation of several bonds in one operation. This chapter deals with the efficacy of nanomaterials for extensive organic reactions ranging from the traditional carbon-heteroatom bond forming reactions to further contemporary transformations. These employed include various varieties such as magnetic nanoparticles (MNPs), core-shell nanoparticles, carbon-based nanocatalyst (CBN), graphene supported nanoparticles and nanocomposite, *etc*. These nanocatalysts could act as a reusable catalyst for the sustainable synthesis of the different heterocyclic compounds. Their intriguing and designable properties and impressive catalytic selectivity open a wide corridor to synthesize diverse heterocycles with maximum molecular complexities. A new era will be highly devoted to achieve skillfullness in the nanoworld with the technological, industrial, and scientific fields.

Keywords: Heterocyclic compounds, Multicomponent reactions, Nanocatalyst.

INTRODUCTION

Multicomponent reactions are a set of prominent processes that thrust for synthetic competence, in which more than two reactions are employed in a reaction to generate a novel compound using a high number of new bond and enormous functionalities [1 - 8] (Fig. **1**). MCRs are idyllically appropriate for the creation of combinatorial libraries inspired from the privileged structures of natural molecules prone to demonstrate biological activity [9 - 13]. Modern synthetic strategies involve organic transformation and conversions in such a logical manner that leads to the building of complex molecular structures [14]. This cascade reaction affords the construction of several bonds during a single operation. MCRs give better economic benefits by producing fewer wastes [15].

* **Corresponding authors Anshu Dandia and Vijay Parewa:** Centre of Advanced Studies, Department of Chemistry, University of Rajasthan, Jaipur-302004, India; Tel: +911412702306; E-mails: parewavijay.parewa@gmail.com, parewavijay@uniraj.ac.in

Fundamentally in multicomponent reactions more than two reactants are combined in a single step either sequential (cascade) or simultaneously (domino) without changing the reaction medium or isolating the intermediate.

Fig. (1). Synthesis of the product *via* multicomponent reaction.

MCRs are seen as synthetic protocols advancing numerous fields such as organic, natural product synthesis, combinatorial, heterocyclic agro, polymer, pharmaceutical chemistry. It is a powerful tool for the construction of various biologically active heterocyclic compounds such as nucleic acid, plant alkaloids, anthocyanins and flavones as well as in heam, chlorophyll, and some protein, vitamin, hormones contain aromatic heterocyclic system. It has enormous potential for the design of new drugs which are used as an antitumor, antibacterial, anticancer, antirheumatic, and antiasthmatic [16 - 19].

Currently, MCRs are considered to be the foremost in terms of the green approach due to their minimalism, cost-effectiveness, high atom economy and efficiency of bond formation, waste avoidance, effortless implementation, reagents, solvents, energy and time saving as well as avoidance of complex purification method [20, 21]. Over the past year, the main challenges of these reactions are to develop organic transformations that are less contaminating in terms of resource, operational simplicity, clean and green, health, and environmental safety. Since its origin, the field of sustainable chemistry has been exclusively planned to congregate such challenges in synthetic transformations.

Classical chemical transformations for the organic synthesis of MCRs usually involve catalysts (homogeneous and/or heterogeneous), harmful/toxic organic solvents, a nasty working environment like elevated temperature and pressure [22]. Inventive efforts are on finding new methodology based on the specific and suitable catalytic system. Considering the significant features of sustainable chemistry, the utilization of nanomaterials as a catalyst in multicomponent reactions is attracting more attention due to their high efficiency under an eco-friendly reaction environment [23]. Nanoparticles have appeared as green substitutes to conventional substances owing to their unique chemical qualities like catalytic activity, enhanced mixing with reactants, easy separation, and physical properties like photoemission, electrical or heat conductivities and also exhibited novel characteristics of quantum size effects [24 - 26]. Nanocatalyst can be effortlessly recycled during organic transformations for numerous cycles without loss of its reactivity. Compared to regular catalysis, nanocatalysis has assured inimitable rewards [27, 28]. Several types of nanocatalyst are used for the encouragement of multi-component reactions (Fig. **2**).

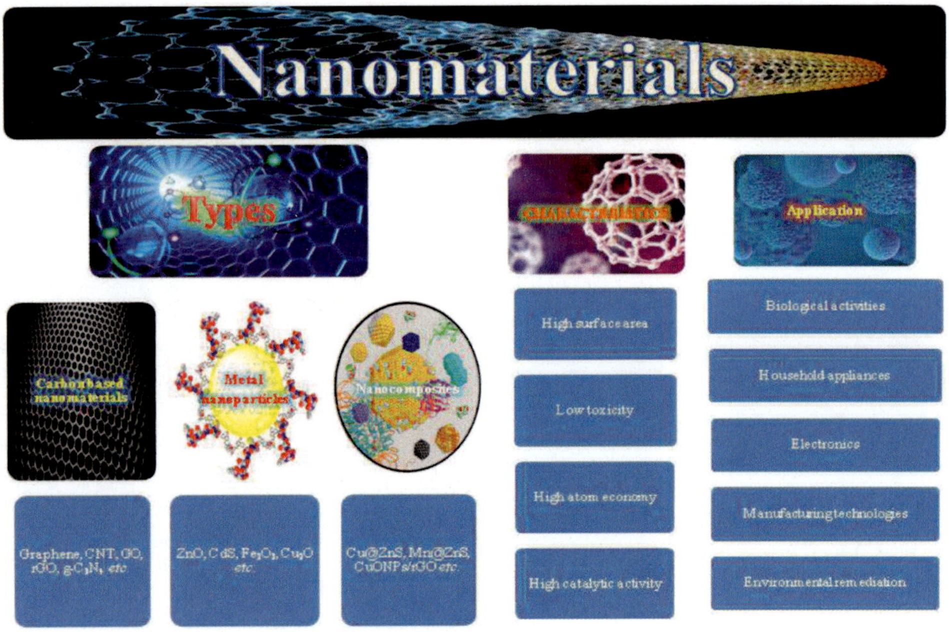

Fig. (2). Types of catalytic nanomaterials and their properties.

CARBON BASED NANOCATALYSTS (CBN)

In the last few years, several carbon-based materials {graphene, fullerenes, carbon nanotubes (CNT), reduced graphene oxide (rGO), graphene oxide (GO) and graphitic carbon nitride (g-C_3N_4), *etc*. have been synthesized and used as a catalyst in many fields such as biomedical, electronics, electrochemistry, metallurgy and some other industries, *etc*. [29 - 31]. Among them, graphene and graphene oxide are measured as auspicious catalyst support substances because of their unusual electronic or conducting properties and high surface area. GO can be a pioneer of graphene through thermal or chemical reduction processes. Carbon-based nanocatalysts have appeared as green substitutes to conventional substances owing to their chemical qualities like catalytic activity, enhanced mixing ability with reactants, easy separation, and physical properties like photoemission, electrical, or heat conductivities. It also exhibits novel characteristics of quantum size effects [32 - 35]. Thus, the employment of CBNis a significant option for one-step synthesis including multicomponent reactions.

To demonstrate the viability study of the "green and sustainable" process, numerous carbon-based nanocatalysts' chemical transformation has been evaluated. Burak and co-worker [36] have described the reaction of aniline, aldehydes, and dimedone *via* multi-component reaction utilizing graphene oxideindimethyl formamide (DMF) as a solvent to give 1,8-dioxoacridine derivatives (Scheme **1**).

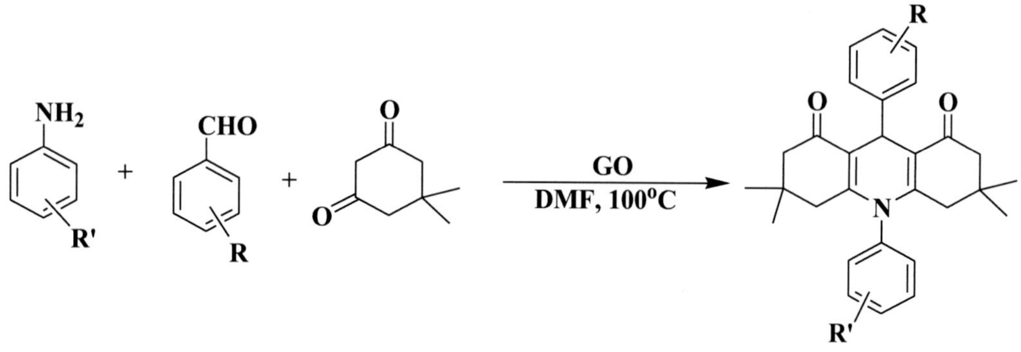

Scheme 1. Synthesis of 1,8-dioxoacridine derivatives.

Behravesh and co-workers [37] have introduced a novel protocol to the construction of 2-Amino-7-hydroxy-4-phenyl-4*H*chromene-3-carbonitrile and its derivatives from aldehydes, phenols, and malononitrile utilizing rGO-SO_3H as a catalyst under the mild condition in water (Scheme **2**).

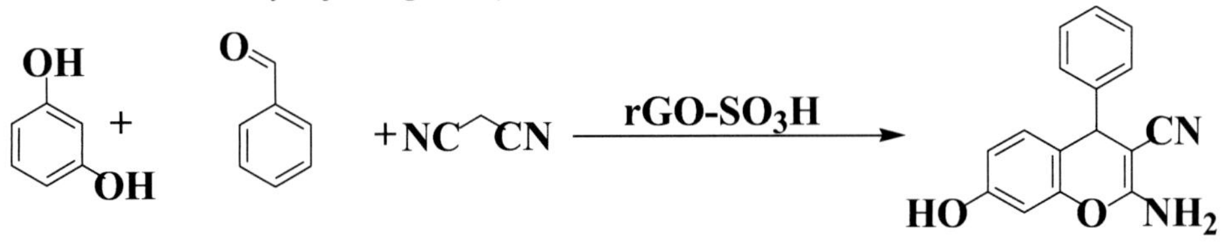

Scheme 2. rGO-SO_3H catalysed synthesis of chromenederivatives.

Das *et al.* [38] have described the reaction of aldehyde, amine, and acetylenic esters for the preparation of 3,4,5-substituted furanones, 2-oxo dihydropyrroles, and 2-pyrrolidinones utilizing graphene oxide (GO) as a catalyst (Scheme **3**).

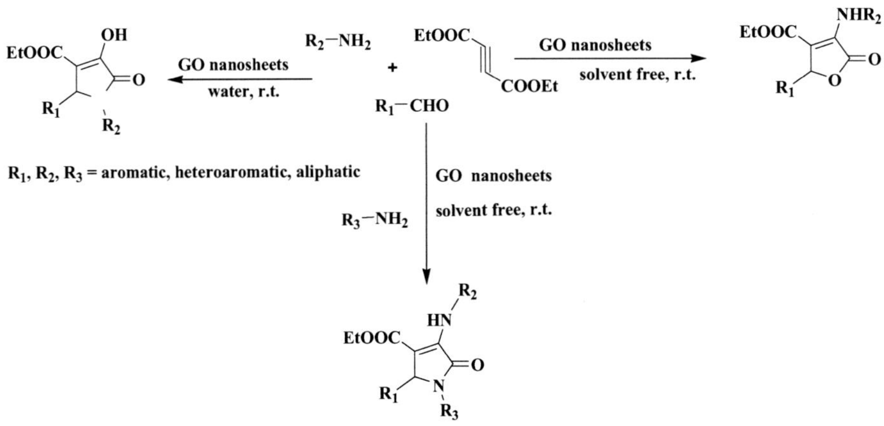

Scheme 3. GO catalysed synthesis of substituted furanones, 2-oxo dihydropyrroles and 2-pyrrolidinones.

Khalili *et al.* [39] have demonstrated 2-amino-6-phenyl-4-p-tolyl-nicotimonitrile and its derivative *via* multicomponent reaction of ammonium acetate, aromatic aldehyde, malononitrile, ketone and catalysed by GO under air in water (Scheme **4**).

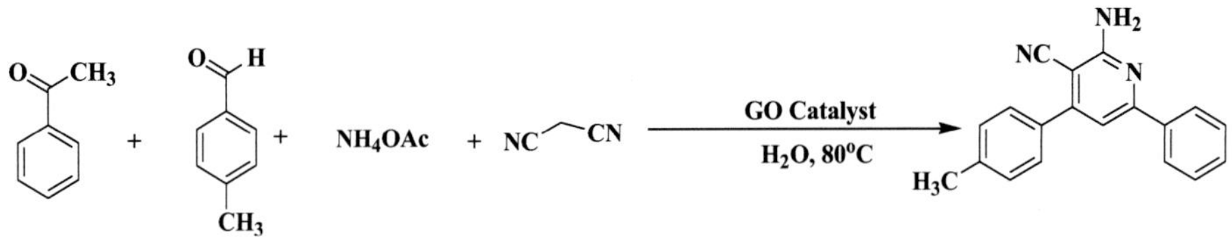

Scheme 4. GO catalysed synthesis of 2-amino-6-phenyl-4-p-tolyl-nicotimonitrile derivatives.

Gupta and co-worker [40] have achieved highly functionalized tetrahydropyridines through single-step multicomponent reaction of aniline, aldehyde, and *β*-ketoester using GO as a catalyst (Scheme **5**).

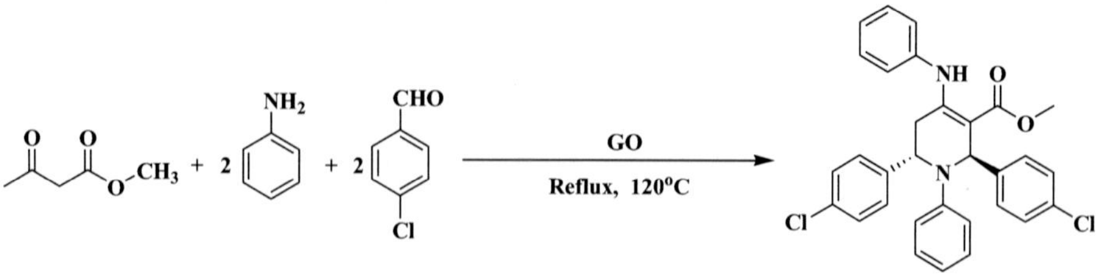

Scheme 5. GO catalysed synthesis of tetrahydropyridines derivatives.

Hassankhani *et al.* [41] have synthesizedtetrazoloquinazolines*via* three-component reactions of 2-aminotetrazole, benzaldehyde, and dimedoneutilizing g-C_3N_4/$NHSO_3H$ as a catalyst in ethanol at 80°C (Scheme **6**).

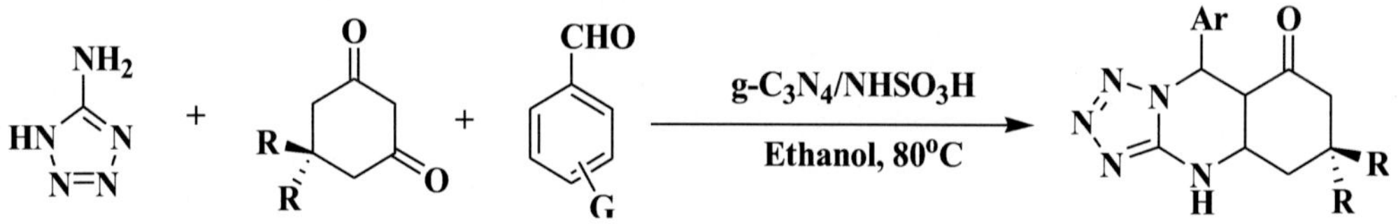

Scheme 6. g-C_3N_4/$NHSO_3H$ catalysed synthesis of tetrazoloquinazolines derivatives.

Shaabani and co-worker [42] have described a new approach for the production of carboxamide-f-multiwall carbon nanotube (MWCNT) *via* a four-component Ugi reaction in ethanol under mild reaction conditions (Scheme **7**).

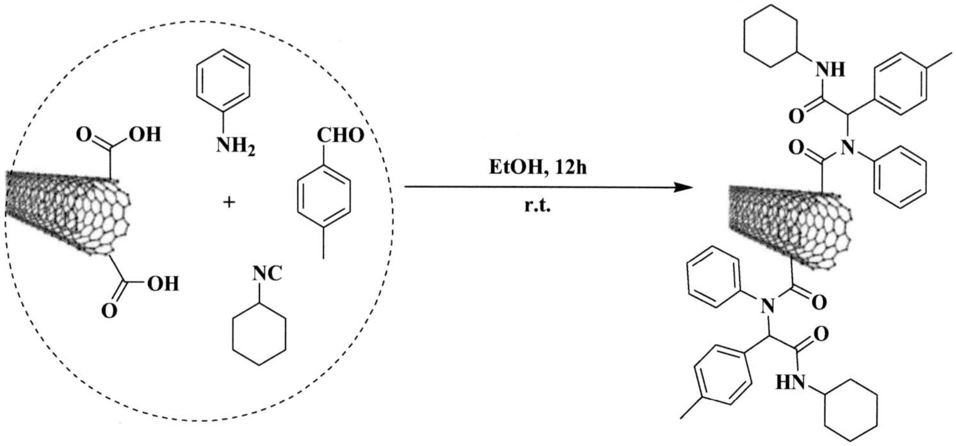

Scheme 7. Ugi reaction in ethanol.

Bahuguna *et al.* [43] have described the reaction of benzaldehyde, 5,5-Dimethyl cyclohexane-1,3-dione, and malononitrile in ethanol resulted in the formation of substituted chromenes catalyzed by potassium-functionalized graphitic carbon nitride supported on reduced graphene oxide (KGCN-rGO) (Scheme **8**).

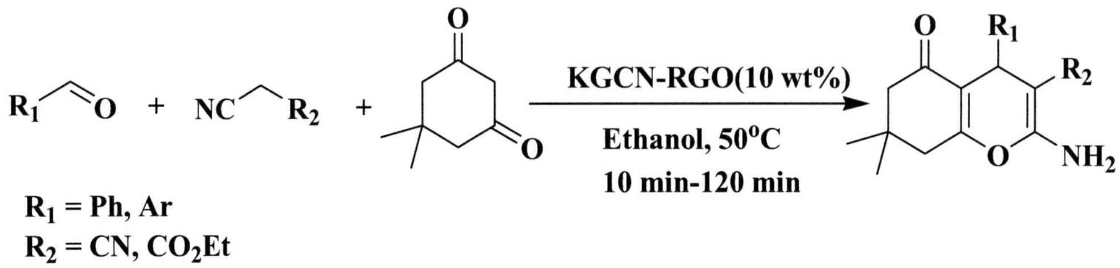

Scheme 8. KGCN-rGO catalysed synthesis of substituted chromenes.

Dehghanpour *et al.* [44] have described a facial pathway for the production of pyrido[2,3-*d*]pyrimidine derivative through the multicomponent reaction of aromatic aldehydes, indane-1,3-dione, and. pyrimidine derivative utilizing GO as a catalyst in water (Scheme **9**).

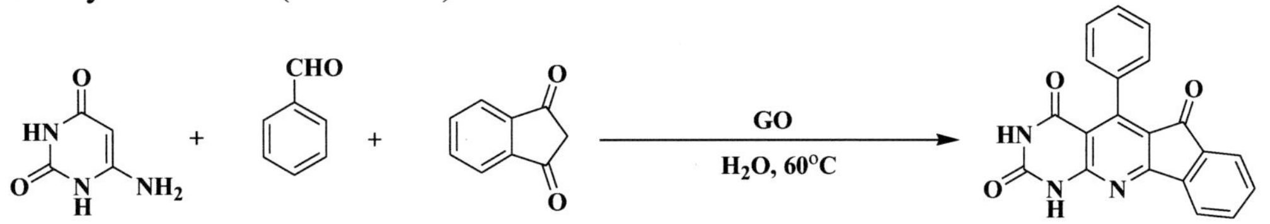

Scheme 9. KGCN-rGO catalysed synthesis of substituted chromenes.

Mohammadipour *et al.* [45] have synthesized polyhydroacridines derivatives with the reaction of aromatic benzaldehyde, dimedone, ammonium acetate using sulfobenzoic acid-functionalized graphene oxide (SBGO) under the solvent-free condition at 80 °C (Scheme **10**).

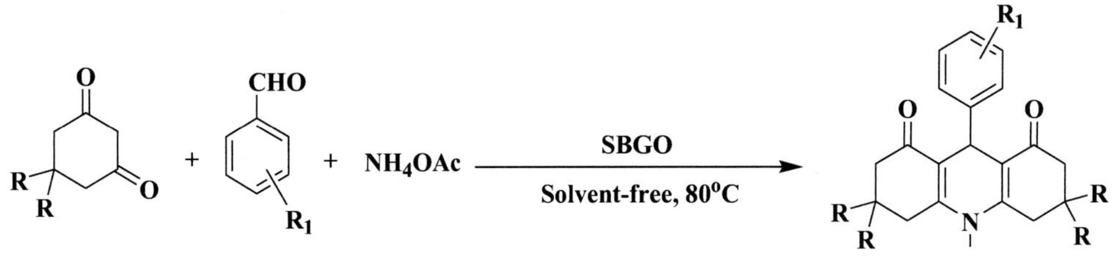

Scehem 10. SBGO catalysed synthesis of polyhydroacridines derivatives.

METAL NANOPARTICLES

Recently, nanotechnology has achieved enormous recognition in science due to its excellently productive uses [46]. In this way, nanoparticles are gaining attention owing to their several remarkable properties like thermal stability, ease of synthesis, low toxicity, environmentally friendly and functionalization, *etc.* Particularly, metal-based nanoparticles are a vital group of nano-catalysts. These nanoparticles (Fe_3O_4, CdS, ZnS, Ni, SiO_2, TiO_2, CuO and Ag_2S, *etc.*) have gained great significance in the past decades, as they are reusable and competent catalysts for the formation of organic moiety and biological activities [47 - 50]. These nanocatalysts have more advantages over the conventional metal-catalyst in terms of high atom economy, shorter reaction times, better yields, low catalyst loading, low corrosion, inexpensive, non-toxicity, and recyclability of the catalyst [51 - 53]. The activity of metal-based nanoparticles as catalysts especially multicomponent cascade reactions has gained attention to the construction of diversely functionalized organic moiety.

The following are some of the organic transformations involving the role of metal nanoparticles.

Dandia *et al.* [54] described the reaction of indole, aromatic aldehyde, and indan-1,3-dione utilizing reusable Cu doped ZnS nanocatalysts in water at room temperature to give pharmacologically important indole derivatives (Scheme **11**).

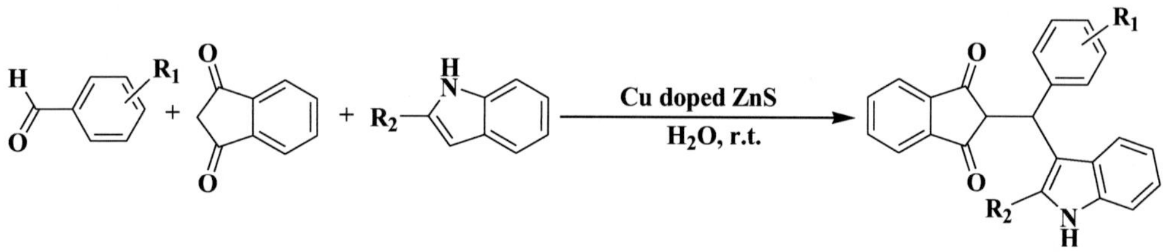

Scheme 11. Cu doped ZnS NPs catalysed synthesis of pharmacologically important indole derivatives.

Dandia*et al.* [55] have synthesized chromeno [1, 6]naphthyridine derivatives *via multicomponent* reaction of 3-phenoxybenzaldehyde, 2-hydroxyacetophenone, and malononitrile using Fe_3O_4NPs in water under microwave (MW) conditions (Scheme **12**).

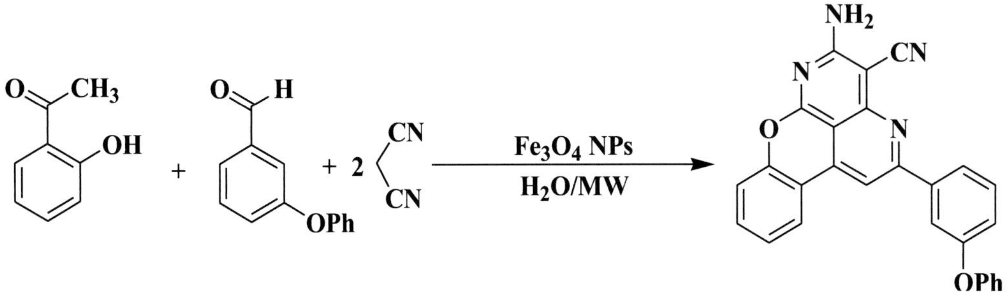

Scheme 12. Fe_3O_4 NPs catalysed synthesis of chromeno[1,6]naphthyridine derivatives under microwave.

Haldmi *et al.* [56] have demonstrated nickel nanoparticles (Ni NPs) mediated prompt and solvent-free reaction of acetamide, aromatic aldehyde, and 2-naphthol to the construction of 1-amidoalkyl-2-naphthols at 100°C (Scheme **13**).

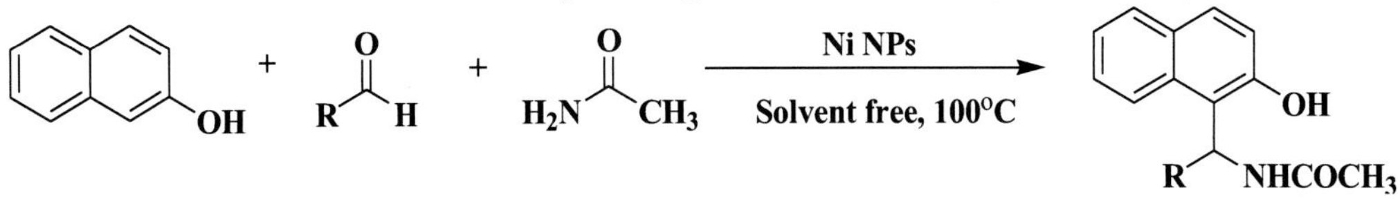

Scheme 13. Ni NPs catalysed synthesis of 1-amidoalkyl-2-naphthols.

An efficient and environmentally benign procedure to afford several pyrazolyl-phosphonatesderivatives benzaldehyde, 3-methyl-1-phenyl-5-pyrazolone, and from using PdNPs was described by Rostamnia *et al.* [57] (Scheme **14**).

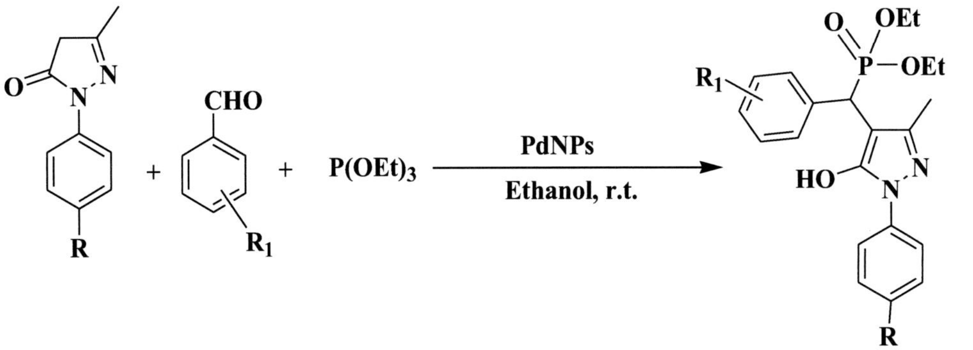

Scheme 14. Pd NPs catalysed synthesis of pyrazolylphosphonates derivatives.

Rostami *et al.* [58] have reported the multicomponent reaction of acid chlorides, ammonium thiocyanate, and α-bromo ketones, using Fe_3O_4 magnetic nanoparticles (MNPs) to give functionalized 1,3-oxazole derivatives in high yields (Scheme **15**).

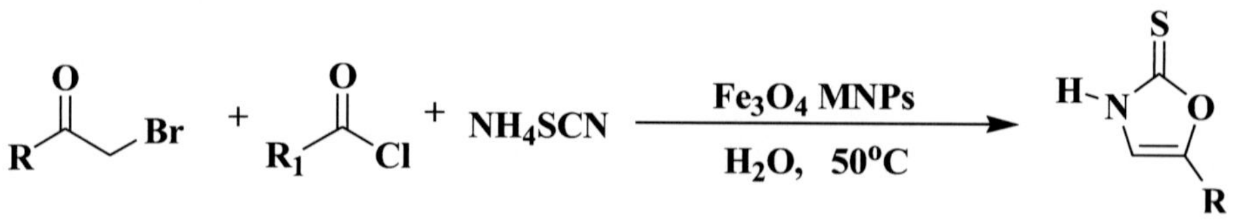

Scheme 15. Fe_3O_4 catalysed synthesis of 1,3-oxazole derivatives.

Shafaei *et al.* [59] have synthesized 1,3-benzoxazole derivatives with α-bromoketones, isothiocyanate, and propiolatecatalyzed by bio-Fe_3O_4 MNPs and NaHin water at room temperature (Scheme **16**).

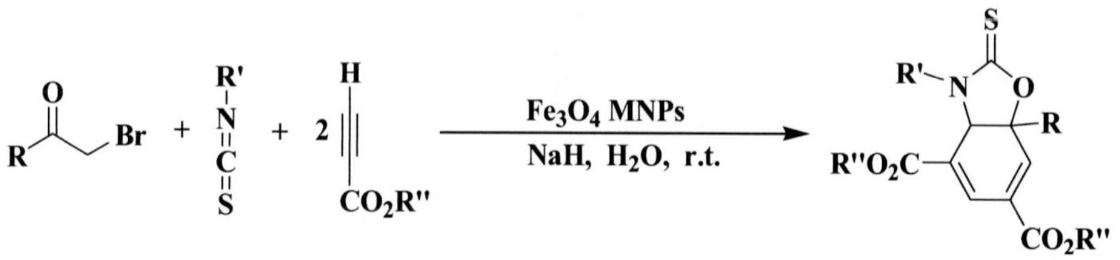

Scheme 16. Bio-Fe_3O_4 MNPs and NaH catalysed synthesis of 1,3-benoxazole derivatives.

Moghaddam *et al.* [60] have demonstrated facile, efficient, environment-friendly protocols for the preparation of substituted pyrroles with a primary amine, aromatic aldehyde, 1,2-dione, and nitromethane using $NiFe_2O_4$ magnetic nanoparticles (MNPs) as a recyclable catalyst (Scheme **17**).

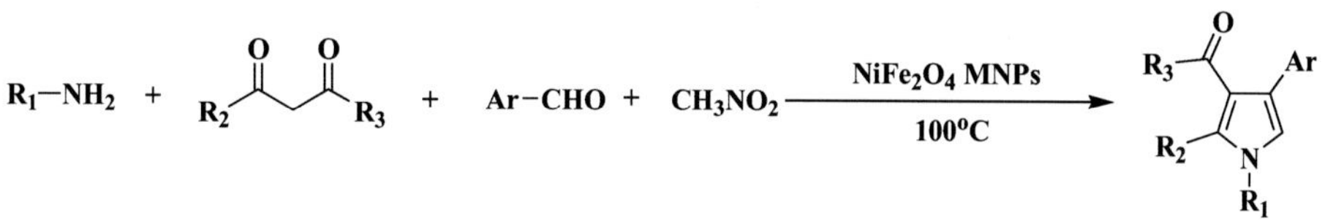

Scheme 17. $NiFe_2O_4$ MNPs scatalysed synthesis of substituted pyrroles.

Mirjafary and co-worker [61] have described ZnO nanoparticles mediated multicomponent reaction to the construction of β-acetamido ketones/esters under mild condition in acetonitrile (Scheme **18**).

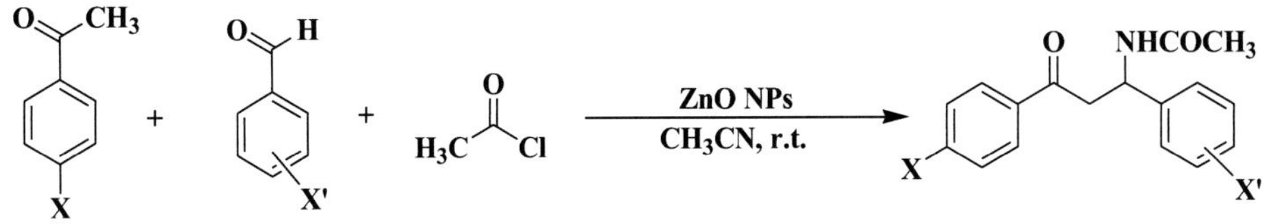

Scheme 18. ZnO nanoparticles catalysed synthesis of β-acetamido ketones/esters.

ZnO catalyzed construction of 4-Aryl-NH-1,2,3-Triazoles utilizing nitroalkane, aromatic aldehydes, and sodium azide at 100 °C was described by Phukan *et al.* [62] (Scheme **19**).

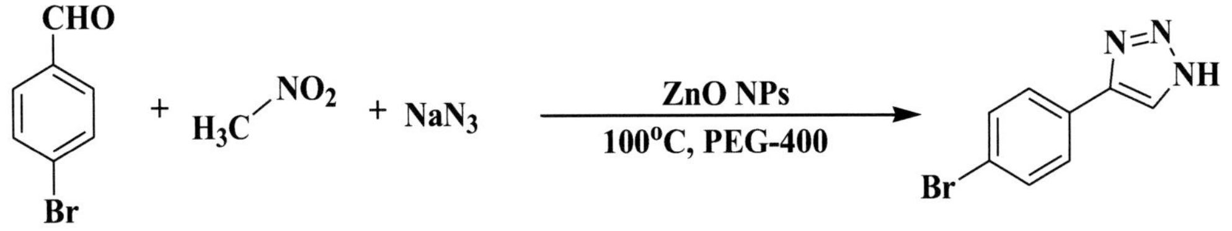

Scheme 19. ZnO nanoparticles catalysed synthesis of Triazoles derivatives.

A broad and proficient process to synthesize Biginelli compounds from the reaction of benzaldehyde, ethyl acetoacetate, and urea catalyzed by the copper oxide nanoparticles in solvent-free condition at 90°C was developed by Alinezhad and co-workers [63] (Scheme **20**).

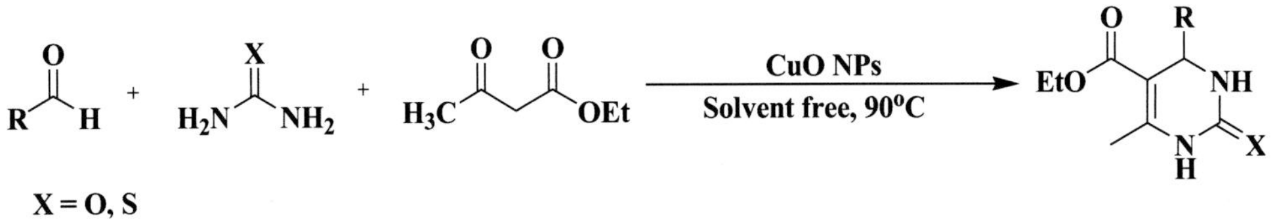

Scheme 20. Copper oxide nanoparticles catalysed synthesis of biginelli compounds.

NANOCOMPOSITES

In recent years, nanotechnology and nanoscience have gained prominence in various fields of science [64]. In the growing field of nanotechnology, nanoparticles has been applied as one of the best ways for different organic transformations towards a greener and highly efficient product [65]. In addition, when nanosize particles are added into a matrix of standard material which is produced as a new material with extraordinary flexibility is called nanocomposites [66]. Incorporation of nanoparticles results in drastic enhancement in properties that may include toughness, mechanical strength, and thermal or electrical conductivity [67]. Nanocomposites (Nano-Fe_3O_4@SiO_2, Fe_2O_3@rGO, mica/Fe_3O_4, CuO NPs/rGO, Ag NPs/rGO, *etc.*) were easily synthesized under mild conditions. Nanocomposites have been utilized in a vast number of catalytic reactions like condensation, photocatalysis, C-C and C-heteroatom coupling, cycloaddition, oxidation, drug delivery, and sensors. Magnetic nanocomposites can be used in many applications such as technical, medical, and catalytic [68].

A few reactions using nanocomposites are given below:

Akocak *et al.* [69] described the reaction of aromatic resorcinol, aromatic aldehydes and malononitrile utilizing reusable Pd@GO composite in ethanol at 80°C to give 2-amino-4*H*-chromene derivatives (Scheme **21**).

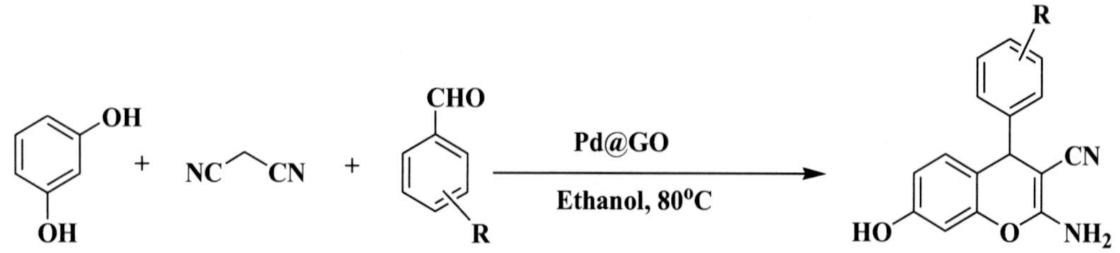

Scheme 21. Pd@GO catalysed synthesis of chromene derivatives.

Dandia *et al.* [70] described a microwave-induced reaction of boronic acids, salicylaldehydes, and amines to synthesized alkylaminophenols catalyzed by CuONPs/rGO composite in dichloromethane (DCM) solvent (Scheme **22**).

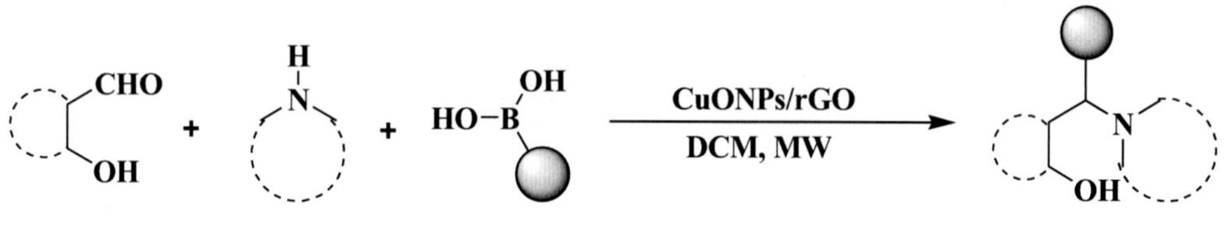

Scheme 22. CuONPs/rGO catalysed synthesis of alkylaminophenols.

Dandia *et al.* [71] introduced a diastereoselective approach for the synthesis of functionalized tetrahydropyridines utilizing recoverable Ag NPs/rGO nano-composite in methanol under ultrasonic irradiation (Scheme **23**).

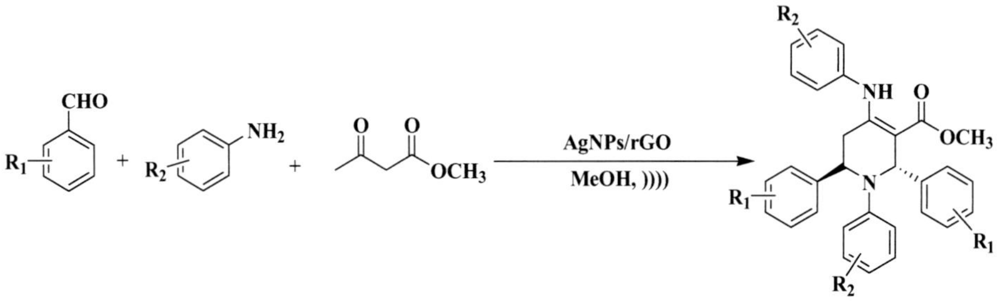

Scheme 23. Ag NPs/rGO catalysed synthesis of tetrahydropyridines derivatives.

A new efficient and environment-friendly microwave assisted chemoselective production of pyrrolo[2,3,4-*kl*]acridin-1-ones form cyclic diketone, tribulin and aniline utilizing Ag NPs/ rGO NPs as a reusable catalyst in ethanol was described by Dandia *et al.* [72] (Scheme **24**).

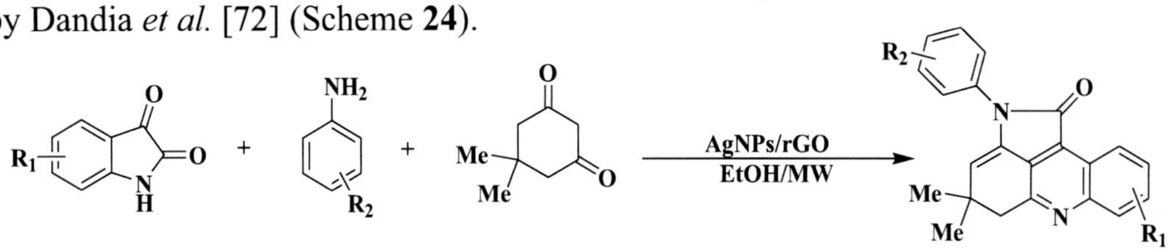

Scheme 24. Ag NPs/rGO NPs catalysed synthesis of acridin derivatives.

Dandia *et al.* [73] described a new microwave-assisted production of 2,8-dioxabicyclo[3.3.1]nonane derivatives utilizing 2-hydroxybenzaldehydes, 4-hydroxycoumarinand acetophenone using CeO_2 NPs/rGO as a recyclable nanocatalyst (Scheme **25**).

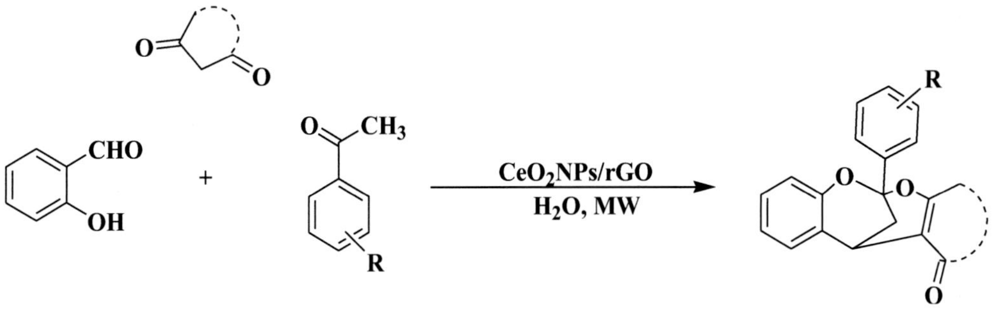

Scheme 25. Microwave assisted production of 2,8-dioxabicyclo[3.3.1]nonane derivatives.

Azgomi *et al.* [74] described the reaction of aromatic aldehydes, thioglycolic acid, and anilines utilizing nano-Fe_3O_4@SiO_2 supported ionic liquid (MNPs@SiO_2-IL) in solvent-free condition to give 1,3-thiazolidin-4-one (Scheme **26**).

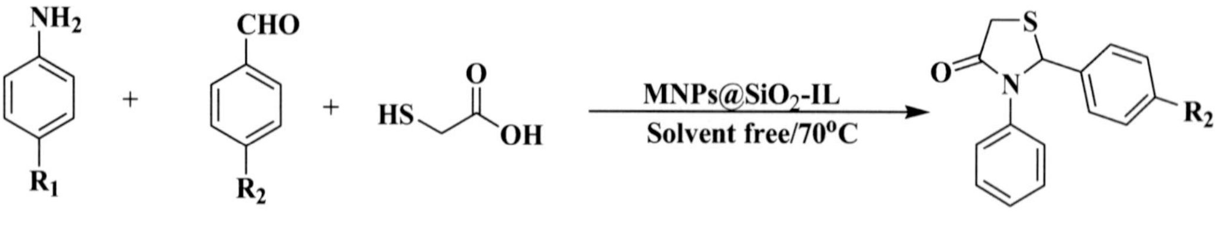

Scheme 26. Fe_3O_4@SiO_2 catalysed synthesis of thiazolidin derivatives.

Tamora *et al.* [75] designed a facile, inexpensive green protocol for the production of polyhydroquinoline and 2,3-dihydroquinazolin-4(1*H*)-one derivative utilizing functionalized $CoFe_2O_4$@Pr magnetic nanoparticles (Scheme **27**).

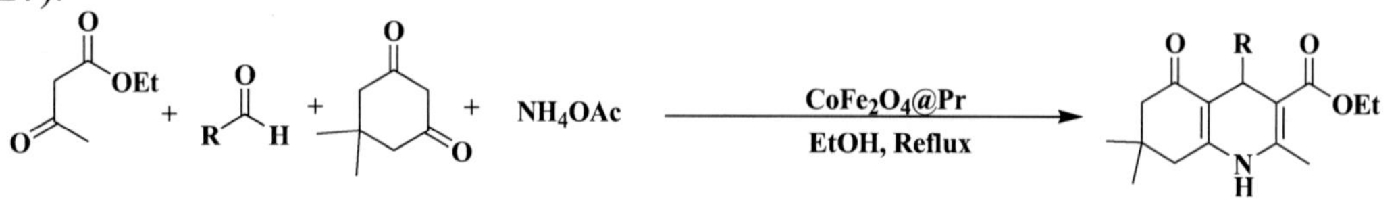

Scheme 27. $CoFe_2O_4$@Pr MNPs catalysed synthesis of polyhydroquinoline and 2,3-dihydroquinazolin derivatives.

Tran *et al.* [76] developed an inventive protocols to the construction of high regioselectivity thieno[2,3-*b*]indole derivative through the three-component reaction from easily accessible acetophenones, indoles, and octasulfur utilizing a magnetically reusable deep eutectic solvent(DES)@MNP catalyst (Scheme **28**).

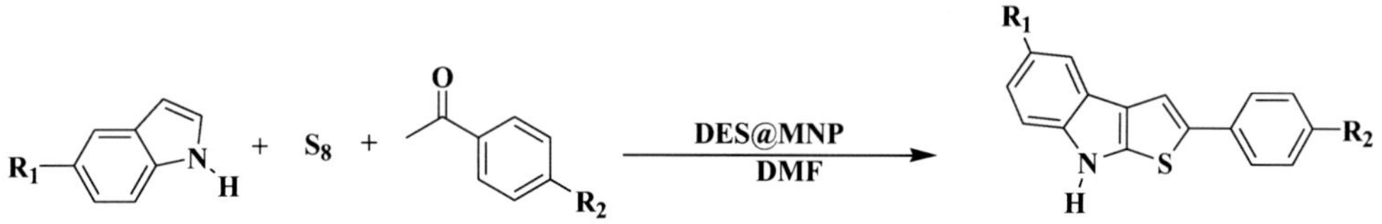

Scheme 28. DES@MNP catalysed synthesis of indole derivative.

Nguyen *et al.* [77] demonstrated that Lewis acid ionic liquid(LAIL)@MNP nanocomposite can work as a productive catalyst for the preparation of substituted polyhydroquinoline *via* Hantzsch condensation under solventless sonication (Scheme **29**).

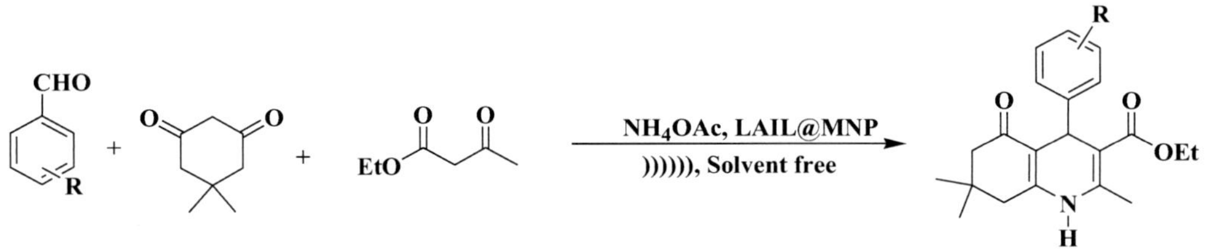

Scheme 29. LAIL@MNP catalysed synthesis of polyhydroquinoline derivative *via* Hantzsch condensation.

Kalhor *et al.* [78] demonstrated that MnO_2@zeolite-Y nanocomposite can act as a green and convenient catalyst for the creation of substituted isoxazole from hydrochloride, ethyl acetoacetate, and aromatic aldehydes (Scheme **30**).

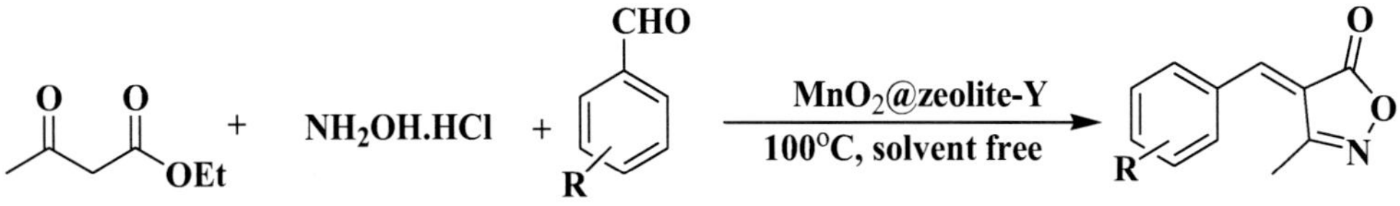

Scheme 30. MnO_2@zeolite-Y nanocomposite catalysed synthesis of isoxazole derivative.

CONCLUDING REMARKS

Multicomponent reactions are useful for the synthesis of a wide range of different heterocyclic complex of biological activities. In synthetic organic chemistry, MCRs act as a powerful tool to synthesize the highly functionalized compound and assess them against diverse areas such as pharmaceutical chemistry and industrial chemistry. Most of the MCRs are one-pot reaction which is eco-friendly and an ideal reaction system. In view of a greener approach, most of the nanocatalysts are used in MCRs. In this chapter, we have summarized many organic syntheses using nanomaterials as a catalyst in multicomponent reactions.

CONSENT FOR PUBLICATION

Not applicable.

CONFLICTS OF INTEREST

The author declares no conflict of interest, financial or otherwise.

ACKNOWLEDGEMENTS

Declared none.

REFERENCES

[1] Reguera, L.; Méndez, Y.; Humpierre, A.R.; Valdés, O.; Rivera, D.G. Multicomponent reactions in ligation and bioconjugation chemistry. *Acc. Chem. Res.,* **2018**, *51*(6), 1475-1486. [http://dx.doi.org/10.1021/acs.accounts.8b00126] [PMID: 29799718]

[2] Wu, S.; Lei, X.; Fan, E.; Sun, Z. Thermolysis-induced two- or multicomponent tandem reactions involving isocyanides and sulfenic-acid-generating sulfoxides: access to diverse sulfur-containing functional scaffolds. *Org. Lett.,* **2018**, *20*(3), 522-525. [http://dx.doi.org/10.1021/acs.orglett.7b03593] [PMID: 29360371]

[3] Chen, S.; Feng, D.F.; Li, D.Y.; Liu, P.N. Radical Cyanotrifluoromethylation of Isocyanides: Step-Economical Access to CF_3-Containing Nitriles, Amines, and Imines. *Org. Lett.,* **2018**, *20*(17), 5418-5422. [http://dx.doi.org/10.1021/acs.orglett.8b02328] [PMID: 30148642]

[4] Saya, J.M.; Berabez, R.; Broersen, P.; Schuringa, I.; Kruithof, A.; Orru, R.V.A.; Ruijter, E. Hexafluoroisopropanol as the acid component in the passerini reaction: one-pot access to β-amino alcohols. *Org. Lett.,* **2018**, *20*(13), 3988-3991. [http://dx.doi.org/10.1021/acs.orglett.8b01561] [PMID: 29906122]

[5] Holmes, M.; Schwartz, L.A.; Krische, M.J. Intermolecular metal-catalyzed reductive coupling of dienes, allenes, and enynes with carbonyl compounds and imines. *Chem. Rev.,* **2018**, *118*(12), 6026-6052. [http://dx.doi.org/10.1021/acs.chemrev.8b00213] [PMID: 29897740]

[6] Hall, D.G.; Rybak, T.; Verdelet, T. Multicomponent hetero-[4 + 2] cycloaddition/allylboration reaction: from natural product synthesis to drug discovery. *Acc. Chem. Res.,* **2016**, *49*(11), 2489-2500. [http://dx.doi.org/10.1021/acs.accounts.6b00403] [PMID: 27753496]

[7] Bhojgude, S.S.; Bhunia, A.; Biju, A.T. Employing arynes in diels-alder reactions and transition-meta--free multicomponent coupling and arylation reactions. *Acc. Chem. Res.,* **2016**, *49*(9), 1658-1670. [http://dx.doi.org/10.1021/acs.accounts.6b00188] [PMID: 27560296]

[8] Yang, L.; Zhang, Z.; Cheng, B.; You, Y.; Wu, D.; Hong, C. Two tandem multicomponent reactions for the synthesis of sequence-defined polymers. *Sci. China Chem.,* **2015**, *58*(11), 1734-1740. [http://dx.doi.org/10.1007/s11426-015-5448-0]

[9] Dömling, A.; Wang, W.; Wang, K. Chemistry and biology of multicomponent reactions. *Chem. Rev.,* **2012**, *112*(6), 3083-3135. [http://dx.doi.org/10.1021/cr100233r] [PMID: 22435608]

[10] Li, L.; Deng, X.X.; Li, Z.L.; Du, F.S.; Li, Z.C. Multifunctional photodegradable polymers for reactive micropatterns. *Macromolecules,* **2014**, *47*(14), 4660-4667. [http://dx.doi.org/10.1021/ma501019c]

[11] Yan, J.J.; Wang, D.; Wu, D.C.; You, Y.Z. Synthesis of sequence-ordered polymers *via* sequential addition of monomers in one pot. *Chem. Commun. (Camb.),* **2013**, *49*(54), 6057-6059. [http://dx.doi.org/10.1039/c3cc42483a] [PMID: 23722596]

[12] Kan, X.W.; Deng, X.X.; Du, F.S.; Li, Z.C. Concurrent oxidation of alcohols and the Passerini three-component polymerization for the synthesis of functional poly(ester/amide)s. *Macromol. Chem. Phys.,* **2014**, *215*(22), 2221-2228. [http://dx.doi.org/10.1002/macp.201400264]

[13] Liu, Y.J.; Gao, M.; Lam, J.W.Y.; Hu, R.R.; Tang, B.Z. Copper-catalyzed poly coupling of diynes, primary amines, andaldehydes: a new one-pot multicomponent polymerization tool to functional polymers. *Macromolecules,* **2014**, *47*(15), 4908-4919. [http://dx.doi.org/10.1021/ma501477w]

[14] Ruijter, E.; Scheffelaar, R.; Orru, R.V.A. Multicomponent reaction design in the quest for molecular complexity and diversity. *Angew. Chem. Int. Ed. Engl.,* **2011**, *50*(28), 6234-6246.

[http://dx.doi.org/10.1002/anie.201006515] [PMID: 21710674]

[15] Da Silva, E.N. Multicomponent reactions as a green approach. *Res. J. Chem. Environ.,* **2007**, *11*, 90-91.

[16] Brauch, S.; van Berkel, S.S.; Westermann, B. Higher-order multicomponent reactions: beyond four reactants. *Chem. Soc. Rev.,* **2013**, *42*(12), 4948-4962. [http://dx.doi.org/10.1039/c3cs35505e] [PMID: 23426583]

[17] Rotstein, B.H.; Zaretsky, S.; Rai, V.; Yudin, A.K. Small heterocycles in multicomponent reactions. *Chem. Rev.,* **2014**, *114*(16), 8323-8359. [http://dx.doi.org/10.1021/cr400615v] [PMID: 25032909]

[18] Touré, B.B.; Hall, D.G. Natural product synthesis using multicomponent reaction strategies. *Chem. Rev.,* **2009**, *109*(9), 4439-4486. [http://dx.doi.org/10.1021/cr800296p] [PMID: 19480390]

[19] Zong, Y.; Lang, Y.; Yang, M.; Li, X.; Fan, X.; Wu, J. Synthesis of β-sulfonyl amides through a multicomponent reaction with the insertion of sulfur dioxide under visible light irradiation. *Org. Lett.,* **2019**, *21*(6), 1935-1938. [http://dx.doi.org/10.1021/acs.orglett.9b00620] [PMID: 30840466]

[20] Garbarino, S.; Ravelli, D.; Protti, S.; Basso, A. Photoinduced multicomponent reactions. *Angew. Chem. Int. Ed. Engl.,* **2016**, *55*(50), 15476-15484. [http://dx.doi.org/10.1002/anie.201605288] [PMID: 27487327]

[21] Zhang, M.; Fu, Q.Y.; Gao, G.; He, H.Y.; Zhang, Y.; Wu, Y.S.; Zhang, Z.H. Catalyst-free, visible-light promoted one-potsynthesis of spirooxindole-pyran derivatives in aqueous ethyl lactate. *ACS Sustain. Chem.& Eng.,* **2017**, *5*(7), 6175-6182. [http://dx.doi.org/10.1021/acssuschemeng.7b01102]

[22] Kumar, G.; Nikolla, E.; Linic, S.; Medlin, J.W.; Janik, M.J. Multicomponent Catalysts: Limitations and Prospects. *ACS Catal.,* **2018**, *8*(4), 3202-3208. [http://dx.doi.org/10.1021/acscatal.8b00145]

[23] Chen, W.; Peng, X.; Zhong, L.; Li, Y.; Sun, R. Lignosulfonic Acid: A Renewable and Effective Biomass-Based Catalyst for Multicomponent Reactions. *ACS Sustain. Chem.& Eng.,* **2015**, *3*(7), 1366-1373. [http://dx.doi.org/10.1021/acssuschemeng.5b00091]

[24] Balu, A.M.; Baruwati, B.; Serrano, E.; Cot, J.; Garcia-Martinez, J.; Varma, R.S.; Luque, R. Magnetically separable nanocomposites with photocatalytic activity under visible light for the selective transformation of biomass-derived platform molecules. *Green Chem.,* **2011**, *13*(10), 2750-2758. [http://dx.doi.org/10.1039/c1gc15692f]

[25] Bhandari, R.; Coppage, R.; Knecht, M.R. Mimicking nature's strategies for the design of nanocatalysts. *Catal. Sci. Technol.,* **2012**, *2*(2), 256-266. [http://dx.doi.org/10.1039/C1CY00350J]

[26] Li, Q.; Guo, B.; Yu, J.; Ran, J.; Zhang, B.; Yan, H.; Gong, J.R. Highly efficient visible-light-driven photocatalytic hydrogen production of CdS-cluster-decorated graphene nanosheets. *J. Am. Chem. Soc.,* **2011**, *133*(28), 10878-10884. [http://dx.doi.org/10.1021/ja2025454] [PMID: 21639097]

[27] Zhang, W.; Chen, P.; Zhao, Z.; Wang, L.; Wang, S.; Tang, Y.; Wang, B.; Wang, Z.; Zhuang, H. Synthesis of functionalized 4-H-Chromenes catalyzed by lipase immobilized on magnetic nanoparticles. *Green Chem. Lett. Rev.,* **2018**, *11*(3), 246-253. [http://dx.doi.org/10.1080/17518253.2018.1474271]

[28] Hajizadeh, Z.; Hassanzadeh-Afruzi, F.; Jelodar, D.F.; Ahghari, M.R.; Maleki, A. Cu(II) immobilized on Fe_3O_4@HNTstetrazole (CFHT) nanocomposite: synthesis, characterization, investigation of its

catalytic role for the 1,3-dipolar cycloaddition reaction, and antibacterial activity. *RSC Advances,* **2020**, *10*(44), 26467-26478.
[http://dx.doi.org/10.1039/D0RA04772D]

[29] Georgakilas, V.; Perman, J.A.; Tucek, J.; Zboril, R. Broad family of carbon nanoallotropes: classification, chemistry, and applications of fullerenes, carbon dots, nanotubes, graphene, nanodiamonds, and combined superstructures. *Chem. Rev.,* **2015**, *115*(11), 4744-4822.
[http://dx.doi.org/10.1021/cr500304f] [PMID: 26012488]

[30] Shaik, S.A.; Goswami, A.; Varma, R.S.; Gawande, M.B. Nitrogen-Doped Nanocarbons (NNCs): Current Status and Future Opportunities. *Curr. Opin. Green Sustain. Chem.,* **2019**, *15*, 67-76.
[http://dx.doi.org/10.1016/j.cogsc.2018.10.001]

[31] Su, D.S.; Perathoner, S.; Centi, G. Nanocarbons for the development of advanced catalysts. *Chem. Rev.,* **2013**, *113*(8), 5782-5816.
[http://dx.doi.org/10.1021/cr300367d] [PMID: 23721498]

[32] Bahuguna, A.; Kumar, A.; Krishnan, V. Carbon support based heterogeneous nanocatalysts: synthesis and applications in organic reactions. *Asian J. Org. Chem.,* **2019**, *8*(8), 1263-1305.
[http://dx.doi.org/10.1002/ajoc.201900259]

[33] Vinayan, B.P.; Ramaprabhu, S. Platinum-TM (TM = Fe, Co) alloy nanoparticles dispersed nitrogen doped (reduced graphene oxide-multiwalled carbon nanotube) hybrid structure cathode electrocatalysts for high performance PEMFC applications. *Nanoscale,* **2013**, *5*(11), 5109-5118.
[http://dx.doi.org/10.1039/c3nr00585b] [PMID: 23644681]

[34] Zhai, Y.; Zhu, Z.; Dong, S. Carbon-based nanostructures for advanced catalysis. *ChemCatChem,* **2015**, *7*(18), 2806-2815.
[http://dx.doi.org/10.1002/cctc.201500323]

[35] Matveev, A.V.; Kaichev, V.V.; Saraev, A.A.; Gorodetskii, V.V.; Knop-Gericke, A.; Bukhtiyarov, V.I.; Nieuwenhuys, B.E. Oxidation of propylene over Pd(5 5 1): Temperature hysteresis induced by carbon deposition and oxygen adsorption. *Catal. Today,* **2015**, *244*, 29-35.
[http://dx.doi.org/10.1016/j.cattod.2014.08.023]

[36] Aday, B.; Pamuk, H.; Kaya, M.; Sen, F. Graphene oxide as highly effective and readily recyclable catalyst using for the one-pot synthesis of 1,8-dioxoacridine derivatives. *J. Nanosci. Nanotechnol.,* **2016**, *16*(6), 6498-6504.
[http://dx.doi.org/10.1166/jnn.2016.12432] [PMID: 27427743]

[37] Behravesh, S.; Fareghi-Alamdari, R.; Badri, R. Sulfonated reduced graphene oxide (rGO-SO_3H): As an efficient nanocatalyst for one-pot synthesis of 2-amino-3-cyano-7-hydroxy-4H-chromenes derivatives in water. *Polycycl. Aromat. Compd.,* **2018**, *38*(1), 51-65.
[http://dx.doi.org/10.1080/10406638.2016.1149080]

[38] Saha, M.; Das, A.R. Access of Diverse 2-Pyrrolidinone, 3, 4, 5-Substituted Furanone and 2-Ox- -dihydropyrroles Applying Graphene Oxide Nanosheet: Unraveling of Solvent Selectivity. *ChemistrySelect,* **2017**, *2*(31), 10249-10260.
[http://dx.doi.org/10.1002/slct.201701989]

[39] Khalili, D. Graphene oxide: a reusable and metal-free carbocatalyst for the one-pot synthesis of 2-amino-3-cyanopyridinesin water. *Tetrahedron Lett.,* **2016**, *57*(15), 1721-1723.
[http://dx.doi.org/10.1016/j.tetlet.2016.03.020]

[40] Gupta, A.; Kaura, R.; Singh, D.; Kapoora, K.K. Graphene oxide: A carbocatalyst for the one-pot multicomponent synthesis of highly functionalized tetrahydropyridines. *Tetrahedron Lett.,* **2017**, *58*(26), 2583-2587.
[http://dx.doi.org/10.1016/j.tetlet.2017.05.067]

[41] Hassankhani, A.; Gholipour, B.; Rostamnia, S. An efficient regioselective three-component synthesis of tetrazoloquinazolines using g-C_3N_4 covalently bonded sulfamic acid. *Polyhedron,* **2020**, *175*, 114217.

[http://dx.doi.org/10.1016/j.poly.2019.114217]

[42] Ahmad, S.; Afshari, R. Synthesis of carboxamide-functionalized multiwall carbon nanotubes *via* Ugi multicomponent reaction: water-dispersible peptidomimetic nanohybrid as controlled drug delivery vehicle. *ChemistrySelect,* **2017**, *2*(18), 5218-5225. [http://dx.doi.org/10.1002/slct.201700615]

[43] Ashish, B.; Kumar, A.; Chhabra, T.; Kumar, A.; Krishnan, V. Potassium-functionalized graphitic carbon nitride supported on reduced graphene oxide as a sustainable catalyst for knoevenagel condensation. ACS Appl. *NanoMate.,* **2018**, *1*, 6711-6723.

[44] Dehghanpour, H.R.; Mosslemin, M.H.; Mohebat, R. Graphene oxide: a carbocatalyst for the one-pot multicomponent synthesis of 5-aryl-1H-indeno[20,10:5,6] pyrido [2,3-d] pyrimidine-2,4,6 (3H)-trione. *J. Chem. Res.,* **2018**, *42*, 35-39. [http://dx.doi.org/10.3184/174751918X15161933697835]

[45] Mohammadipour, M.; Amoozadeh, A. The synthesis of polyhydroacridines by covalent 5-sulfobenzoic acid-functionalized graphene oxide as a novel, green, efficient, and heterogeneous catalyst. *Monatsh. Chem.,* **2017**, *148*(6), 1075-1084. [http://dx.doi.org/10.1007/s00706-016-1876-6]

[46] Miyazaki, K.; Islam, N. Nanotechnology systems of innovation-An analysis of industry and academia research activities. *Technovation,* **2007**, *27*(11), 661-675. [http://dx.doi.org/10.1016/j.technovation.2007.05.009]

[47] D'Souza, D.M.; Müller, T.J.J. Multi-component syntheses of heterocycles by transition-metal catalysis. *Chem. Soc. Rev.,* **2007**, *36*(7), 1095-1108. [http://dx.doi.org/10.1039/B608235C] [PMID: 17576477]

[48] Cioc, R.C.; Ruijter, E.; Orru, R.V.A. Multicomponent reactions: advanced tools for sustainable organic synthesis. *Green Chem.,* **2014**, *16*(6), 2958-2975. [http://dx.doi.org/10.1039/C4GC00013G]

[49] Sunderhaus, J.D.; Martin, S.F. Applications of multicomponent reactions to the synthesis of diverse heterocyclic scaffolds. *Chemistry,* **2009**, *15*(6), 1300-1308. [http://dx.doi.org/10.1002/chem.200802140] [PMID: 19132705]

[50] Kakuchi, R. Metal-catalyzed multicomponent reactions for the synthesis of polymers, Multi-Component and Sequential Reactions in Polymer Synthesis. *Springer,* **2015**.

[51] Lu, A.H.; Schmidt, W.; Matoussevitch, N.; Bönnemann, H.; Spliethoff, B.; Tesche, B.; Bill, E.; Kiefer, W.; Schüth, F. Nanoengineering of a magnetically separable hydrogenation catalyst. *Angew. Chem. Int. Ed.,* **2004**, *43*(33), 4303-4306. [http://dx.doi.org/10.1002/anie.200454222] [PMID: 15368378]

[52] Tsang, S.C.; Caps, V.; Paraskevas, I.; Chadwick, D.; Thompsett, D. Magnetically separable, carbon-supported nanocatalysts for the manufacture of fine chemicals. *Angew. Chem. Int. Ed.,* **2004**, *43*(42), 5645-5649. [http://dx.doi.org/10.1002/anie.200460552] [PMID: 15495206]

[53] Jiang, Y.; Gao, Q. Heterogeneous hydrogenation catalyses over recyclable Pd(0) nanoparticle catalysts stabilized by PAMAM-SBA-15 organic-inorganic hybrid composites. *J. Am. Chem. Soc.,* **2006**, *128*(3), 716-717. [http://dx.doi.org/10.1021/ja056424g] [PMID: 16417350]

[54] Dandia, A.; Gupta, S.L.; Parewa, V.; Sharma, A.; Rathore, K.S.; Sharma, A. "On-water" synthesis of 3-substituted indoles *via* Knoevenagel/Michael addition sequence catalyzed by Cu doped ZnS NPs. *Tetrahedron Lett.,* **2013**, *54*(42), 5711-5717. [http://dx.doi.org/10.1016/j.tetlet.2013.08.013]

[55] Dandia, A.; Parewa, V.; Gupta, S.L.; Sharma, A.; Rathore, K.S.; Jain, A. Microwave-assisted Fe_3O_4 nanoparticles catalyzed synthesis of chromeno [1, 6] naphthyridines in aqueous media. *Catal.*

Commun., **2015**, *61*, 88-91.
[http://dx.doi.org/10.1016/j.catcom.2014.12.015]

[56] Hakimi, F.; Oreyzi, F.S.; Fatemeh, F.B. Golrasan. E. A new application of nickel nanoparticles as a heterogeneous catalyst for synthesis of 1-amidoalkyl-2-naphthols according to green chemistry principles. *Asian J. Green Chem.*, **2020**, *4*, 134-141.

[57] Basavegowda, N.; Mishra, K.; Lee, Y.R. Ultrasonic-assisted green synthesis of palladium nanoparticles and their nanocatalytic application in multicomponent reaction. *New J. Chem.*, **2015**, *39*(2), 972-977.
[http://dx.doi.org/10.1039/C4NJ01543F]

[58] Rostami-Charati, F.; Akbaria, R. Biosynthesis of Fe_3O_4-magnetic nanoparticles using Clover Leaf aqueous extract: Green synthesis of oxazole derivatives. *Iran J. Org. Chem.*, **2019**, *11*, 2751-2755.

[59] Shafaei, F.; Babaei, S.E.; Shahvelayati, A.S.; Honarmand Janatabadi, F. HonarmandJanatabadi, F. Biosynthesis of Fe_3O_4-magnetic nanoparticles using clover leaf aqueous extract: Green synthesis of 1, 3-benzoxazole derivatives. *J. Chin. Chem. Soc. (Taipei)*, **2020**, *67*(5), 891-897.
[http://dx.doi.org/10.1002/jccs.201800489]

[60] Moghaddam, F.M.; Foroushani, B.K.; Rezvani, H.R. Nickel ferrite nanoparticles: an efficient and reusable nanocatalyst for a neat, one-pot and four-component synthesis of pyrroles. *RSC Advances*, **2015**, *5*(23), 18092-18096.
[http://dx.doi.org/10.1039/C4RA09348H]

[61] Mirjafary, Z.; Saeidian, H.; Sadeghi, A.; Moghaddam, F.M. ZnO nanoparticles: an efficient nanocatalyst for the synthesis of β-acetamido ketones/esters *via* a multi-component reaction. *Catal. Commun.*, **2008**, *9*(2), 299-306.
[http://dx.doi.org/10.1016/j.catcom.2007.06.018]

[62] Phukan, P.; Agarwal, S.; Deori, K. Sarma. D. Zinc Oxide Nanoparticles Catalysed One-Pot Three-Component Reaction: A Facile Synthesis of 4-Aryl-NH-1, 2, 3-Triazoles. *Catal. Lett.*, **2020**, 1-12.

[63] Alinezhad, H.; Pakzad, K. Green Synthesis of Copper Oxide Nanoparticles with an Extract of Euphorbia maculate and their use in the Biginelli Reaction. *Org. Prep. Proced. Int.*, **2020**, *52*(4), 319-327.
[http://dx.doi.org/10.1080/00304948.2020.1764819]

[64] Roco, M.C. Nanotechnology: convergence with modern biology and medicine. *Curr. Opin. Biotechnol.*, **2003**, *14*(3), 337-346.
[http://dx.doi.org/10.1016/S0958-1669(03)00068-5] [PMID: 12849790]

[65] Lagowski, J.J. Macmillan Encyclopedia of Chemistry, Simon and Schuster Macmillan. New York, **1997**; 4, pp. 1492-1495.

[66] Ates, B.; Koytepe, S.; Ulu, A.; Gurses, C.; Thakur, V.K. Chemistry, Structures, and Advanced Applications of Nanocomposites from Biorenewable Resources. *Chem. Rev.*, **2020**, *120*(17), 9304-9362.
[http://dx.doi.org/10.1021/acs.chemrev.9b00553] [PMID: 32786427]

[67] Choi, J.; Shin, H.; Yang, S.; Cho, M. The influence of nanoparticle size on the mechanical properties of polymer nanocomposites and the associated interphase region: a multiscale approach. *Compos. Struct.*, **2015**, *119*, 365-376.
[http://dx.doi.org/10.1016/j.compstruct.2014.09.014]

[68] Marcelo, L.R.; Santos de Gois, J.; Araujo da Silva, A.; Cesar, D.V. Synthesis of iron-based magnetic nanocomposites and applications in adsorption processes for water treatment: a review. *Environ. Chem. Lett.*, **2020**, 1-46.

[69] Akocak, S.; Şen, B.; Lolak, N.; Şavk, A.; Koca, M.; Kuzu, S.; Şen, F. One-pot three-component synthesis of 2-Amino-4H-Chromene derivatives by using monodisperse Pd nanomaterials anchored graphene oxide as highly efficient and recyclable catalyst. Nano-Struct. *Nano-Objects*, **2017**, *11*, 25-

31.
[http://dx.doi.org/10.1016/j.nanoso.2017.06.002]

[70] Dandia, A.; Bansal, S.; Sharma, R.; Rathore, K.S.; Parewa, V. Microwave-assisted nanocatalysis: A CuO NPs/rGO composite as an efficient and recyclable catalyst for the Petasis-borono-Mannich reaction. *RSC Advances,* **2018**, *8*(53), 30280-30288.
[http://dx.doi.org/10.1039/C8RA05203D]

[71] Dandia, A.; Indora, A.; Parewa, V.; Rathore, K.S. An Ultrasound-assisted Nanocatalysis: Ag NPs/rGO Nanocomposite Catalyzed One-pot Diastereoselective Synthesis of Functionalized Tetrahydropyridines. *Indian J. Heterocycl. Chem.,* **2018**, *28*, 103-111.

[72] Dandia, A.; Sharma, A.; Parewa, V.; Kumawat, B.; Rathore, K.S.; Sharma, A. Amidic C-N bond cleavage of isatin: chemoselective synthesis of pyrrolo [2, 3, 4-kl] acridin-1-ones using Ag NPs decorated rGO composite as an efficient and recoverable catalyst under microwave irradiation. *RSC Advances,* **2015**, *5*(111), 91888-91902.
[http://dx.doi.org/10.1039/C5RA11747J]

[73] Dandia, A.; Sharma, A.; Indora, A.; Rathore, K.S.; Jain, A.; Parewa, V. CeO_2 NPs/rGO composite catalyzed chemoselective synthesis of 2, 8-dioxabicyclo [3.3.1] nonanes in aqueous medium *via* aldol condensation/Michael addition/bicyclization reaction sequence. *Mol. Catal.,* **2018**, *459*, 97-105.
[http://dx.doi.org/10.1016/j.mcat.2018.08.024]

[74] Azgomi, N.; Mokhtary, M. Nano-Fe_3O_4@SiO_2 supported ionic liquid as an efficient catalyst for the synthesis of 1, 3thiazolidin-4-ones under solvent-free conditions. *J. Mol. Catal. Chem.,* **2015**, *398*, 58-64.
[http://dx.doi.org/10.1016/j.molcata.2014.11.018]

[75] Tamoradi, T.; Mousavi, S.M.; Mohammadi, M. Praseodymium (iii) anchored on $CoFe_2O_4$MNPs: an efficient heterogeneous magnetic nanocatalyst for one-pot, multi-component domino synthesis of polyhydroquinoline and 2, 3dihydroquinazolin-4 (1 H)-one derivatives. *New J. Chem.,* **2020**, *44*(7), 3012-3020.
[http://dx.doi.org/10.1039/C9NJ05468E]

[76] Tran, P.H.; Nguyen, T. One-pot multicomponent synthesis of thieno [2, 3-b] indoles catalyzed by a magnetic nanoparticle supported [Urea] 4 [$ZnCl_2$] deep eutectic solvent. *RSC Advances,* **2020**, *10*(16), 9663-9671.
[http://dx.doi.org/10.1039/D0RA00773K]

[77] Nguyen, H.T.; Truong, V.A.; Tran, P.H. Synthesis of polyhydroquinolines and propargylamines through one-pot multicomponent reactions using an acidic ionic liquid immobilized onto magnetic Fe_3O_4 as an efficient heterogeneous catalyst under solvent-free sonication. *RSC Advances,* **2020**, *10*(42), 25358-25363.
[http://dx.doi.org/10.1039/D0RA04008H]

[78] Mehdi, K.; Samiei, S.; Mirshokraie, S.A. MnO_2@Zeolite-Y Nanoporous: Preparation and Application as a High Efficient Catalyst for Multi-Component Synthesis of 4-Arylidene-Isoxazolidinones. *Silicon,* **2021**, *13*(1), 201-210.
[http://dx.doi.org/10.1007/s12633-020-00413-5]

CHAPTER 10

Synthesis of Heterocycles Involving Nanomaterials as Heterogeneous Catalysts

Tanvi Jandial[1], Muthu Karuppasamy[1] and **Vellaisamy Sridharan[1,*]**

[1] *Department of Chemistry and Chemical Sciences, Central University of Jammu, Rahya-Suchani (Bagla), District-Samba, Jammu-181143, J & K, India*

Abstract: Heterocycles are the most significant class of organic compounds comprising a wide range of biological and pharmacological activities. In particular, nitrogen and oxygen heterocycles are the important core structures, frequently found in a large number of natural products and medicinally important molecules. Consequently, great efforts have been made to develop direct methodologies for the synthesis of these heterocycles including nanomaterial-catalyzed approaches. In this chapter, we summarize the important synthetic methodologies for the synthesis of nitrogen and oxygen-containing heterocycles involving nanomaterial catalyzed reactions. The catalytic activity and the advantages of a variety of nanomaterials are described to understand their importance as heterogeneous catalysts towards the synthesis of various heterocyclic compounds.

Keywords: Heterogeneous catalysis, Nanocatalysts, Nanomaterials, Nitrogen heterocycles, Oxygen heterocycles.

INTRODUCTION

Heterocyclic compounds are the largest class of organic compounds that exhibit a variety of interesting pharmacological activities. Among the different classes of heterocycles, nitrogen and oxygen heterocycles are part of various natural products and compounds of biological significance [1]. Consequently, considerable effort has been devoted to the development of synthetic methodologies for these heterocycles, especially the five- and six-membered nitrogen and oxygen heterocyclic compounds [2, 3]. Some of the well-known examples of medicinally important heterocycles include ciprofloxacin **1**, which is a wide spectrum antibiotic drug used for the treatment of respiratory tract infections and urinary tract infections [4], and the antibacterial drug sulfamethox-

* **Corresponding author Vellaisamy Sridharan:** Department of Chemistry and Chemical Sciences, Central University of Jammu, Rahya-Suchani (Bagla), District-Samba, Jammu-181143, J & K, India. E-mail: vesridharan@gmail.com; sridharan.che@cujammu.ac.in

azole **2**, which is used to treat urinary tract infections in patients with a kidney transplant (Fig. **1**) [5].

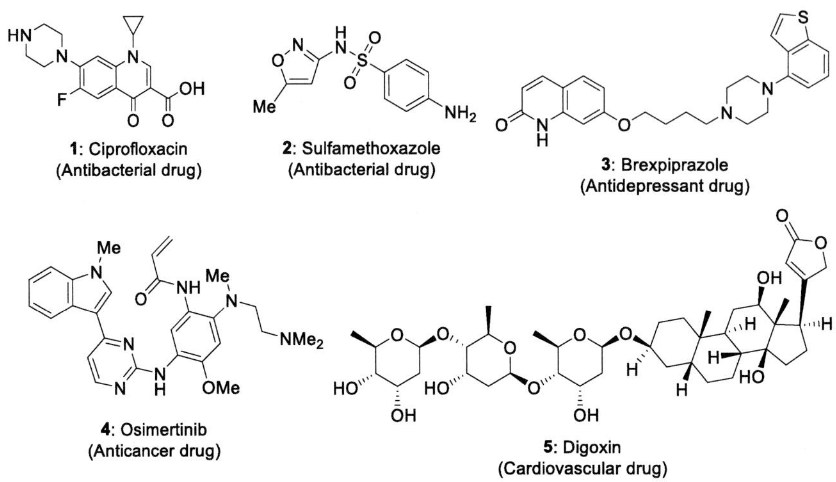

Fig. (1). Representative examples of biologically active heterocycles.

Furthermore, brexipiprazole **3** is an example of antipsychotic drug used for the treatment of the major depressive disorder (MDD) [6] and osimertinib **4** is widely used for the treatment of non-small cell lung cancer [7]. The oxygen heterocycle digoxin **5** is an effective drug for the treatment of irregular heartbeats and heart failure [8].

Numerous conventional methods are available in the literature for the synthesis of these biologically relevant heterocycles and some of the downsides of many of these approaches involve expensive catalysts, harsh reaction conditions, toxic solvent, high temperature, and multistep synthetic sequence. Consequently, in recent years, synthetic chemists are paying more attention to the development of greener strategies involving nanocatalysts for the synthesis of these valuable heterocycles [9]. This chapter summarizes the heterogeneous nanomaterial catalyzed synthesis of various heterocyclic compounds.

SYNTHESIS OF NITROGEN HETEROCYCLES

A variety of nanomaterials have been extensively used as heterogeneous catalysts to access various nitrogen heterocycles. In this section, we describe heterogeneous nanocatalyzed strategies for the synthesis of five- and six-membered nitrogen heterocycles.

Synthesis of Five-Membered Nitrogen Heterocycles

2H-Indazoles

Recently, Pathak and co-workers developed a nanocomposite CuO@CB [6] involving CuO nanoparticles immobilized on cucurbit[6]uril, which showed effective catalytic activity for the synthesis of a series of 2*H*-indazoles **8** starting from 2-bromobenzaldehydes **6**, primary amines **7**, and sodium azide under base- and ligand-free conditions in high yields (Fig. **2**) [10]. It was observed that the nanocatalyst was an air-stable green powder and reusable up to five runs with no remarkable loss of catalytic property.

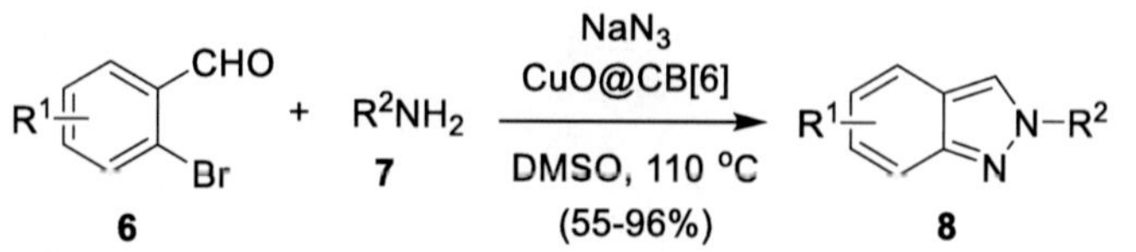

Fig. (2). Copper nanoparticles-catalyzed synthesis of 2*H*-indazoles.

An ultrasonic-mediated one-pot, the three-component reaction between 2-bromobenzaldehydes **9**, primary amines **10**, and tetrabutylammonium azide (TBAA) **11** was established for the synthesis of 2*H*-indazole derivatives **12** in good to excellent yields (Fig. **3**) [11]. The reaction was achieved using copper-doped silica cuprous sulphate (CDSCS) as a heterogeneous nanocatalyst and was reused for several sequential runs without any significant loss in catalytic activity.

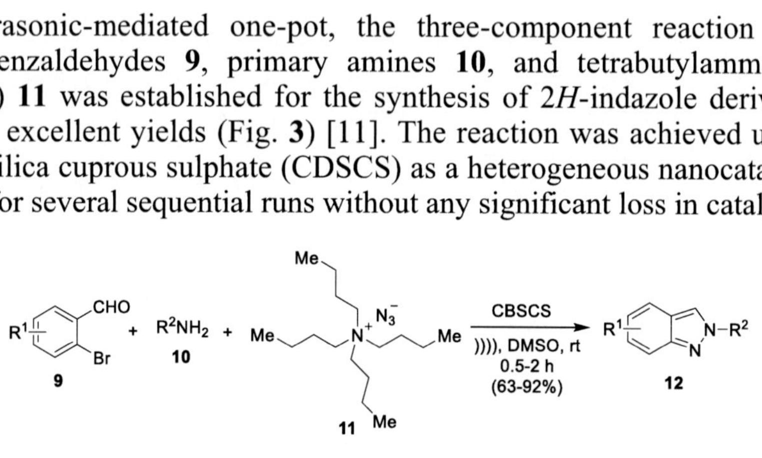

Fig. (3). Copper-doped silica cuprous sulphate-catalyzed ultrasonic-mediated synthesis of 2*H*-indazoles..

Imidazoles and Benzimidazoles

Aberi and co-workers established a one-pot approach for the synthesis of 2,4,5-triarylimidazoles **16** *via* a three-component reaction of benzil **13** or benzoin **14**, aryl aldehydes **15**, and ammonium acetate under mild conditions (Fig. **4**) [12]. The Mo^{IV}-salen complex was found to be an efficient and reusable nanocatalyst to access the desired products in high yields.

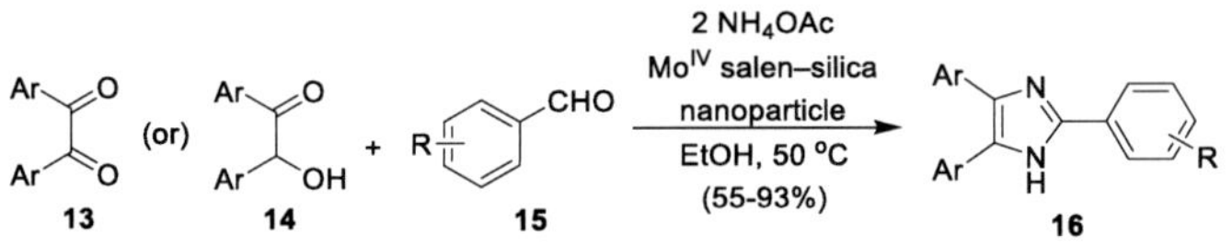

Fig. (4). Mo^{IV}-catalyzed three-component synthesis of 2,4,5-triarylimidazoles.

The catalytic activity of $NiFe_2O_4$/geopolymer nanocomposite was investigated for the one-pot synthesis of 2,4,5-triaryl-1*H*-imidazole derivatives **19** involving a three-component condensation between benzil **17**, substituted aldehydes **18**, and ammonium acetate under ultrasonic irradiation (Fig. **5**) [13]. The catalyst was reusable and found to be effective to furnish the products in high yields.

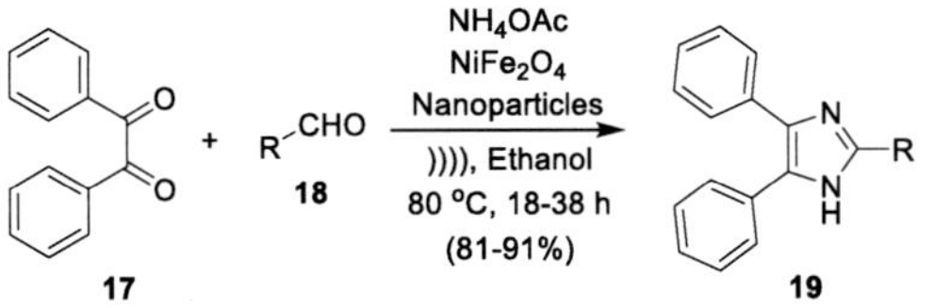

Fig. (5). $NiFe_2O_4$/geopolymer nanocomposite-catalyzed three-component synthesis of imidazoles.

Alemi and co-workers synthesized nano-structured $Bi_{1.5}(Lu, Er)_{0.5}O_3$ and utilized it as recyclable heterogeneous catalyst to access tetrasubstituted imidazole derivatives. The four-component reaction between benzil **20**, primary amines **21**, aryl aldehydes **22**, and ammonium acetate delivered imidazoles **23** in good yields in water under ultrasonic irradiation (Fig. **6**) [14].

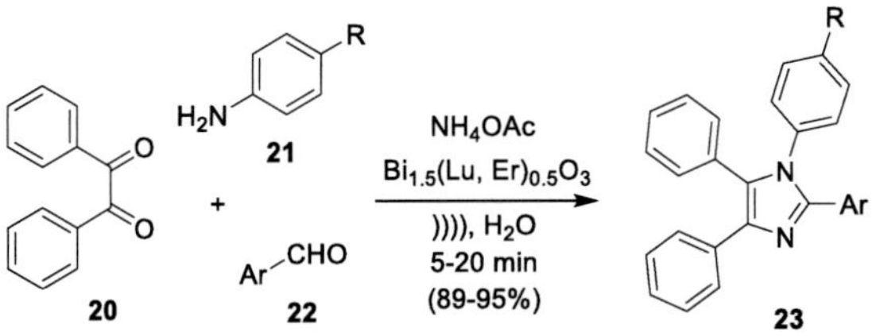

Fig. (6). Nanocatalyzed four-component synthesis of tetrasubstituted imidazole derivatives..

In addition, chitosan-coated Fe_3O_4 nanoparticles (Fe_3O_4@CS) were also utilized as a renewable heterogeneous magnetic biocatalyst to give access to 2,4,5-trisubstituted imidazoles **26** in appreciable yields. This methodology involves the cyclization reaction between substituted benzils **24**, aryl aldehydes **25**, and ammonium acetate in ethanol under reflux conditions (Fig. **7**) [15].

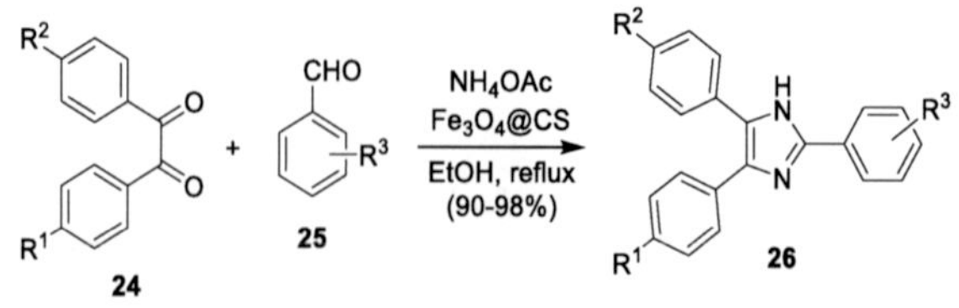

Fig. (7). Chitosan-coated Fe_3O_4 nanoparticles-catalyzed synthesis of 2,4,5-trisubstituted imidazoles.

A solvent-free approach was developed by Bardajee and co-workers for the synthesis of benzimidazoles **29** starting from *o*-phenylenediamines **27** (X = NH_2) and aryl aldehydes **28** in good to excellent yields using Cu^{II}-DiAmSar/SBA-15 as a non-leaching heterogeneous nanocatalyst (Fig. **8**) [16]. This reusable catalyst was found to be efficient and used for the synthesis of other related heterocycles including benzoxazoles and benzothiazoles.

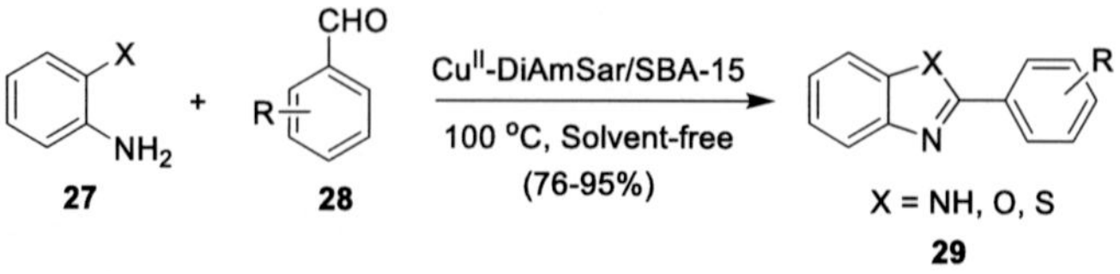

Fig. (8). Cu^{II}-catalyzed synthesis of benzimidazoles, benzoxazoles, and benzothiazoles.

1H-1,2,3-Triazoles

Amini and co-workers employed a copper nanoparticle immobilized on metal oxide (Cu/WO_3) as a heterogeneous catalyst to access 1,2,3-triazole derivatives **32** in good to excellent yields involving azide-alkyne cycloaddition reaction (click reaction) of alkynes **30** with various benzyl halides **31** in the presence of sodium azide (Fig. **9**) [17]. It was noted that the catalyst was reused for three runs without considerable loss of catalytic activity.

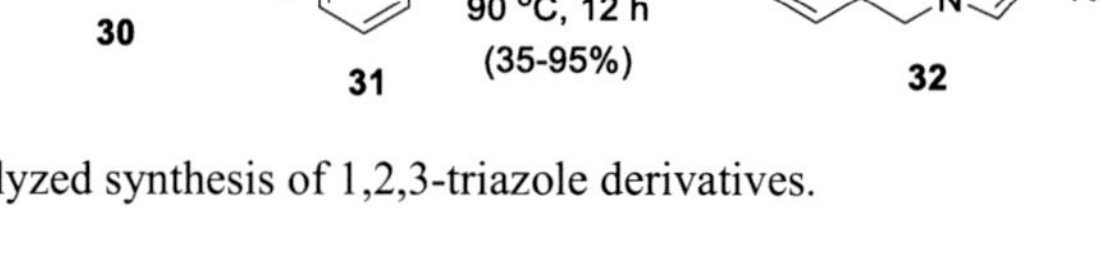

Fig. (9). Cu/WO_3-catalyzed synthesis of 1,2,3-triazole derivatives.

Hamzehloueian and co-workers illustrated the condensation reaction of primary halides **33**, terminal alkynes **34**, and NaN_3 for the one-pot synthesis of 1,4-disubstituted 1,2,3-triazoles **35** in the presence of heterogeneous nanocatalyst (Fig. **10**) [18]. This three-component transformation was achieved involving $MnFe_2O4$@GO@CS/Cu as a heterogeneous reusable nanocatalyst under mild conditions. The catalyst was successfully recovered and reused for nine runs with insignificant loss in the catalytic behaviour. The same three-component synthesis of 1,2,3-triazoles was also achieved by employing a heterogeneous copper catalyst (MNP@ImAc/Cu) based on poly(ionic liquid)-coated magnetic nanoparticles [19] and a copper immobilized ferromagnetic nanoparticle triazine dendrimer (FMNP@TD–Cu(II)) [20].

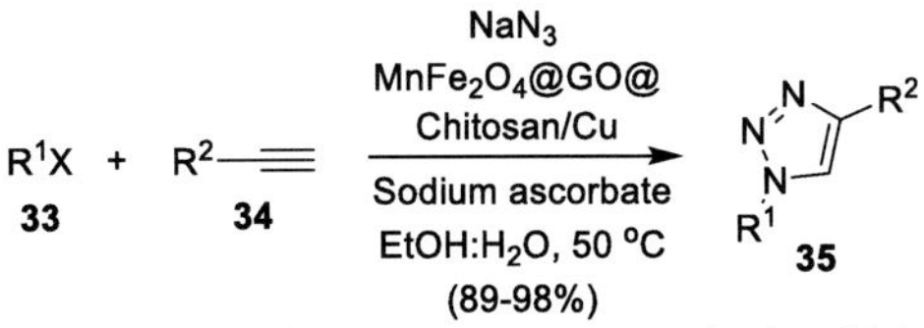

Fig. (10). $MnFe_2O4$@GO@CS/Cu-catalyzed three-component synthesis of 1,2,3-triazoles.

1H-Tetrazoles

Nikoorazm and co-workers tested the efficiency of a palladium complex immobilized onto MCM-41 pores as a nanocatalyst for the homoselective synthesis of 5-substituted 1*H*-tetrazole derivatives **37** starting from aryl nitriles **36** and sodium azide (Fig. **11**) [21]. The desired products were obtained in high yields and the catalyst was recyclable up to six runs with no remarkable decrease in catalytic activity and product selectivity.

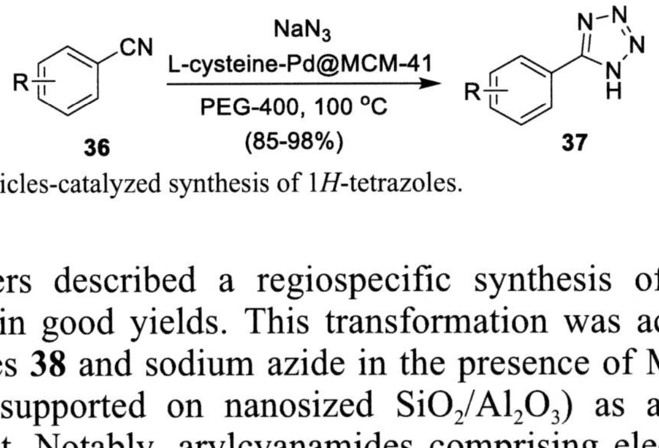

Fig. (11). Palladium nanoparticles-catalyzed synthesis of 1*H*-tetrazoles.

Habibi and co-workers described a regiospecific synthesis of 5-amino-1*H*-tetrazoles **39** and **40** in good yields. This transformation was achieved by the reaction of cyanamides **38** and sodium azide in the presence of M@Si/Al (M = Mn^{III}, Co^{II} and Fe^{III} supported on nanosized SiO_2/Al_2O_3) as a reusable and heterogeneous catalyst. Notably, arylcyanamides comprising electron-releasing substituents on the aryl ring afforded the isomer **40**. The other isomer **39** can be accessed with the influence of electron-withdrawing substituents on the aryl ring of arylcyanamides (Fig. **12**) [22].

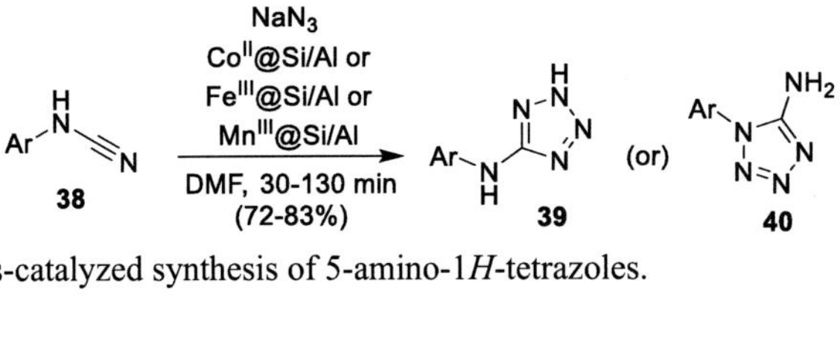

Fig. (12). Metal nanoparticles-catalyzed synthesis of 5-amino-1*H*-tetrazoles.

Synthesis of Five-Membered Fused Nitrogen Heterocycles

The three-component synthesis of indole-fused imidazoles **43** was achieved starting from isatin **41**, aldehydes **42**, and ammonium acetate in excellent yields under nanocatalyzed conditions (Fig. **13**) [23]. A cerium-immobilized silicotungstic acid nanoparticle-impregnated zirconia (Ce@STANPs/ZrO_2) was utilized as a nanocatalyst to activate the C–O double bond under microwave irradiation. The advantages of this synthetic approach include reusability of the catalyst up to seven cycles, excellent yields, shorter reaction times, and green reaction conditions.

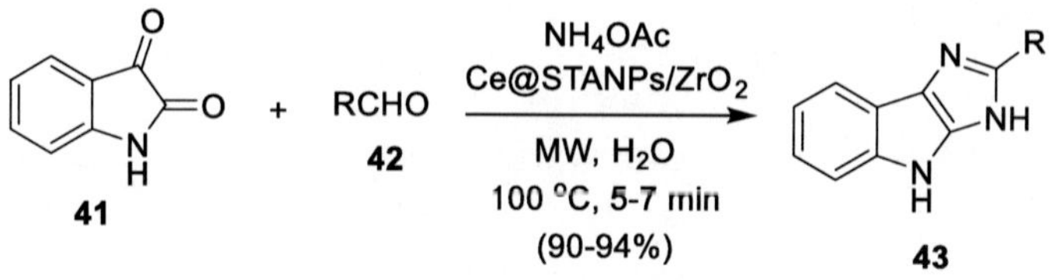

Fig. (13). Three-component synthesis of 3,4-dihydroimidazo[4,5-*b*]indoles.

A recyclable and superparamagnetic nano-Fe_3O_4@SiO_2 anchored sulfuric acid-catalyzed one-pot, the three-component reaction between benzene-1,2-dicarbaldehyde **44**, pyridin-2-amines **45** and trimethylsilylcyanide was developed for the synthesis of pyrido[2',1':2,3]imidazo[4,5-*c*]isoquinolines **46** in good yields under mild conditions (Fig. **14**) [24].

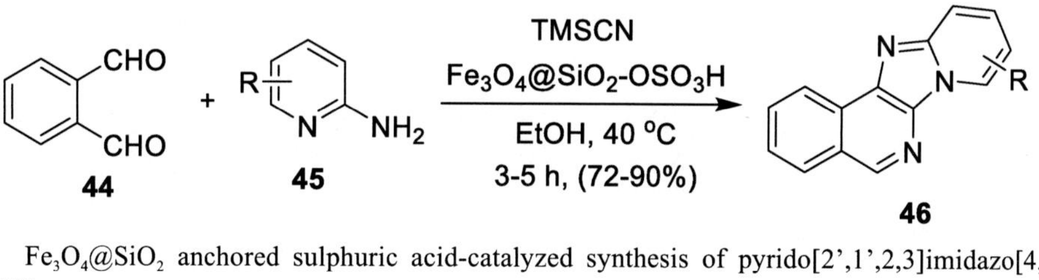

Fig. (14). Fe_3O_4@SiO_2 anchored sulphuric acid-catalyzed synthesis of pyrido[2',1',2,3]imidazo[4,5-*c*] isoquinolines.

Javanshir and co-workers described the reusable SO_3H-functionalized mesoporous silica material (SO_3H-FMSM) catalyzed one-pot, three-component condensation reaction of 2,3-dihydriphthalazine-1,4-dione **47**, dimedone **48**, and aryl aldehydes **49** for the direct synthesis of 2*H*-indazolo[2,1-*b*]phthalazine-1,6,11-trione derivatives **50** in excellent yields under solvent-free reaction conditions (Fig. **15**) [25].

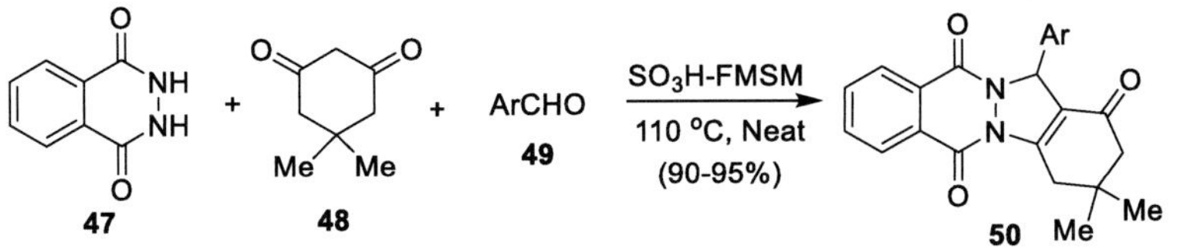

Fig. (15). Mesoporous silica nanomaterial-catalyzed synthesis of 2*H*-indazolo[2,1-*b*]phthalazine-1,6,11-triones.

Synthesis of Six-Membered Nitrogen Heterocycles

Heterocycles Bearing One Nitrogen Atom

The Hantzsch 1,4-dihydropyridine synthesis was established involving $CoFe_2O_4@SiO_2$-NH_2-Co^{II} nanoparticles as heterogeneous and reusable catalysts. A series of 1,4-dihydropyridine derivatives **53** were synthesized in good to excellent yields *via* a four-component reaction of aryl aldehydes **51**, β-ketoesters **52**, and ammonium acetate in EtOH/Water solvent system under reflux conditions (Fig. **16**) [26].

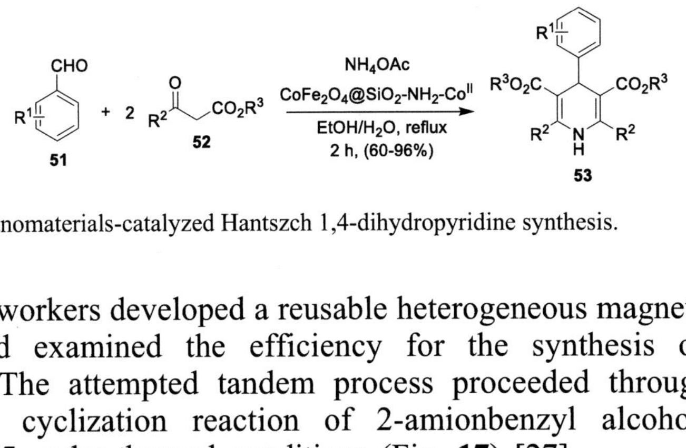

Fig. (16). Co-based nanomaterials-catalyzed Hantszch 1,4-dihydropyridine synthesis.

Bhanage and co-workers developed a reusable heterogeneous magnetic CuNiFeO nanocatalyst and examined the efficiency for the synthesis of quinoline derivatives **56**. The attempted tandem process proceeded through a double dehydrogenation cyclization reaction of 2-amionbenzyl alcohols **54** with benzylalcohols **55** under thermal conditions (Fig. **17**) [27].

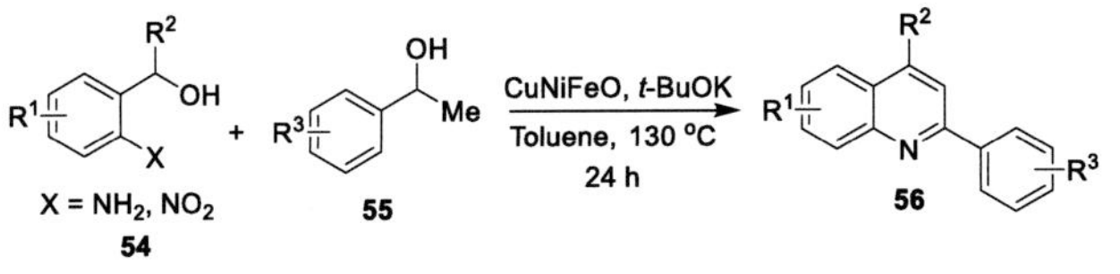

Fig. (17). CuNiFeO-catalyzed synthesis of quinolines.

On the other side, a green and one-pot, three-component synthesis of functionalized quinolines **60** was established starting from isatin **57**, activated acetylenic compounds **58**, and alcohols **59** in the presence of KF/CP (NPs) as a heterogeneous basic nanocatalyst in water at room temperature (Fig. **18**) [28].

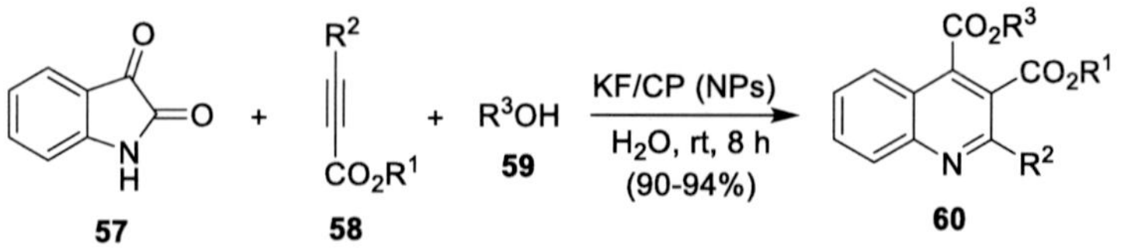

Fig. (18). Nanocatalyzed three-component synthesis of quinolines.

Heterocycles Bearing Two Nitrogen Atoms

Tayebee and co-workers demonstrated a one-pot, three-component Biginelli condensation of substituted benzaldehydes **61**, ethyl acetoacetate **62**, and urea **63** to access 3,4-dihydropyrimidin-2-(1*H*)-ones **64** in the presence of phosphotungstic acid ($H_3PW_{12}O_{40}$, PTA) supported on ZIF-9(NH_2) as catalyst under solvent-free conditions (Fig. **19**) [29]. It was observed that the nanocatalyst could be easily separated from the reaction mixture and reused at least five times with a 14% reduction in yield after the fifth run.

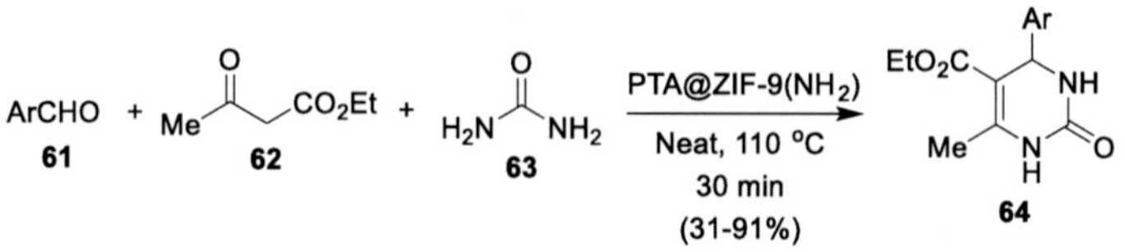

Fig. (19). Nanocatalyzed Biginelli 3,4-dihydropyrimidin-2-(1*H*)-one synthesis.

A magnetic silica-based recyclable nanocomposite (Fe_3O_4/SBA-15) was synthesized and its catalytic activity was investigated for the one-pot, three-component reaction between isatoic anhydride **65**, a variety of aldehydes **66**, and ammonium acetate to access 2,3-dihydroquinazolin-4(1*H*)-ones **67** in good yields in ethanol under reflux conditions (Fig. **20**) [30].

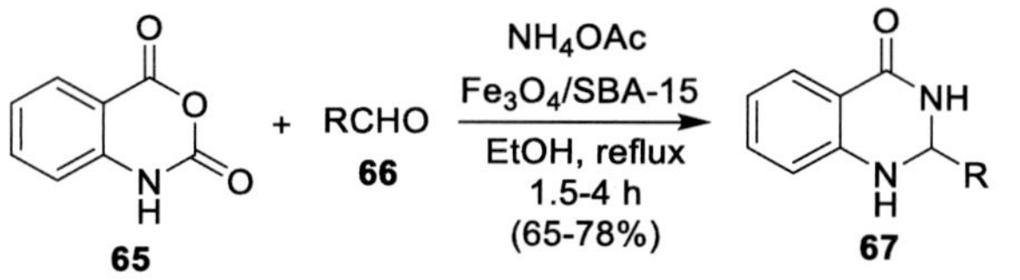

Fig. (20). Fe_3O_4/SBA-15 nanocomposite-catalyzed synthesis of dihydroquinazolin-4(1*H*)-ones.

A heterogeneous catalyst, MCM-41-dtz-Ni, was prepared by anchoring a nickel complex in functionalized mesoporous MCM-41 and used as a thermally stable nanocatalyst for the one-pot synthesis of 2,3-dihydroquinazolin-4(1*H*)-ones **70** starting from 2-aminobenzamide **68** and aryl aldehydes **69** in PEG-400 under mild conditions (Fig. **21**) [31]. The heterogeneous catalyst showed recyclability for five consecutive runs with no remarkable loss of catalytic behaviour.

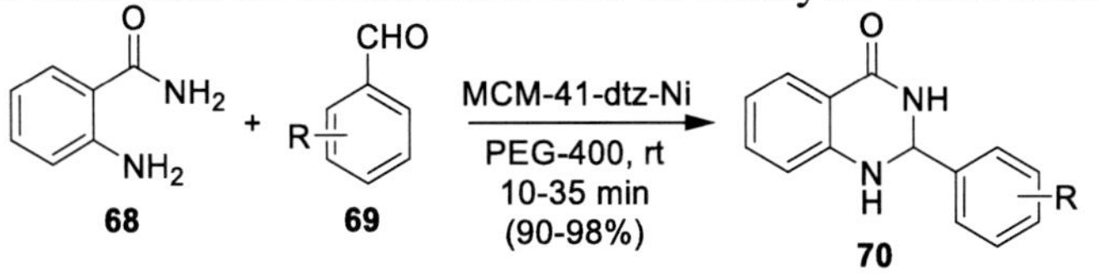

Fig. (21). Synthesis of 2,3-dihydroquinazolin-4(1*H*)-ones using Ni-nanoparticles.

Bhanage and co-workers employed Cu_2O nanocubes as a recyclable heterogeneous nanocatalyst for the synthesis of quinazoline derivatives **73** in excellent yields in a very short reaction time. The methodology involves a tandem cyclization of 2-bromobenzaldehydes **71** and amidines **72** in ethylene glycol under microwave irradiation conditions (Fig. **22**) [32].

Fig. (22). Cu_2O nanocubes-catalyzed synthesis of quinazolines.

A green protocol for the synthesis of medicinally important spiropyrimidine derivatives **76** was developed using a magnetically separable and reusable heterogeneous copper ferrite nanocatalyst under mild conditions. The three-component cyclization reaction between cyclic ketones **74**, aromatic amines **75**, and formaldehyde allowed the generation of six new bonds in a single synthetic operation to deliver spiro compounds **76** in good yields (Fig. **23**) [33].

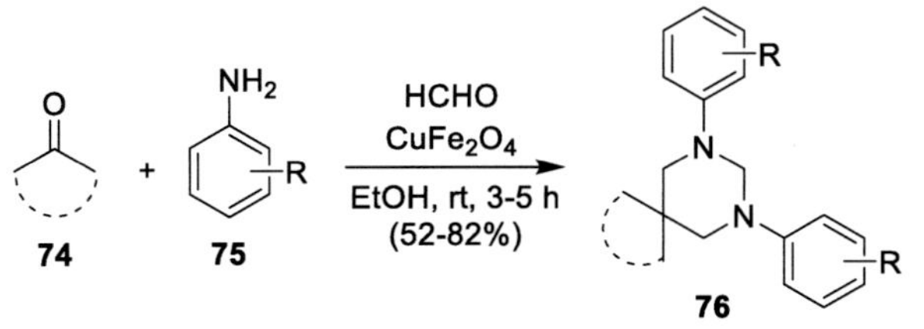

Fig. (23). Heterogeneous copper ferrite-catalyzed synthesis of spiropyrimidine derivatives.

A simple protocol for the synthesis of *N*-cyclohexyl-3-aryl-quinoxaline-2-amines **80** was established utilizing ZnO nanoparticles as a reusable catalyst. The methodology involves a three-component reaction of *o*-phenylenediamine **77**, aryl aldehydes **78**, and cyclohexyl isocyanide **79** under mild conditions to deliver the target compounds in high yields (Fig. **24**) [34].

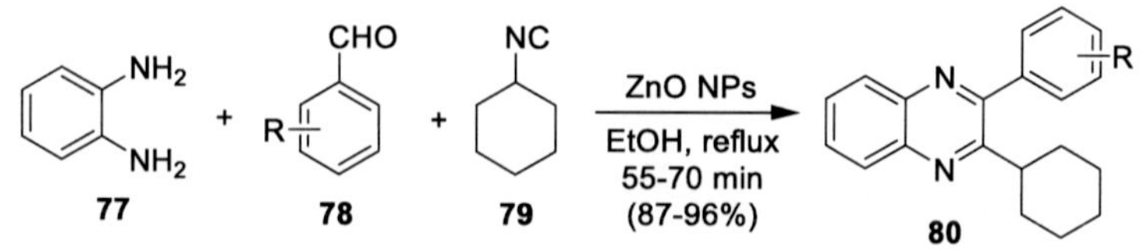

Fig. (24). Mesoporous silica phenylsulfonic acid-catalyzed synthesis of quinoxalines.

Synthesis of Six-Membered Fused Nitrogen Heterocycles

Foroughifar and co-workers described *N*-propyl-benzoguanamine-SO_3H-stabilized magnetic nanoparticles catalyzed approach for the synthesis of 1,8-diox--decahydroacridine derivatives **84** involving a four-component condensation between dimedone **81**, aromatic aldehydes **82**, and arylamines **83** or ammonium acetate in good to excellent yields in shorter reaction times (Fig. **25**) [35].

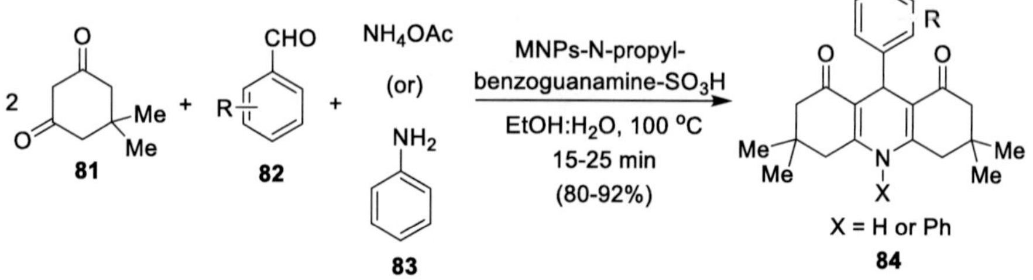

Fig. (25). *N*-propyl-benzoguanamine-SO_3H-stabilized magnetic nanoparticles catalyzed synthesis of 1,8-dioxo-decahydroacridines.

Recently, Saeidiroshan and Moradi synthesized MWCNTs@TEPA/Co^{II} as a new heterogeneous nanocatalyst and used it for the four-component synthesis of fused dihydropyridines. The reaction between dimedone **85**, aryl aldehydes **86**, barbituric acid/thiobarbituric acid **87** (X = O, S), and ammonium acetate or arylamines **89** in the presence of MWCNTs@TEPA/Co^{II} catalyst delivered dihydropyrimido [4,5-*b*]quinolinetrione derivatives **88** and **90** in high yields under reflux conditions in ethanol (Fig. **26**) [36]. The nanocatalyst was recovered by filtration and reused for several runs without noticeable loss of catalytic activity.

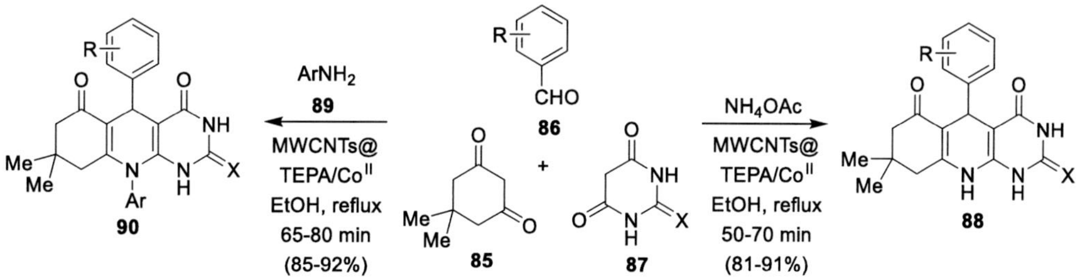

Fig. (26). Co-nanoparticles-catalyzed four-component synthesis of dihydropyrimido[4,5-*b*]quinolinetriones.

A nickel complex immobilized on superparamagnetic nanoparticles $MnFe_2O_4$ @NH_2@2AB-Ni was employed as a recyclable nanocatalyst for the green synthesis of spirooxindoles **95** and **96** *via* a three-component reaction between isatins **91**, cyclohexane-1,3-diones **92**, and 2,6-diaminopyrimidin- 4(3*H*)-one **93** or barbituric acid/thiobarbituric acid **94** in water at 90 °C (Fig. **27**) [37]. The catalyst was recovered simply by applying a permanent external magnet and reused for subsequent runs without significant loss of catalytic activity.

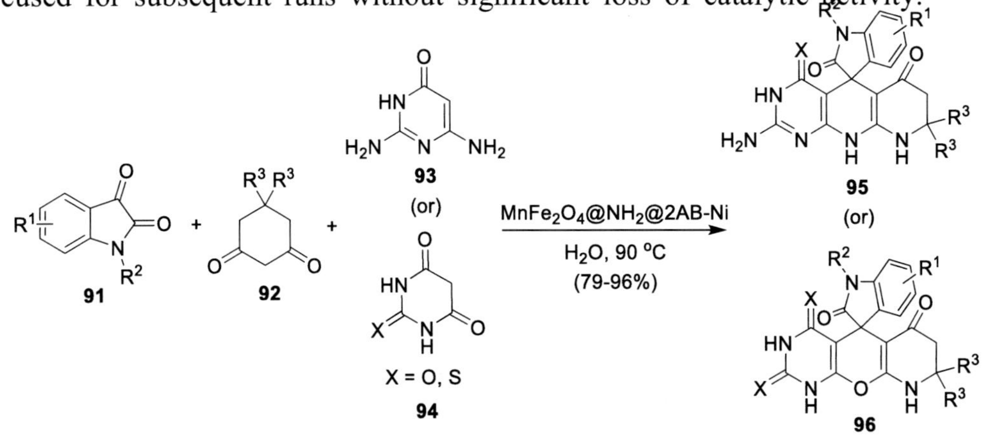

Fig. (27). Nickel complex immobilized superparamagnetic nanoparticles-catalyzed synthesis of spirooxindoles.

Debnath and Pramanik established a nanocatalyzed four-component reaction between isatin derivatives **97**, arylamines **98**, dialkyl acetylenedicarboxylates **99**, and cyclohexane-1,3-diones **100** for the synthesis of functionalized tetrahydrospiro[indoline-3,2′-quinoline] derivatives **101** (Fig. **28**) [38]. The bimetallic $ZnFe_2O_4$ nanopowder, a dual Lewis acid-base combined catalyst was found to be an efficient catalyst for this transformation in water under mild conditions.

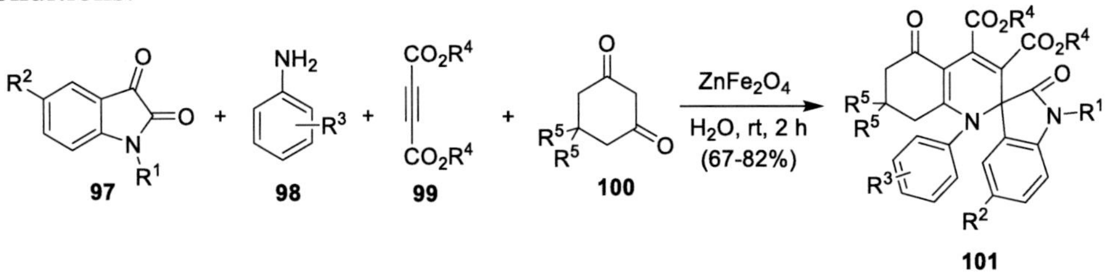

Fig. (28). Bimetallic $ZnFe_2O_4$ nanopowder-catalyzed four-component synthesis of spirooxindoles.

A one-pot, three-component assembly of 5-methylpyrazol-3-amine **102**, aryl aldehydes **103**, and Meldrum's acid **104** was developed to obtain 3-methyl-4-

aryl-2,4,5,7-tetrahydropyrazolo[3,4-*b*]pyridine-6-ones **105** using mesoporous silica phenylsulfonic acid (SBA-15-Ph-SO_3H) as heterogonous solid acid nanocatalyst (Fig. **29**) [39]. The catalyst was found to be reusable and afforded the desired products in good to excellent yields.

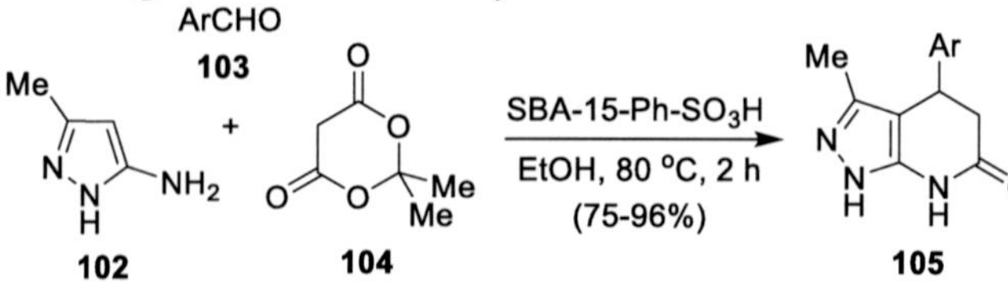

Fig. (29). Mesoporous silica phenylsulfonic acid-catalyzed synthesis of fused pyrazoles.

Recently, Riadi and co-workers developed a simple methodology to access 2-arylpyridopyrimidines **108** in good yields involving Ni-doped TiO_2 nanoparticles catalyzed cyclization reaction of (2-amino-pyridin-3-yl)-methanol **106** and benzylamines **107** in toluene under reflux conditions (Fig. **30**) [40].

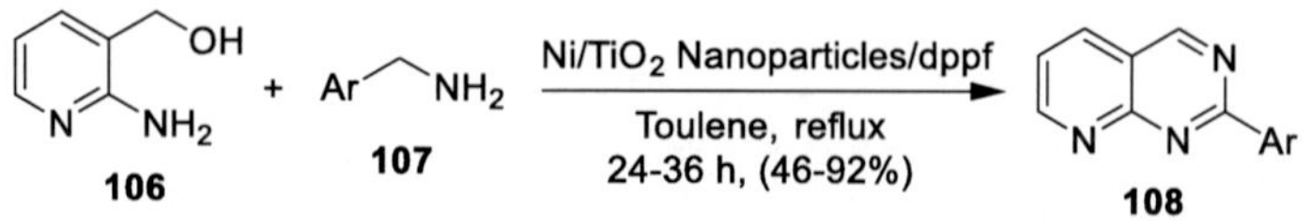

Fig. (30). Ni-doped TiO_2 nanoparticles-catalyzed synthesis of pyridopyrimidines.

Malakooti and co-workers encapsulated an iron-Schiff base complex in SBA-15 mesoporous silica to furnish a Fe^{III}-Schiff base/SBA-15 heterogeneous nanocatalyst for the construction of functionalized pyridopyrazines and quinoxalines. The cyclization reaction of 1,2-diamines **109** and 1,2-diketones **110** in the presence of the nanocatalyst delivered the corresponding fused heterocycles **111** in excellent yields (up to 99%) in water (Fig. **31**) [41].

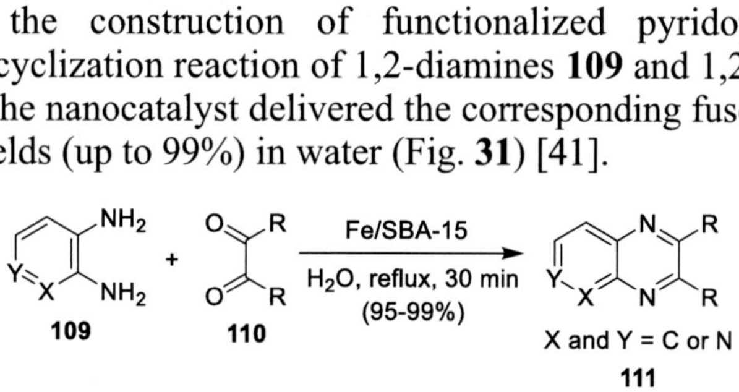

Fig. (31). Fe/SBA-15-catalyzed synthesis of quinoxaline derivatives.

Mohammadi and Shaterian demonstrated a reusable (3-oxo- [1, 2, 4]triazolidin-1-yl)bis(butane-1-sulfonic acid) supported on γ-Fe_2O_3 catalyzed synthesis of spiro[indeno[1,2-*b*]quinoxaline derivatives **116** in excellent yields *via* a four-component condensation between isatoic anhydride **112**, primary amines **113**, ninhydrin **114**, and *o*-phenylenediamines **115** in ethanol at 70 °C in excellent yields (Fig. **32**) [42].

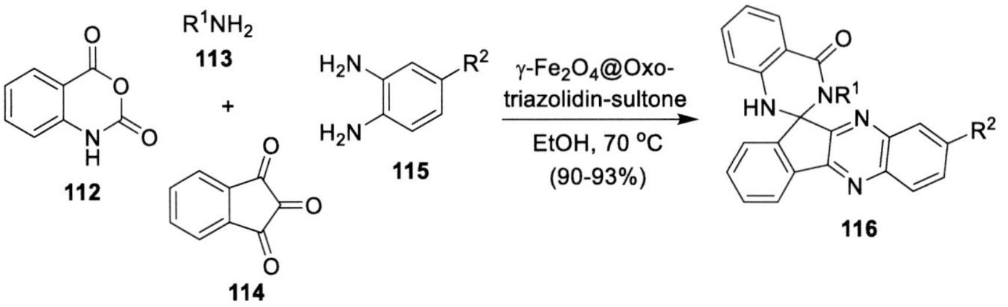

Fig. (32). Fe-nanoparticle catalyzed four-component synthesis of spiro[indeno[1,2-*b*]quinoxaline derivatives.

SYNTHESIS OF OXYGEN HETEROCYCLES

The synthesis of a variety of oxygen heterocycles has also been achieved involving heterogeneous nanocatalysts. This section summarizes the methodologies established for the direct synthesis of five- and six-membered oxygen heterocycles under nanocatalyzed heterogeneous reaction conditions.

Synthesis of Five-Membered Oxygen Heterocycles

An efficient tandem approach was established for the synthesis of 2-phenylbenzofuran **119** using Cu-doped Pd-Fe_3O_4 nanocatalyst starting from 2-iodophenol **117** and phenylpropiolic acid **118** (Fig. **33**) [43]. The magnetically recyclable catalyst was synthesised using thermal decomposition of $Fe(CO)_5$ and reduction of $Pd(OAc)_2$ and $Cu(acac)_2$. The advantages of this catalyst include high selectivity, physical and chemical stability, and reusability over the conventional non-metallic catalytic system.

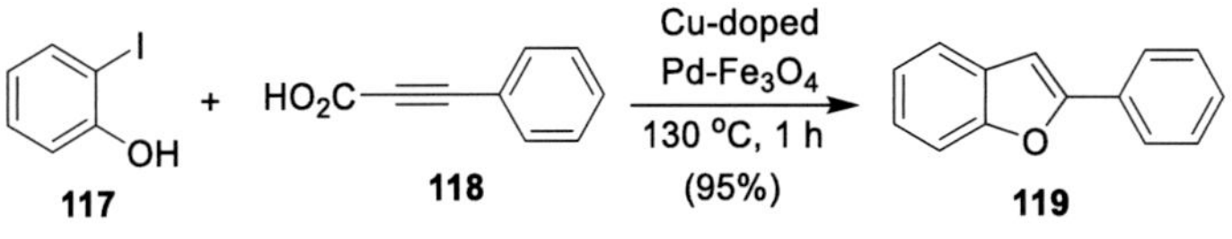

Fig. (33). Cu-doped Pd-Fe_3O_4-catalyzed synthesis of 2-phenybenzofuran.

Dehnavi and Ghashang devised a three-component, one-pot, solvent-free approach for the synthesis of furo[3,2-*c*]chromen-4-one derivatives **123** starting from arylamines **120**, α,β-epoxy ketones **121**, and 4-hydroxycoumarin **122** using ZnO-$ZnAl_2O_4$ nanocomposites as heterogeneous catalyst (Fig. **34**) [44]. The catalyst was recovered by filtration and reused for five consecutive runs without any loss in catalytic property.

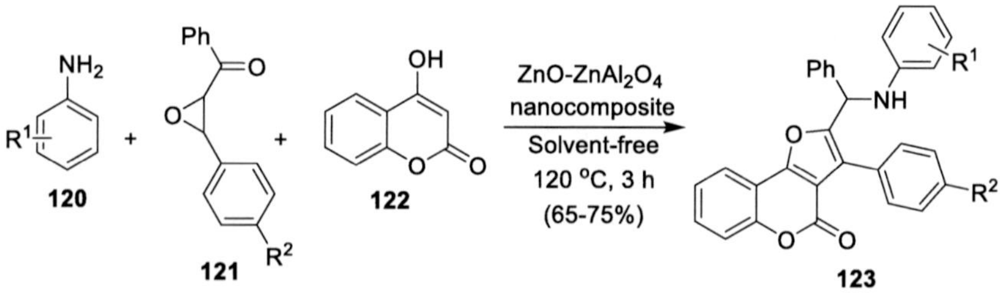

Fig. (34). ZnO-$ZnAl_2O_4$-catalysed synthesis of furo[3,2-*c*]chromen-4-ones.

Recently, Liu and Hu synthesised a Cu-based green and recyclable nanocatalyst, which was employed for the synthesis of cyclic carbonates **125** through fixation of CO_2 to epoxides **124** under solvent-free conditions (Fig. **35**) [45]. The authors prepared the heterogeneous catalyst $PMO@ILCu_2(OH)_3Cl_2$ in combination with ionic liquid and periodic mesoporous organosilicons (PMOs).

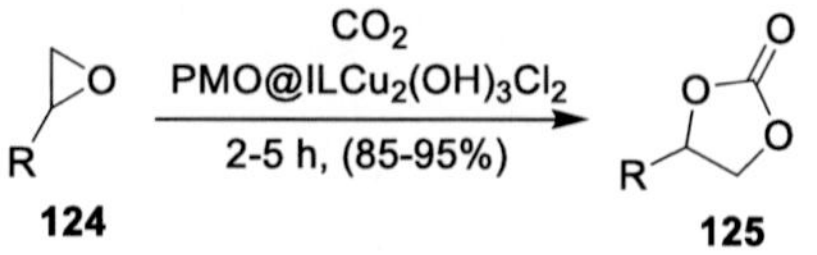

Fig. (35). $PMO@ILCu_2(OH)_3Cl_2$-catalyzed chemical fixation of CO_2 with epoxides.

Synthesis of Six-Membered Oxygen Heterocycles

Shaabani and co-workers demonstrated a nickel-Schiff-base complex immobilised on magnetic nanoparticles for the synthesis of tetrahydrobenzo[*b*]pyran derivatives **128** starting from cyclic 1,3-diketones **126**, aryl aldehydes **127**, and malononitrile (Fig. **36**) [46]. This one-pot, three-component approach offers high selectivity, reusability of the catalyst, easy handling and high yield of the product than that of the conventional methods. The same three-component reaction was also achieved in the presence of $CuFe_3O_4$ catalyst in ethanol at 60 °C [47] and reusable *n*-butylsulfonated nano-titania (*n*-TiO_2-NH-$(CH_2)_4$-SO_3H) catalyst under solvent-free conditions [48].

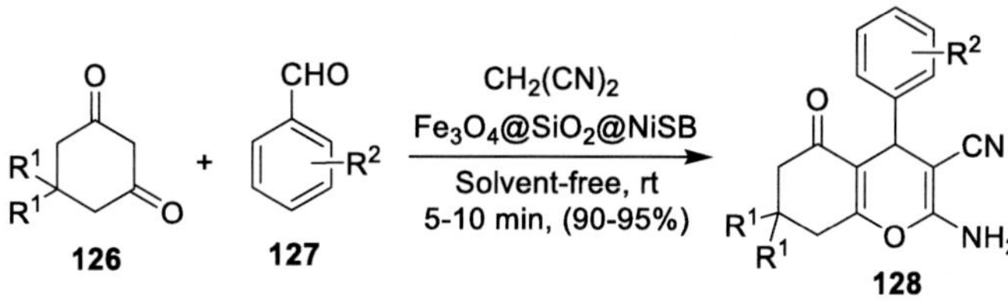

Fig. (36). Ni-based nanoparticles-catalyzed synthesis of tetrahydrobenzo[*b*]pyran derivatives.

A $CuFe_3O_4$@starch catalyzed three-component reaction between aryl aldehydes **129**, enols **130**, and malononitrile was developed for the synthesis of highly functionalised 4*H*-pyran derivatives **131** (Fig. **37**) [49]. In this green protocol, the products were isolated in high yields, and other advantages of this approach include low cost and reusability of the catalyst, easy work-up, and mild reaction conditions.

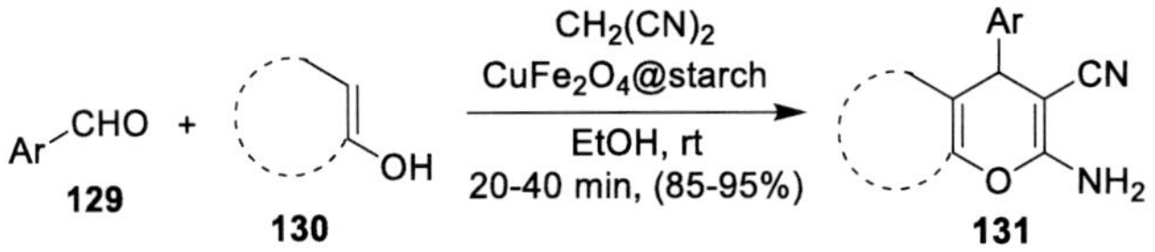

Fig. (37). $CuFe_2O_4$-catalysed synthesis of 4*H*-pyran derivatives.

An (α-Fe_2O_3)-MCM-41-supported dual acidic ionic liquid (DAIL) catalytic system was developed for the synthesis of 2-amino-4*H*-chromenes **134** in excellent yields (Fig. **38**) [50]. The one-pot, three-component reaction between resorcinol **132**, aryl aldehydes **133**, and malononitrile in the presence of the nanocatalyst in water delivered the desired products **134** in excellent yields.

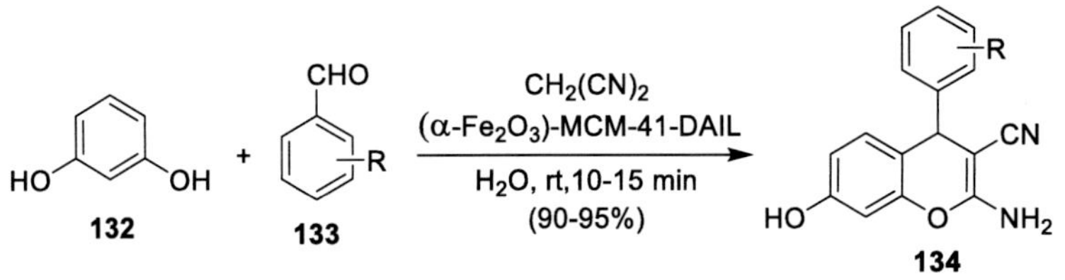

Fig. (38). (α-Fe_2O_3)-MCM-41-DAIL-catalysed synthesis of 2-amino-4*H*-chromenes.

Hazeri and co-workers established a Fe_3O_4@THAM-piperazine catalyzed synthesis of hexahydro-1*H*-xanthene-1,8(2*H*)-diones **137** involving a AB_2 three-component condensation of aryl aldehydes **135** and two equivalents of dimedone **136** (Fig. **39**) [51]. The methodology was also extended to the synthesis of tetrahydro-5*H*-chromen-5-ones **138** using malononitrile as one of the reactants.

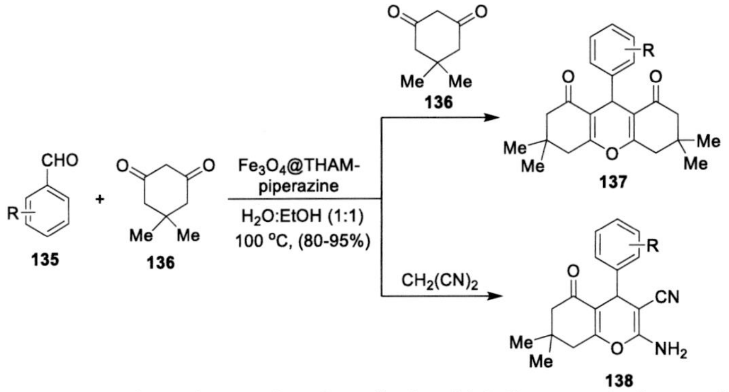

Fig. (39). Fe_3O_4@THAM-piperazine-catalyzed synthesis of 1,8-dioxo-octahydro-xanthenes and tetrahydro-benzo[*b*]pyrans.

Recently, the related Fe_3O_4-supported propane-1-sulfonic acid-grafted graphene oxide quantum dots (Fe_3O_4@GOQD-*O*-(propane-1-sulfonic acid)) catalyzed synthesis of 4*H*-chromene derivatives **141** and dihydropyrano[2,3-*c*]pyrazole **143** was established in water under mild reaction conditions (Fig. **40**) [52]. The reusability of the catalyst for five consecutive cycles without significant loss of activity, use of water as a solvent, and easy work-up are some of the advantages of the methodology.

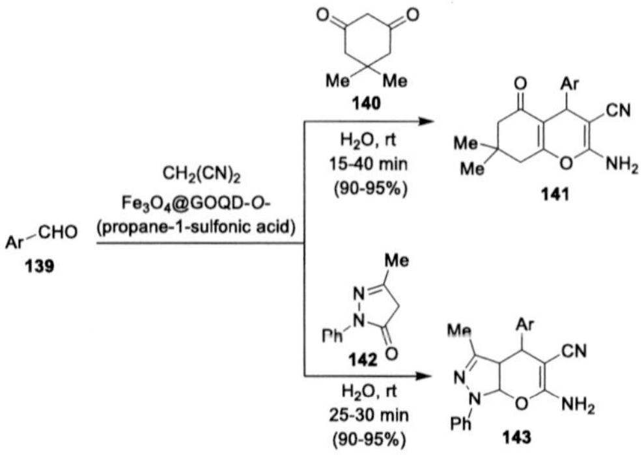

Fig. (40). Synthesis of 4*H*-chromenes and pyrano[2,3-*c*]pyrazoles involving Fe_3O_4@GOQD-*O*-(propane-1-sulfonic acid) nanocatalyst.

Dekamin and co-workers reported the synthesis of pyranopyrazole derivatives **146** involving a four-component reaction between ethyl acetoacetate **144**, aryl aldehydes **145**, malononitrile, and hydrazine using Fe_3O_4@SiO_2@PTS-DABA as a heterogeneous nanocatalyst under mild reaction conditions in shorter reaction times (Fig. **41**) [53]. Almost simultaneously, Mohtasham and Gholizadeh also reported the same four-component reaction in the presence of Fe_3O_4@SiO_2-E-NH-HPA catalyst in high yields [54].

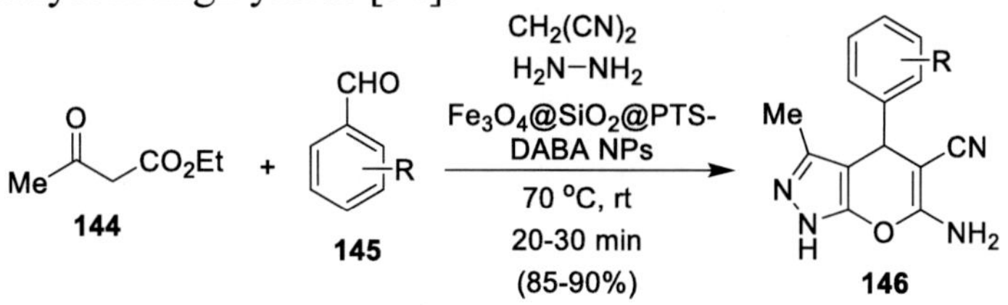

Fig. (41). Fe_3O_4@SiO_2@PTS-DABA-catalyzed four-component synthesis of pyrano[2,3-*c*]pyrazoles.

Tayebee and co-workers established a simple approach for the synthesis of 2-amino-4*H*-chromenes **149** starting from aryl aldehydes **147**, 4-hydroxycoumarin **148**, and malononitrile in the presence of $SiO_2/H_3PW_{12}O_{40}$ nanocatalyst (Fig. **42**) [55]. This heterogeneous catalyst offered several advantages including high yield, low cost, and easy preparation under solvent-free conditions.

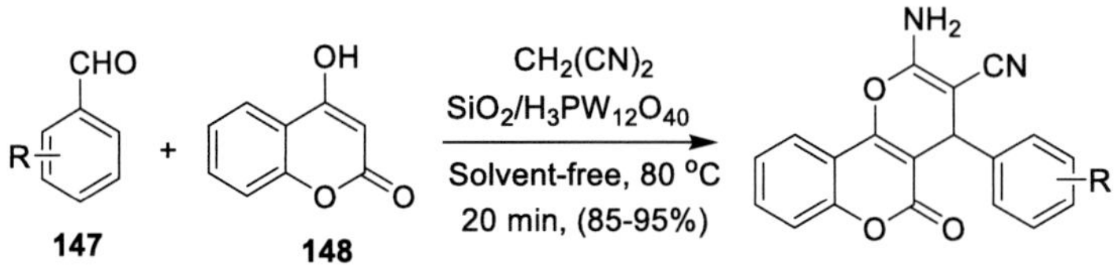

Fig. (42). $SiO_2/H_3PW_{12}O_{40}$-catalyzed synthesis of 2-amino-4*H*-chromenes.

Heydari and co-workers developed a three-component reaction involving aryl aldehydes **150**, barbituric acid **151**, and malononitrile for the synthesis of pyrano[2,3-*d*]pyrimidinone derivatives **152** using formamidine sulfonic acid (FSA) stabilized on silica-coated Fe_3O_4 magnetic nanoparticles as a catalyst in water (Fig. **43**) [56].

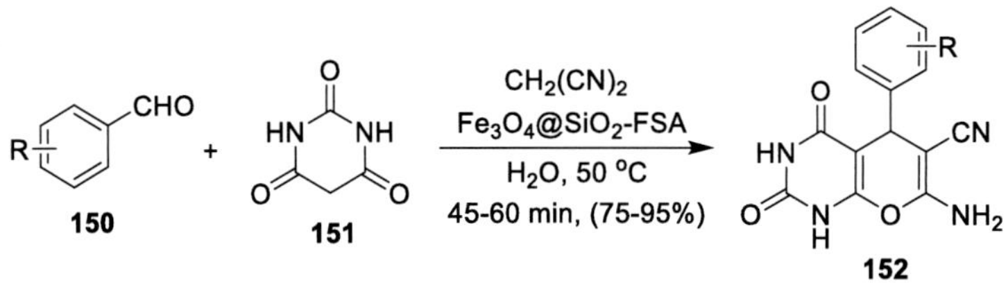

Fig. (43). Nanocatalyzed synthesis of pyrano[2,3-*d*]pyrimidinones.

A simple approach was established for the synthesis of pyrano[2,3-*d*] pyrimidinones **155** and pyrano[2,3-*d*]pyrimidine-2,4,7-trione derivatives **157** using SCMNPs@Urea/Py-$CuCl_2$ as catalyst (Fig. **44**) [57]. The reaction involves the use of aldehydes **153**, barbituric acid **154** and malononitrile for the synthesis of pyrano[2,3-*d*]pyrimidinone derivatives **155**, and the use of meldrum's acid **156** instead of malononitrile allowed access to pyrano[2,3-*d*]pyrimidine-2,4,7-trione derivatives **157**. The catalyst was separated by magnetic separation and reused for six consecutive runs without any significant loss in catalytic activity.

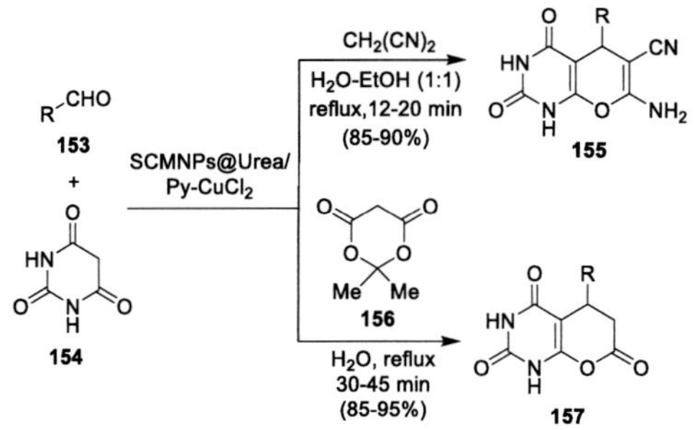

Fig. (44). SCMNPs@Urea/Py-$CuCl_2$-catalysed synthesis of pyrano[2,3-*d*]pyrimidinones and pyrano[2,3-*d*] pyrimidine-2,4,7-triones.

The synthesis of pyrano[3,2-*b*]pyran derivatives **160** was achieved using ferrocene functionalized ionic liquid supported on magnetic nanoparticle (Fe_3O_4@SiO_2-I--Fc) as catalyst (Fig. **45**) [58]. The reaction was performed using aryl aldehydes **158**, kojic acid **159** and malononitrile to access the desired products **160** in 85-95% yields in ethanol under ultrasonic irradiation. This nanocatalyst offered the advantages of shorter reaction times, high yields, and reusability for at least six consecutive cycles.

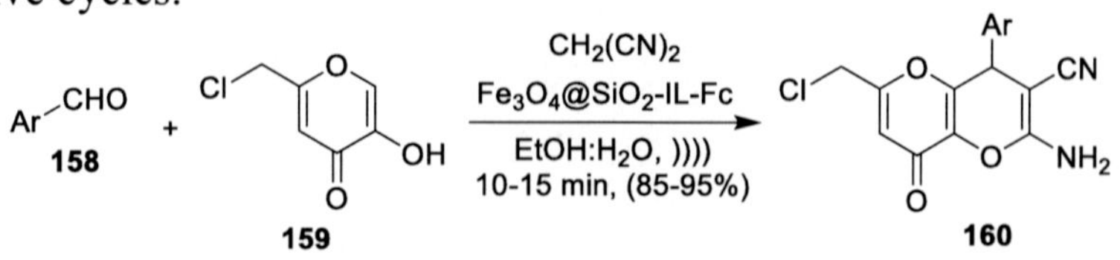

Fig. (45). Fe_3O_4@SiO_2-IL-Fc-catalyzed synthesis of pyrano[3,2-*b*]pyrans.

Multicomponent synthesis of pyrano[2,3-*b*]pyridine-3-carboxamide derivatives **164** was established involving magnetic Fe_3O_4 nanoparticle as a catalyst starting from aryl aldehydes **161**, 6-hydroxy-4-methyl-2-oxo-1,2-dihydropy-idine-3-carbonitrile **162**, and 2-cyanoacetamide **163** under solvent-free conditions (Fig. **46**) [59]. The heterogeneous catalyst has gained considerable importance due to its high specificity, high surface area, non-toxicity and ease of preparation. The catalyst was also reused for up to six catalytic cycles without any loss in catalytic activity.

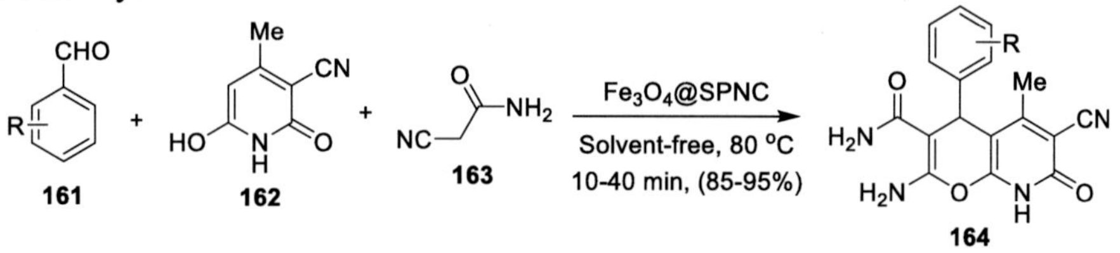

Fig. (46). Fe_3O_4@SPNC-catalyzed synthesis of pyrano[2,3-*b*]pyridine-3-carboxamides.

Mamaghani and co-workers established an efficient green methodology for the synthesis of benzo[*a*]xanthenone derivatives **168** using [Fe_2O_3@HAp]-supported nanocatalyst starting from 2-naphthol **165**, aryl aldehydes **166**, and dimedone **167** (Fig. **47**) [60]. This multi-component reaction involves a Knoevenagel condensation-initiated cascade mechanism, which proceeded *via* an *o*-quinonemethide intermediate.

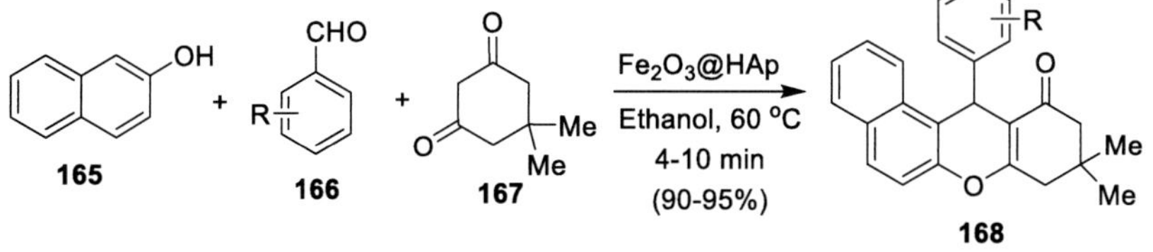

Fig. (47). Fe_2O_3@HAp-catalyzed synthesis of benzoxanthenones.

Azarifar, Keypour and co-workers illustrated a one-pot, three-component reaction between aromatic aldehydes **169**, 1,3-dimethylbarbituric acid **170**, and 3-cyano-6-hydroxy-4-methyl-pyridine-2(1*H*)-one **171** for the synthesis of 2*H*-pyrido [3',2':6,7]pyrano[2,3-*d*]pyrimidine-7-carbonitrile derivatives **172** using GO/Fe_3O_4 @SPNC as nanocatalyst (Fig. **48**) [61]. The reaction showed high yields and lower reaction times for aldehydes bearing electron-withdrawing groups.

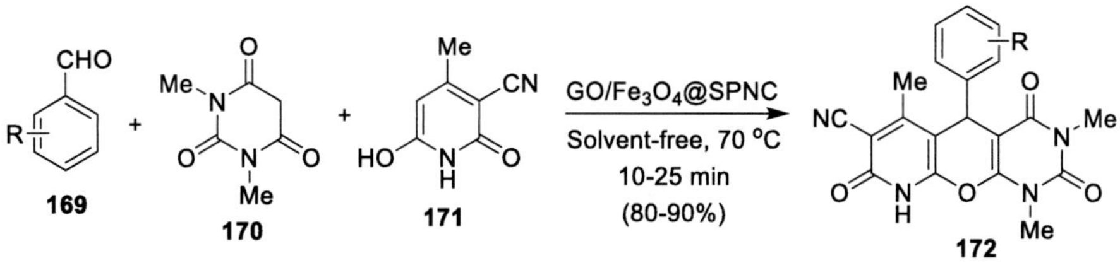

Fig. (48). GO/Fe_3O_4@SPNC-catalyzed synthesis of pyrido[2,3-*d*]pyrimidine-7-carbonitriles.

The Fe_2O_3-based nanomaterial γ-Fe_2O_3@PAMAM-SO_3H was synthesized and used as a nanocatalyst for the synthesis of a set of xanthene derivatives **175**, **177**, **178**, and **180** (Fig. **49**) [62]. This methodology involves a one-pot, the three-component reaction between aromatic aldehydes **173** and β-naphthol **174** or cyclic 1,3-dicarbonyl compounds **176** or β-naphthol **174** and cyclic 1,3-dicarbonyl compounds **176** or 2-hydroxy-1,4-naphthoquinone **179**.

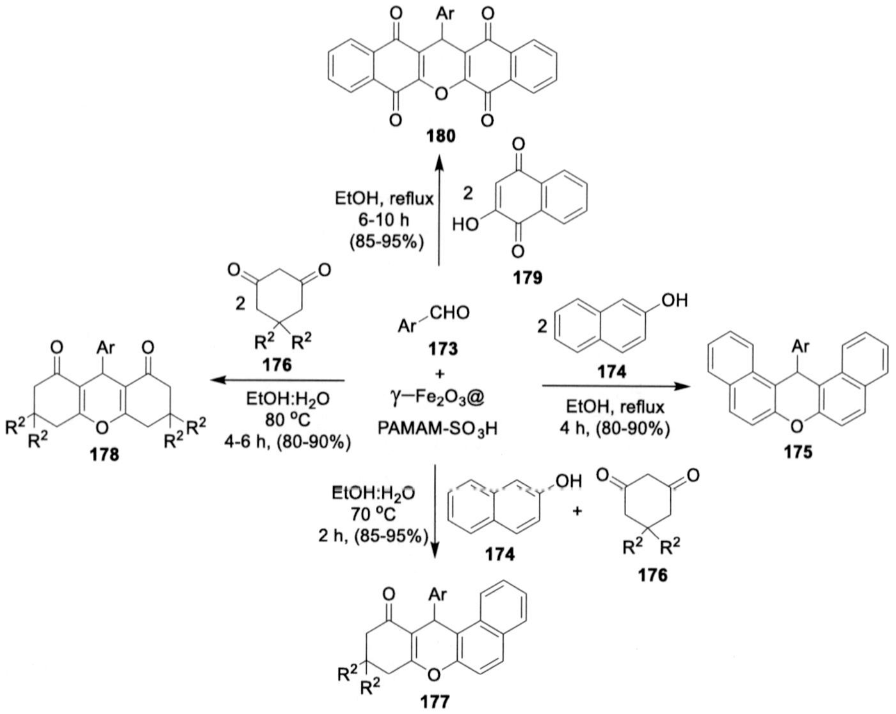

Fig. (49). γ–Fe_2O_3@PAMAM–SO_3H-catalyzed synthesis of xanthene derivatives.

Recently, Moradi and co-workers developed Fe_3O_4@SiO_2/PEtOx as a recyclable heterogeneous catalyst for the synthesis of benzo[*b*]xanthene-triones **184** and tetrahydrochromeno[2,3-*b*] xanthene-tetraones **186** from aryl aldehydes **181** cyclic 1,3-diketones **182**, and quinones **183** or **185** in ethanol under reflux conditions (Fig. **50**) [63]. The mechanism of the reaction involves sequential Knoevenagel condensation, deprotonation, intermolecular cyclization and dehydration steps.

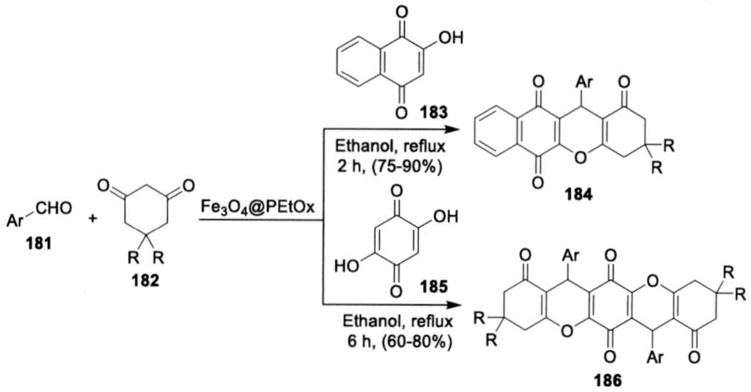

Fig. (50). Synthesis of benzo[*b*]xanthene-triones and tetrahydrochromeno[2,3-*b*]xanthene-tetraones.

The synthesis of xanthene derivatives **189** was achieved starting from aryl aldehydes **187** and β-naphthol **188** in the presence of calix [4]resorcinarene based *N*-propyl sulfamic acid-functionalized nanocatalyst under solvent-free conditions (Fig. **51**) [64]. The advantages of this heterogeneous catalyst include lesser reaction times, high yields, reusability without any loss of catalytic activity and easy workup. Previously, Rezayati and co-workers also employed γ-Fe_2O_3-HAp-Fe^{2+} nanocatalyst for the same reaction under solvent-free conditions to access compounds **189** in 60-85% yields [65].

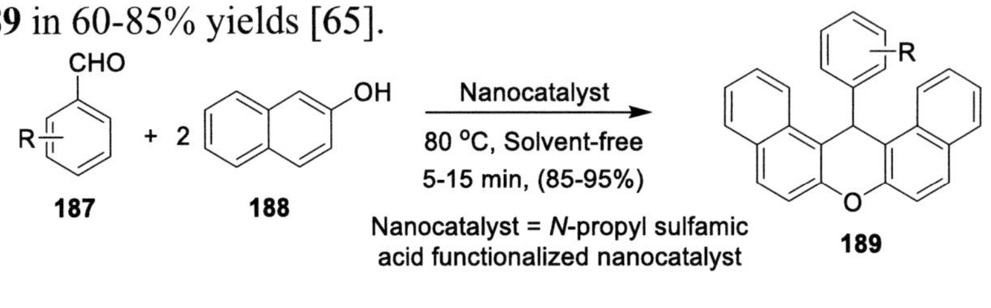

Fig. (51). Nanocatalyzed-synthesis of 14-aryl-14*H*-dibenzo[*a,j*]xanthenes.

The three-component synthesis of xanthene derivatives **193** was established starting from 2-naphthol **190**, aryl aldehydes **191**, and dimedone **192** using silica-bonded imidazolium-sulfonic acid chloride (SBISAC) as a catalyst (Fig. **52**) [66]. The catalyst was found to be effective to deliver the desired products in high yields in shorter reaction times under solvent-free conditions.

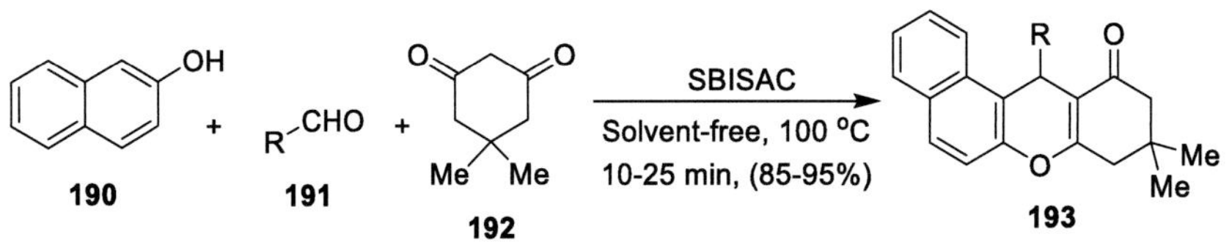

Fig. (52). SBISAC-catalyzed synthesis of xanthenes.

CONCLUDING REMARKS

Heterocyclic compounds, in particular, nitrogen and oxygen heterocycles are widespread and possess a wide range of biological activities. Consequently, enormous effort has been devoted to the development of synthetic approaches to access these heterocyclic compounds. In recent years, a large number of heterogeneous nanocatalysts were developed and employed for the synthesis of heterocyclic compounds owing to the advantages of these nanocatalysts including better reactivity, selectivity, and reusability. In this chapter, we have summarized the important heterogeneous nanocatalyzed reactions for the synthesis of various nitrogen- and oxygen-containing heterocycles. We believe that this chapter confers a clear insight to the readers about the catalytic activity of various nanomaterials for the direct synthesis of an important class of heterocycles.

CONSENT FOR PUBLICATION

Not Applicable.

CONFLICT OF INTEREST

The author declares no conflict of interest, financial or otherwise.

ACKNOWLEDGEMENTS

Financial support from the Department of Science and Technology-Science and Engineering Research Board, DST-SERB (No. CRG/2021/000760) is gratefully acknowledged.

REFERENCES

[1] Katritzky, A.R.; Ramsden, C.A.; Scriven, E.F.V.; Taylor, R.J.K., Eds. Elsevier: Oxford, U.K., **2008**. Katritzky, A.R.; Scriven, E.F.V.; Rees, C.W., Eds. Elsevier: Oxford, U.K., **1996**.

[2] aLiu, J.; Jiang, J.; Zheng, L.; Liu, Z-Q. Recent advances in the synthesis of nitrogen heterocycles using arenediazonium salts as nitrogen sources. *Adv. Synth. Catal.,* **2020**, *362*(22), 4876-4895. [http://dx.doi.org/10.1002/adsc.202000700] bVachan, B.S.; Karuppasamy, M.; Vinoth, P.; Kumar, S.V.; Perumal, S.; Sridharan, V.; Menéndez, J.C. Proline and its derivatives as organocatalysts for multicomponent reactions in aqueous media: synergic pathways to the green synthesis of heterocycles. *Adv. Synth. Catal.,* **2020**, *362*(1), 87-110. [http://dx.doi.org/10.1002/adsc.201900558] cMuthukrishnan, I.; Sridharan, V.; Menéndez, J.C. Progress in the chemistry of tetrahydroquinolines. *Chem. Rev.,* **2019**, *119*(8), 5057-5191. [http://dx.doi.org/10.1021/acs.chemrev.8b00567] [PMID: 30963764] dZhang, B.; Studer, A. Recent advances in the synthesis of nitrogen heterocycles *via* radical cascade reactions using isonitriles as radical acceptors. *Chem. Soc. Rev.,* **2015**, *44*(11), 3505-3521. [http://dx.doi.org/10.1039/C5CS00083A] [PMID: 25882084] eSridharan, V.; Suryavanshi, P.A.; Menéndez, J.C. Advances in the chemistry of tetrahydroquinolines. *Chem. Rev.,* **2011**, *111*(11), 7157-7259. [http://dx.doi.org/10.1021/cr100307m] [PMID: 21830756] fCandeias, N.R.; Branco, L.C.; Gois, P.M.P.; Afonso, C.A.M.; Trindade, A.F. More sustainable approaches for the synthesis of *N*-based heterocycles. *Chem. Rev.,* **2009**, *109*(6), 2703-2802. [http://dx.doi.org/10.1021/cr800462w] [PMID: 19385653]

[3] aAhmad, T.; Ullah, N. The oxa-Michael reaction in the synthesis of 5- and 6-membered oxygen-containing heterocycles. *Org. Chem. Front.,* **2021**, *8*(6), 1329-1344. [http://dx.doi.org/10.1039/D0QO01312A] bKaur, N.; Kishore, D. Solid-phase synthetic approach toward the synthesis of oxygen-containing heterocycles. *Synth. Commun.,* **2014**, *44*(8), 1019-1042. [http://dx.doi.org/10.1080/00397911.2012.760131] cDeiters, A.; Martin, S.F. Synthesis of oxygen- and nitrogen-containing heterocycles by ring-closing metathesis. *Chem. Rev.,* **2004**, *104*(5), 2199-2238. [http://dx.doi.org/10.1021/cr0200872] [PMID: 15137789]

[4] Sharma, P.C.; Jain, A.; Jain, S.; Pahwa, R.; Yar, M.S. Ciprofloxacin: review on developments in synthetic, analytical, and medicinal aspects. *J. Enzyme Inhib. Med. Chem.,* **2010**, *25*(4), 577-589. [http://dx.doi.org/10.3109/14756360903373350] [PMID: 20235755]

[5] de Moura Oliveria, C.H.; de Melo, C.C.; Doriguetto, A.C. Sulfamethoxazole salts: crystal structures, conformations and solubility. *New J. Chem.,* **2019**, *43*, 10250-10258. [http://dx.doi.org/10.1039/C9NJ00586B]

[6] Markovic, M.; Gallipani, A.; Patel, K.H.; Maroney, M. Brexipiprazole: a new treatment option for

schizophrenia and major depressive disorder. *Ann. Pharmacother.*, **2017**, *51*, 315-322. [http://dx.doi.org/10.1177/1060028016678262] [PMID: 28228056]

[7] Sun, J.M.; Lee, S.H.; Ahn, J.S.; Park, K.; Ahn, M.J. Osimertinib for the treatment of non-small cell lung cancer. *Expert Opin. Pharmacother.*, **2017**, *18*(2), 225-231. [http://dx.doi.org/10.1080/14656566.2017.1285283] [PMID: 28116908]

[8] Gheorghiade, M.; Adams, K.F., Jr; Colucci, W.S. Digoxin in the management of cardiovascular disorders. *Circulation,* **2004**, *109*(24), 2959-2964. [http://dx.doi.org/10.1161/01.CIR.0000132482.95686.87] [PMID: 15210613]

[9] aHemalatha, K.; Madhumita, G.; Kajbafvala, A.; Anupama, N.; Sompalle, R.; Roopan, S.M. Function of nanocatalyst in chemistry of organic compounds revolution: an overview. *J. Nanomater.*, **2013**, *2013*: 341015. [http://dx.doi.org/10.1155/2013/341015] bMadan, Y.A. Mini review study on the generation of bioactive heterocyclic motifs by using sustainable metal nanocatalysed organic synthetic routes. *Mater. Today Proc.*, **2020**, *30*, 197-202. [http://dx.doi.org/10.1016/j.matpr.2020.05.742]

[10] Verma, S.; Kujur, S.; Agrahari, B.; Layek, S.; Pathak, D.D. Synthesis and characterization of cucurbit[6]uril supported copper oxide nanoparticles, CuO@CB[6]: application as nanocatalyst for the synthesis of 2*H*-indazoles. *ChemistrySelect,* **2019**, *4*(35), 10408-10416. [http://dx.doi.org/10.1002/slct.201902957]

[11] Soltani Rad, M.N. Ultrasound promoted mild and facile one-pot, three component synthesis of 2*H*-indazoles by consecutive condensation, CN and NN bond formations catalysed by copper-doped silica cuprous sulphate (CDSCS) as an efficient heterogeneous nano-catalyst. *Ultrason. Sonochem.*, **2017**, *34*, 865-872. [http://dx.doi.org/10.1016/j.ultsonch.2016.07.026] [PMID: 27773314]

[12] Sharghi, H.; Aberi, M.; Doroodmand, M.M. A mild, three-component one-pot synthesis of 2,4,5-trisubstituted imidazoles using Mo(IV) salen complex in homogeneous catalytic system and Mo(IV) salen complex nanoparticles onto silica as a highly active, efficient, and reusable heterogeneous nanocatalyst. *Mol. Divers.*, **2015**, *19*(1), 77-85. [http://dx.doi.org/10.1007/s11030-014-9558-y] [PMID: 25515148]

[13] Hajizadeh, Z.; Radinekiyan, F.; Eivazzadeh-Keihan, R.; Maleki, A. Development of novel and green $NiFe_2O_4$/geopolymer nanocatalyst based on bentonite for synthesis of imidazole heterocycles by ultrasonic irradiations. *Sci. Rep.*, **2020**, *10*(1), 11671-11681. [http://dx.doi.org/10.1038/s41598-020-68426-z] [PMID: 32669578]

[14] Kohan, E.; Gholamhosseini-Nazari, M.; Allahvirdinesbat, M.; Alemi, A.A. Green and efficiently synthesized tetrasubstituted imidazole: introduced bismuth oxide co-doped Lu^{3+}, Er^{3+} as a novel reusable heterogeneous nanocatalyst. Inorg. Nano-Met. *Chem,* **2020**. [http://dx.doi.org/10.1080/24701556.2020.1814327]

[15] Zarnegar, Z.; Safari, J. Fe_3O_4@chitosan nanoparticles: a valuable heterogeneous nanocatalyst for the synthesis of 2,4,5-trisubstituted imidazoles. *RSC Advances,* **2014**, *4*(40), 20932-20939. [http://dx.doi.org/10.1039/C4RA03176H]

[16] Bardajee, G.R.; Mohammadi, M.; Kakavand, N. Copper(II)-diaminosarcophagine functionalized SBA-15: a heterogeneous nanocatalyst for the synthesis of benzimidazole, benzoxazole and benzothiazole derivatives under solvent-free conditions. *Appl. Organomet. Chem.*, **2016**, *30*(1), 51-58. [http://dx.doi.org/10.1002/aoc.3400]

[17] Amini, M.; Hajipour, E.; Akbari, A.; Chae, K.H. Immobilization of copper nanoparticles on WO_3 with enhanced catalytic activity for the synthesis of 1,2,3-triazoles. *Appl. Organomet. Chem.*, **2020**, *3*: e5959.

[18] Mahdavinasab, M.; Hamzehloueian, M.; Sarrafi, Y. Preparation and application of magnetic chitosan/graphene oxide composite supported copper as a recyclable heterogeneous nanocatalyst in the

synthesis of triazoles. *Int. J. Biol. Macromol.*, **2019**, *138*, 764-772.
[http://dx.doi.org/10.1016/j.ijbiomac.2019.07.013] [PMID: 31284011]

[19] Pourjavadi, A.; Tajbakhsh, M.; Farhangb, M.; Hosseinia, S.H. Copper-loaded polymeric magnetic nanocatalysts as retrievable and robust heterogeneous catalysts for click reactions. *New J. Chem.*, **2015**, *39*(6), 4591-4600.
[http://dx.doi.org/10.1039/C4NJ02134G]

[20] Bahrami, K.; Arabia, M.S. Copper immobilized ferromagnetic nanoparticle triazine dendrimer (FMNP@TD-Cu(II))-catalyzed regioselective synthesis of 1,4-disubstituted 1,2,3-triazoles. *New J. Chem.*, **2016**, *40*(4), 3447-3455.
[http://dx.doi.org/10.1039/C5NJ03219A]

[21] Nikoorazm, M.; Moradi, P.; Noori, N. L-Cysteine complex of palladium onto mesoporous channels of MCM-41 as reusable, homoselective and organic-inorganic hybrid nanocatalyst for the synthesis of tetrazoles. *J. Porous Mater.*, **2020**, *27*(4), 1159-1169.
[http://dx.doi.org/10.1007/s10934-020-00894-0]

[22] Habibi, D.; Faraji, R.; Sheikh, D.; Sheikhid, M.; Abedi, S. Application of supported Mn(III), Fe(III) and Co(II) as heterogeneous, selective and highly reusable nano catalysts for synthesis of arylaminotetrazoles, and DFT studies of the products. *RSC Advances*, **2014**, *4*(88), 47625-47636.
[http://dx.doi.org/10.1039/C4RA06463A]

[23] Khan, M.U.; Siddiqui, Z.N. U.; Siddiqui, Z. N. Ce@STANPs/ZrO_2 as nanocatalyst for multicomponent synthesis of isatin-derived imidazoles under green reaction conditions. *ACS Omega*, **2018**, *3*(8), 10357-10364.
[http://dx.doi.org/10.1021/acsomega.8b01043] [PMID: 31459163]

[24] Maleki, A. One-pot three-component synthesis of pyrido[20,10:2,3]imidazo[4,5-*c*]isoquinolines using Fe_3O_4@SiO_2-OSO_3H as an efficient heterogeneous nanocatalyst. *RSC Advances*, **2014**, *4*, 64169-64173.
[http://dx.doi.org/10.1039/C4RA10856F]

[25] Amiri, A.A.; Javanshir, S.; Dolatkhah, Z.; Dekamin, M.G. SO_3H-functionalized mesoporous silica materials as solid acid catalyst for facile and solvent-free synthesis of 2*H*-indazolo[2,1-*b*]phthalazine-1,6,11-trione derivatives. *New J. Chem.*, **2015**, *39*(12), 9665-9671.
[http://dx.doi.org/10.1039/C5NJ01733E]

[26] Allahresani, A.; Sangani, M.M.; Nasseri, M.A. $CoFe_2O_4$@SiO_2-NH_2-CoII NPs catalyzed Hantzsch reaction as an efficient, reusable catalyst for the facile, green, one-pot synthesis of novel functionalized 1,4-dihydropyridine derivatives. *Appl. Organomet. Chem.*, **2020**, *34*(9): e5759.
[http://dx.doi.org/10.1002/aoc.5759]

[27] Chaurasia, S.R.; Tiwari, A.R.; Bhanage, B.M. Synthesis of quinolines *via* acceptorless dehydrogenative tandem cyclization of 2-amionbenzyl alcohol with alcohols using magnetic CuNiFeO nanocatalyst. *Mol. Catal.*, **2018**, *478*, 110565-110571.
[http://dx.doi.org/10.1016/j.mcat.2019.110565]

[28] Sajjadi-Ghotbabadi, H.; Javanshir, S.; Rostami-Charati, F. Nano KF/clinoptilolite: an effective heterogeneous base nanocatalyst for synthesis of substituted quinolines in water. *Catal. Lett.*, **2016**, *146*(2), 338-344.
[http://dx.doi.org/10.1007/s10562-015-1652-y]

[29] Tayebee, R.; Abdizadeh, M.F.; Erfaninia, N.; Amiri, A.; Baghayeri, M.; Kakhki, R.M.; Maleki, B.; Esmaili, E. Phosphotungstic acid grafted zeolite imidazolate framework as an effective heterogeneous nanocatalyst for the one-pot solvent-free synthesis of 3,4-dihydropyrimidinones. *Appl. Organomet. Chem.*, **2019**, *33*(8): e4959.
[http://dx.doi.org/10.1002/aoc.4959]

[30] Mohammadi, H.; Shaterian, H.R. Preparation and characterization of a silica-based magnetic nanocomposite and its application as a recoverable catalyst for the one-pot multicomponent synthesis

of quinazolinone derivatives. *Appl. Organomet. Chem.*, **2015**, *29*(12), 809-814.
[http://dx.doi.org/10.1002/aoc.3373]

[31] Havasia, F.; Ghorbani-Choghamarani, A.; Nikpour, F. Synthesis and characterization of nickel complex anchored onto MCM-41 as a novel and reusable nanocatalyst for the efficient synthesis of 2,3-dihydroquinazolin-4(1*H*)-ones. *Microporous Mesoporous Mater.*, **2016**, *224*, 26-35.
[http://dx.doi.org/10.1016/j.micromeso.2015.10.036]

[32] Raut, A.B.; Tiwari, A.R.; Bhanage, B.M. Ultrasonic irradiation assisted preparation of Cu_2O-nanocubes and their high catalytic activity in synthesis of quinazolines. *ChemCatChem*, **2017**, *9*, 1292-1297.
[http://dx.doi.org/10.1002/cctc.201601330]

[33] Dandia, A.; Jain, A.K.; Sharma, S. $CuFe_2O_4$ Nanoparticles as a highly efficient and magnetically recoverable catalyst for the synthesis of medicinally privileged spiropyrimidine scaffolds. *RSC Advances*, **2013**, *3*(9), 2924-2934.
[http://dx.doi.org/10.1039/c2ra22477a]

[34] Ziarati, A.; Safaei-Ghomi, J.; Rohani, S. A One-pot multicomponent synthesis of *N*-cyclohexyl-3-arylquinoxaline-2-amines using ZnO nanoparticles as a heterogeneous reusable catalyst. *Lett. Org. Chem.*, **2013**, *10*, 47-52.

[35] Dehbalaei, M.G.; Foroughifar, N.; Pasdara, H.; Khajeh-Amirib, A. *N*-propyl benzoguanamine sulfonic acid supported on magnetic Fe_3O_4 nanoparticles: a novel and efficient magnetically heterogeneous catalyst for the synthesis of 1,8-dioxo-decahydroacridine derivatives. *New J. Chem.*, **2018**, *42*(1), 327-335.
[http://dx.doi.org/10.1039/C7NJ03508J]

[36] Saeidiroshan, H.; Moradi, L. Efficient and green synthesis of dihydropyrimido[4,5-*b*] guinolinetriones using MWCNTs@TEPA/Co(II) as a novel and eco-friendly catalyst. *Appl. Organomet. Chem.*, **2020**, *34*(9): e5732.
[http://dx.doi.org/10.1002/aoc.5732]

[37] Naeimi, H.; Rashid, Z.; Zarnani, A-H.; Ghahremanzadeh, R. $MnFe_2O_4$@NH_2@2AB-Ni: a novel, highly active, stable and magnetically recoverable nanocatalyst and use of this heterogeneous catalyst in green synthesis of spirooxindoles in water. *New J. Chem.*, **2014**, *38*(11), 5527-5535.
[http://dx.doi.org/10.1039/C4NJ01182A]

[38] Debnath, K.; Pramanik, A. Heterogeneous bimetallic $ZnFe_2O_4$ nanopowder catalysed facile four-component reaction for the synthesis of spiro[indoline-3,2′-quinoline] derivatives from isatins in water medium. *Tetrahedron Lett.*, **2015**, *56*(13), 1654-1660.
[http://dx.doi.org/10.1016/j.tetlet.2015.02.030]

[39] Veisi, H.; Hekmati, M.; Hemmatic, S. Mesoporous SBA-15 silica phenylsulfonic acid (SBA-15-Ph-SO_3H) as efficient nanocatalyst for one-pot three-component synthesis of 3-methyl-4-aryl-2,4,5,7-tetrahydropyrazolo[3,4-*b*]pyridine-6-ones. *J. Heterocycl. Chem.*, **2017**, *54*(2), 1630-1635.
[http://dx.doi.org/10.1002/jhet.2689]

[40] Kaiba, A.; Ouerghi, O.; Geesi, M.H.; Elsanousi, A.; Belkacem, A.; Dehbi, O.; Alharthi, A.I.; Alotaibi, M.A.; Riadi, Y. Characterization and catalytic performance of Ni-doped TiO_2 as a potential heterogeneous nanocatalyst for the preparation of substituted pyridopyrimidines. *J. Mol. Struct.*, **2020**, *1203*, 127376-127384.
[http://dx.doi.org/10.1016/j.molstruc.2019.127376]

[41] Malakooti, R.; Bardajee, G.R.; Hadizadeh, S.; Atashin, H.; Khanjari, H. An iron Schiff base complex loaded mesoporous silica nanoreactor as a catalyst for the synthesis of pyrazine-based heterocycles. *Trans. Met. Chem. (Weinh.)*, **2014**, *39*(1), 47-54.
[http://dx.doi.org/10.1007/s11243-013-9772-y]

[42] Mohammadi, H.; Shaterian, H.R. (3-Oxo-[1,2,4]triazolidin-1-yl)bis (butane-1-sulfonic acid)

functionalized magnetic γ-Fe_2O_3 nanoparticles: a novel and heterogeneous nanocatalyst for one-pot and efficient four-component synthesis of novel spiro[indeno[1,2-*b*]quinoxaline derivatives. *Appl. Organomet. Chem.*, **2019**, *33*(6): e4901.
[http://dx.doi.org/10.1002/aoc.4901]

[43] Woo, H.; Kim, D.; Park, J.C.; Kim, J-W.; Park, S.; Lee, J.M.; Park, K.H. New hybrid nanocatalyst based on Cu-doped Pd-Fe_3O_4 for tandem synthesis of 2-phenylbenzofurans. *J. Mater. Chem. A Mater. Energy Sustain.*, **2015**, *3*(42), 20992-20998.
[http://dx.doi.org/10.1039/C5TA05111H]

[44] Dehnavi, H.A.; Ghashang, M. Solvent-free preparation of 3-aryl-2-[(aryl)(arylamino)]methyl-4*H*-furo[3,2-*c*]chromen-4-one derivatives using ZnO-$ZnAl_2O_4$ nanocomposite as a heterogeneous catalyst. *Heterocycl. Commun*, **2018**, *24*(1), 19-22.
[http://dx.doi.org/10.1515/hc-2017-0141]

[45] Liu, X.B.; Hu, Y.L. Novel and highly efficient conversion of carbon dioxide to cyclic carbonates using benzotriazolium ionic liquid-modified periodic mesoporous organosilica as a heterogeneous and recyclable nanocatalyst. *Silicon*, **2020**, *13*(4), 1-12.
[http://dx.doi.org/10.1007/s12633-020-00693-x]

[46] Maleki, H.; Rakhtshah, J.; Shaabani, B. Effective one-pot synthesis of tetrahydrobenzo[*b*]pyranderivatives using nickel Schiff-base complex immobilized on iron oxide nanoparticles. *Appl. Organomet. Chem.*, **2020**, *34*(8): e5683.
[http://dx.doi.org/10.1002/aoc.5683]

[47] Rajput, J.K.; Arora, P.; Kaur, G.; Kaur, M. $CuFe_2O_4$ magnetic heterogeneous nanocatalyst: Low power sonochemical-coprecipitation preparation and applications in synthesis of 4*H*-chromene-3-carbonitrile scaffolds. *Ultrason. Sonochem.*, **2015**, *26*, 229-240.
[http://dx.doi.org/10.1016/j.ultsonch.2015.01.008] [PMID: 25649833]

[48] Shamsi, T.; Amoozadeh, A.; Sajjadi, S.M.; Tabrizian, E. Novel type of SO_3H-functionalized nano-titanium dioxide as a highly efficient and recyclable heterogeneous nanocatalyst for the synthesis of tetrahydrobenzo[*b*]pyrans. *Appl. Organomet. Chem.*, **2017**, *31*(7): e3636.
[http://dx.doi.org/10.1002/aoc.3636]

[49] Kamalzare, M.; Bayat, M.; Maleki, A. Green and efficient three-component synthesis of 4H-pyran catalysed by $CuFe_2O_4$@starch as a magnetically recyclable bionanocatalyst. *R. Soc. Open Sci.*, **2020**, *7*(7), 200385-200397.
[http://dx.doi.org/10.1098/rsos.200385] [PMID: 32874634]

[50] Rostamizadeh, S.; Zekri, N. An eficient, one-pot synthesis of 2-amino-4*H*-chromenes catalyzed by (α-Fe_2O_3)-MCM-41-supported dual acidic ionic liquid as a novel and recyclable magnetic nanocatalyst. *Res. Chem. Intermed.*, **2016**, *42*(3), 2329-2341.
[http://dx.doi.org/10.1007/s11164-015-2152-9]

[51] Niya, H.F.; Hazeri, N.; Fatahpour, M.; Maghsoodlou, M.T. Fe_3O_4@THAM-piperazine: a novel and highly reusable nanocatalyst for one-pot synthesis of 1,8-dioxo-octahydro-xanthenes and benzopyrans. *Res. Chem. Intermed.*, **2020**, *46*(7), 3651-3666.
[http://dx.doi.org/10.1007/s11164-020-04166-z]

[52] Abbasabadi, M.K.; Azarifar, D.; Zand, H.R.E. Sulfonic acid-functionalized Fe_3O_4-supported magnetized graphene oxide quantum dots: a novel organic-inorganic nanocomposite as an efficient and recyclable nanocatalyst for the synthesis of dihydropyrano[2,3-*c*]pyrazole and 4*H*-chromene derivatives. *Appl. Organomet. Chem.*, **2020**, *34*: e6004.

[53] Karami, S.; Dekamin, M.G.; Valiey, E.; Shakib, P. DABA MNPs: a new and efficient magnetic bifunctional nanocatalyst for green synthesis of biologically active pyrano[2,3-*c*]pyrazole and benzylpyrazolyl coumarin derivatives. *New J. Chem.*, **2020**, *44*(33), 13952-13961.
[http://dx.doi.org/10.1039/D0NJ02666B]

[54] Mohtasham, N.H.; Gholizadeh, M. Nano silica extracted from horsetail plant as a natural silica support

for the synthesis of $H_3PW_{12}O_{40}$ immobilized on aminated magnetic nanoparticles (Fe_3O_4@SiO_2-E--NH-HPA): a novel and efficient heterogeneous nanocatalyst for the green one-pot synthesis of pyrano[2,3-*c*]pyrazole derivatives. *Res. Chem. Intermed.,* **2020**, *46*(6), 3037-3066. [http://dx.doi.org/10.1007/s11164-020-04133-8]

[55] Tayebee, R.; Pejhan, A.; Ramshini, H.; Maleki, B.; Erfaninia, N.; Tabatabaie, Z.; Esmaeili, E. Equisetum arvense as an abundant source of silica nanoparticles. $SiO_2/H_3PW_{12}O_{40}$ nanohybrid material as an efficient and environmental benign catalyst in the synthesis of 2-amino-4*H*-chromenes under solvent-free conditions. *Appl. Organomet. Chem.,* **2018**, *32*(1): e3924. [http://dx.doi.org/10.1002/aoc.3924]

[56] Ghandi, L.; Miraki, M.K.; Radfar, I.; Yazdani, E.; Heydari, A. Formamidinesulfinic acid-functionalized Fe_3O_4@SiO_2 as a green and magnetic recyclable catalyst for synthesis of pyrano[2, 3-*d*]pyrimidinone derivatives. *ChemistrySelect,* **2018**, *3*(6), 1787-1792. [http://dx.doi.org/10.1002/slct.201702887]

[57] Zhang, J.; Song, H.; Cui, R.; Deng, C.; Yousif, Q.A. SCMNPs@Urea/Py-$CuCl_2$: a recyclable catalyst for the synthesis of pyrano[2,3-*d*]pyrimidinone and pyrano[2,3-*d*]pyrimidine-2,4,7-trione derivatives. *J. Coord. Chem.,* **2020**, *73*(4), 558-578. [http://dx.doi.org/10.1080/00958972.2020.1737681]

[58] Teimuri-Mofrad, R.; Esmati, S.; Rabiei, M.; Gholamhosseini-Nazari, M. Efficient synthesis of new pyrano[3,2-*b*]pyran derivatives *via* Fe_3O_4@SiO_2-IL-Fc catalyzed three component reaction. *Heterocycl. Commun.,* **2017**, *23*(6), 439-444. [http://dx.doi.org/10.1515/hc-2017-0140]

[59] Mahmoudi-GomYek, S.; Azarifar, D.; Ghaemi, M.; Keypour, H.; Mahmoudabadi, M. M.; Azarifar, D.; Ghaemi, M.; Keypour, H.; Mahmoudabadi, M. Fe_3O_4-Supported Schiff-base copper-(II) complex: a valuable heterogeneous nanocatalyst for one-pot synthesis of new pyrano[2,3-*b*]pyridine-3-arboxamide derivatives. *Appl. Organomet. Chem.,* **2019**, *33*(6): e4918. [http://dx.doi.org/10.1002/aoc.4918]

[60] Kheirkhah, L.; Mamaghani, M.; Yahyazadeh, A.; Mahmoodi, N.O. HAp-encapsulated γ-Fe_2O_3-supported dual acidic heterogeneous catalyst for highly efficient one-pot synthesis of benzoxanthenones and 3-pyranylindoles. *Appl. Organomet. Chem.,* **2018**, *32*(2): e4072. [http://dx.doi.org/10.1002/aoc.4072]

[61] Yek, S.M-G.; Azarifar, D.; Khaleghi-Abbasabadi, M.; Keypour, H.; Mahmoudabadi, M. Heterogenized magnetic graphene oxide-supported N_6-Schiff base Cu(II) complex as an exclusive nanocatalyst for synthesis of new pyrido[2,3-*d*]pyrimidine-7-carbonitrile derivatives. *Appl. Organomet. Chem.,* **2020**, *34*: e5989.

[62] Sheikh, S.; Nasseri, M.A.; Chahkandi, M.; Allahresani, A.; Reiser, O. Functionalized magnetic PAMAM dendrimer as an efficient nanocatalyst for a new synthetic strategy of xanthene pigments. *J. Hazard. Mater.,* **2020**, *400*, 122985-123024. [http://dx.doi.org/10.1016/j.jhazmat.2020.122985] [PMID: 32580092]

[63] Rahnamafar, R.; Moradi, L.; Mehdi Khoobi, M. Synthesis of benzo[*b*]xanthene-triones and tetrahydrochromeno[2,3-*b*]xanthene tetraones *via* three- or pseudo-five-component reactions using Fe_3O_4@SiO_2/PEtOx as a novel, magnetically recyclable, and eco-friendly nanocatalyst. *J. Heterocycl. Chem.,* **2020**, *57*(4), 1825-1837. [http://dx.doi.org/10.1002/jhet.3911]

[64] Mouradzadegun, A.; Mostafavi, M.A.; Ganjali, M.R. Sulfamic acid functionalized 3D-network nanoporous polymer based on calix[4]resorcinarene: a recyclable heterogeneous nanocatalyst for the efficient synthesis of 14-aryl-14*H*-dibenzo[*a,j*]xanthenes under thermal neat conditions. *J. Incl. Phenom. Macrocycl. Chem.,* **2018**, *91*(1-2), 25-36. [http://dx.doi.org/10.1007/s10847-018-0792-2]

[65] Haeri, H.S.; Rezayati, S.; Nezhad, E.R.; Darvishi, H. Fe^{2+} supported on hydroxyapatite-core–shell-c-Fe_2O_3 nanoparticles: efficient and recyclable green catalyst for the

synthesis of 14-aryl-14*H*-dibenzo[*a,j*]xanthenes derivatives. *Res. Chem. Intermed.,* **2016**, *42*(5), 4773-4784.
[http://dx.doi.org/10.1007/s11164-015-2318-5]

[66] Moosavi-Zare, A.R.; Zolfigol, M.A.; Zarei, M.; Zare, A.; Khakyzadeh, V. Application of silica-bonded imidazolium-sulfonic acid chloride (SBISAC) as a heterogeneous nanocatalyst for the domino condensation of aryl aldehydes with 2-naphthol and dimedone. *J. Mol. Liq.,* **2015**, *211*, 373-380.
[http://dx.doi.org/10.1016/j.molliq.2015.07.049]

CHAPTER 11

Role of Nanocatalyst in Electrochemical Sensing of Heavy Metal Ions

H.C. Ananda Murthy[1,*], **Gezahegn Tadesse Ayanie**[1], **Tegene Desalegn**[1], **Suresh Ghotekar**[2] and **Dhanalakshmi Muniswamy**[3]

[1] *Department of Applied Chemistry, School of Applied Natural Science, Adama Science and Technology University, Adama, P.O. Box.1888, Ethiopia*

[2] *Department of Chemistry, Smt. Devkiba Mohansinhji Chauhan College of Commerce and Science College, Silvassa- 396 230, Dadra and Nagar Haveli (UT), India*

[3] *Department of Physics, Government Science College (Nrupathunga University), Bengaluru 560 001, India*

Abstract: This book chapter introduces the based nanocatalysts-based electrochemical sensors for the detection of heavy metal ions (HMIs) from their different sources. A serious challenge in the field of detection of HMIs is the innovation of simple and fast electrochemical sensor techniques with the potential ability to detect at low concentrations. The electrochemical sensors for the assay of heavy metal ions were designed and developed by many researchers in the past. To achieve repeatability, reproducibility, and stability, the fabrication of electrodes for electrochemical sensing applications has been given much emphasis in recent days. The selectivity of the modified electrode towards the detection of transition metal ions (Cd^{2+}, Zn^{2+}, Co^{2+}, Fe^{2+}, Mn^{2+}, Ni^{2+}, and Ag^{+}) was investigated by many researchers and was discussed in this chapter. In addition, the applications of 3 dimensional (3-D) nanomaterials and functionalized nanomaterials of metal and metal oxides supported on graphene, graphene oxide, and carbon nanotubes as nanocatalysts have been presented in this chapter. The modified and functionalized electrodes are anticipated to play a decent role in the latest electrochemical sensing of HMIs to demonstrate quantum confinement effects.

Keywords: Electrochemical sensor, Electrodes, Functionalization, Heavy metal ions, Nanocatalyst.

INTRODUCTION

Nowdays, with the advancement in the industrial, mining, and agricultural activities, a large amount of heavy metal ions were released into the environment,

* **Corresponding Author H.C. Ananda Murthy:** Department of Applied Chemistry, School of Applied Natural Science, Adama Science and Technology University, P O Box 1888, Adama, Ethiopia; Tel: +251 988683640; Fax: +251-221-100038; E-mail: anandkps350@gmail.com.

causing serious global problems. Heavy metals are chemical elements with a specific gravity that is at least 5 times the specific gravity of water. They are toxic or poisonous even at low concentrations [1]. The most common heavy metals, such as chromium (Cr), zinc (Zn), cadmium (Cd), copper (Cu), mercury (Hg), and nickel (Ni) are recognized as a great threat [2] in the earth due to their toxicity and even in small concentration, they seriously affect human beings [3].

The movement and toxicity of heavy metal ions are based on their speciations in terrestrial and aquatic environments, especially interactions with the solid fractions such as various (hydro) oxides [4] clay minerals, organic matters, and microorganisms. These biotic and abiotic constituents are naturally interacting to form varieties of compounds, composites and complexes by the physicochemical processes such as adsorption, coating, co-precipitation, *etc.* [5]. The mining activities by man have introduced these heavy metals into the environment during the extraction of different metals from different sources [6].

HEAVY METAL DETECTION

In the environment, heavy metals can remain for decades or centuries due to their non-biodegradability [7]. They are dispersed throughout the environment by several natural processes such as volcanic eruptions, anthropogenic activities, bacterial activities, industrial processes, an agricultural activities [8]. Heavy metals will normally cause acute diseases such as nausea, vomiting, diarrhea, or allergy after their entry into the human body. They can also cause chronic diseases, such as cancers, lung disease, stroke, Alzheimer's, diabetes, kidney problem, or nerve-related damages and even death [9]. In the human body, these heavy metals are believed to destroy the macromolecules and disrupt their cellular functions [10]. The detection and determination of these heavy metal ions in the soil, food, and water are very significant and therefore, serious efforts have been made in this direction by the researchers. Therefore, a sensitive and selective methodology for the detection of heavy metal ions is the need of the hour. World Health Organization (WHO) and the Centre for Disease Control (CDC) have declared heavy metals as the prominent species to be monitored and have a set of permissible limits as per the environmental quality standards (EQS) [11]. Therefore, a quick, extremely sensitive and efficient tracking of heavy metal ions from all types of samples is highly desirable. The most conventional methods applied for the detection of trace metals at lower concentrations include atomic absorption spectroscopy (AAS), inductively coupled plasma/atomic emission spectrometry (ICP-AES), neutron activation analysis (NAA), ion chromatography ultraviolet-visible spectroscopy (IC-UV-vis), inductively coupled plasma-mass spectroscopy (ICP-MS), microprobes (MP) and X-ray fluorescence spectroscopy (XFS) [12, 13]. Although, all these methods have been being proved to be sensitive and selective, but laborious and time-consuming pretreatment process and expensive analytical instrumentation remain their major drawbacks [3]. In addition to, application of the optical chemical sensors (OCSs) by utilizing electromagnetic (EM) radiation generated an analytical signal which exhibits great potential for the detection of heavy metals ions. But, OCS has also some limitations such as time-consuming sample preparation procedures, being costly, with no portability for the applications in the field [14].

For these reasons, material scientists have been researching for many years to develop a rapid, cheap, and eco-friendly technique to trace heavy metal ions . At present, electrochemical sensors have been proved to be a better alternative to conventional methods, in terms of samples handing, separation, and detection of heavy metal ions [15]. Electrochemical stripping analysis has been used as a powerful technique for a long time to trace metals and now it is the most sensitive electro-analytical technique for the analysis of trace metals in the environmental, clinical, and industrial samples [16]. The process of electrochemical sensing relies on the electrocatalytic oxidation of preconcentrated target analyte on the surface of a specially designed electrode [17].

ELECTROCHEMICAL SENSOR

Electrochemical sensors are very much useful in chemistry and medicine to detect and determine the presence of chemical species. Electrochemical sensors are special type of sensors in which the electrode acts as a transducer element and provides quantitative information about analyte chemical species. Electrochemical sensing has gained significance as a promising analytical tool with applications in the fields of energy, medicine, environment, and pharmaceutical industry [18]. Electrochemical sensors were found to exhibit superior sensitivity and selectivity during the quantitative evaluation of analyte species, in addition to their ability to undergo functionalization [19]. The chapter will only focus and discuss the electrochemical sensing activities for heavy metal ions. Electrochemical sensors are operated by reacting with the chemical solutions and by producing an electrical signal. A sensor transforms electrochemical information into an applicable qualitative or quantitative signal. The reactions which occur at the interface of the surface of an electrode between the recognition element, and the target/binding analyte generate an electrical double layer and thus this potential is measured after transforming these chemical reactions into this measurable electrochemical signal by a recognition element, and a transducer of the sensor (Fig. **1**). An electrochemical sensor mainly consists of two main components, one is a chemical recognition system which is responsible for recognizing the analyte species and the other one is a physicochemical transducer that converts chemical interactions into electrical signals that could be detected and displayed easily by the modern electrical instruments [20]. One of the recent trends in the development of electrochemical sensors is to use nanomaterials leading to substantial enhancement of their performance concerning sensitivity, selectivity and rate of detection [21].

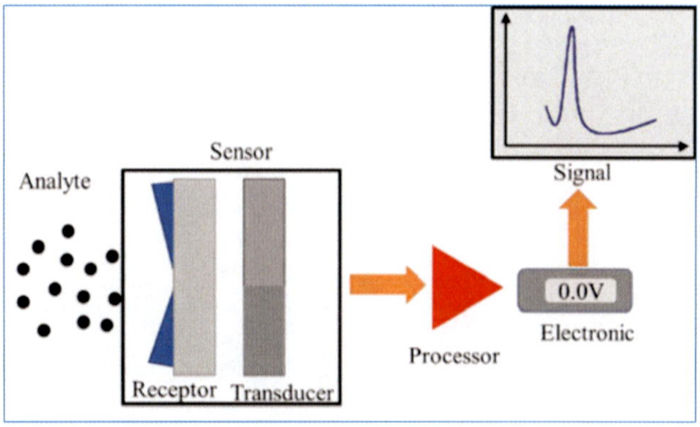

Fig. (1). A schematic representation of the response principle of electrochemical sensor [3].

Types of Electrochemical Sensors

Electrochemical sensors can be classified, based on the electroanalytical technique employed to measure the chemical interactions, into the main classes such as conductometric, voltammetric, and potentiometric. The electrochemical sensors are mainly divided into three types, as shown in Fig. (**2**).

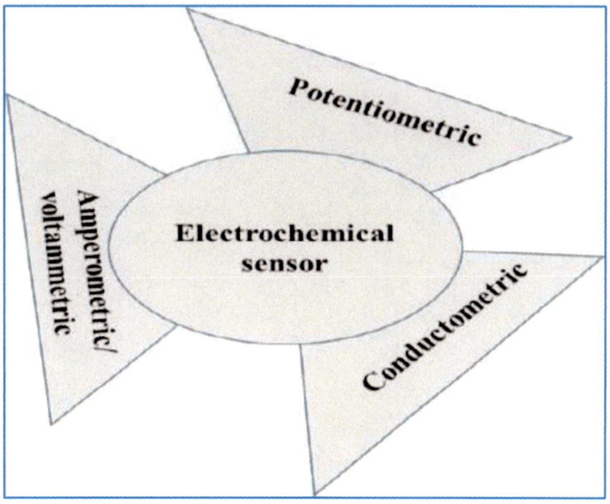

Fig. (2). Show the classification of electrochemical sensor [22].

Working Principle of Electrochemical Sensor

Electrochemical sensor's working mechanism involves the interaction of the target analyte material with the electrode surface and bringing the desired change as a consequence to a redox reaction, which generates an electrical signal that can be transformed to explore the nature of the analyte species (Fig. **3**). Electrochemical sensing can be applied for solid, liquid, or gaseous analytes, while the most common analytes are liquid or gas [23].

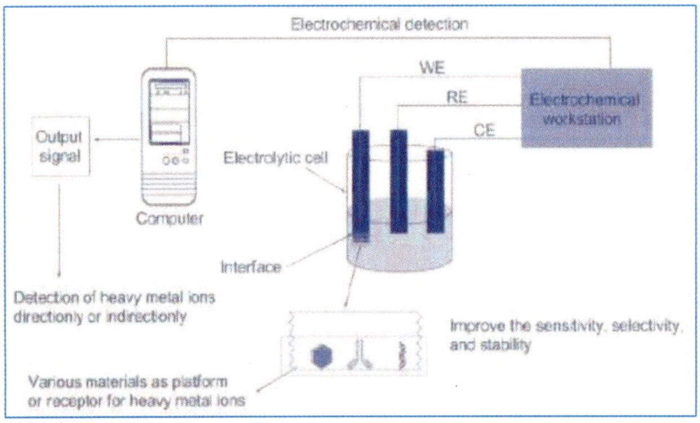

Fig. (3). General working principle of electrochemical sensing of heavy metal ions [23].

Potentiometric Sensors

In the case of potentiometric sensors, a potential difference across a pair of electrodes is determined for the analyte species of interest when no current flows in the system (Fig. **4**). There are three main classes of potentiometric devices: ion-selective electrodes (ISE), field-effect transistors (FETS), and coated wire electrodes (CWES) [21].

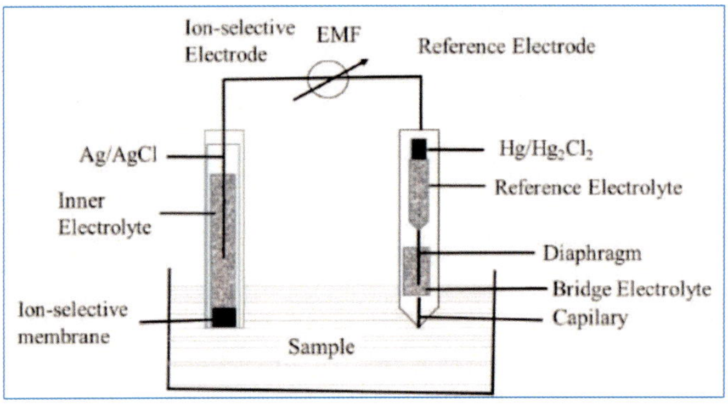

Fig. (4). Shows the electrode in an electrochemical cell used for potentiometry with ion-selective electrodes [25].

The ion-selective electrode is the most common electrode in the environmental analysis and representative potentiometric sensor. ISEs are membrane-based devices with a membrane that separates the sample from the electrode [24]. The ions of interest are measured based on the potential developed across the membrane between the analyte super resolution and the reference electrolyte.

Each electrode in a potentiometric sensor, involving two half-cell reactions. One of these two reactions is utilized in sensing the analytes species of choice. The other should be a well-understood reversible and non-interfering reaction [26 - 28].

Conductometric Sensors

In the case of a conductometric sensor, changes in the electrical conductivity of the selective layer on the electrode are utilized to determine the concentration of the analyte species. The conductometric sensors are nonselective in nature [28]. The past studies have revealed the successful application of copper doped oxides Ga_2O_3 and CdS for the detection of H_2S, CH_4, and O_2 molecules [29].

Amperometric Sensors

An amperometric sensor involves the application of small currents with a measurement at a fixed potential using a non-mercury working electrode (Fig. **5**). Thus, amperometric measurements were conducted by recording the current flow in the cell at a single applied potential [30]. In this case, the analyte involves a faradaic reaction at a desired polarity and magnitude of the potential applied. moreover, the faradaic reaction is incomplete due to the availability of less surface area of the working electrode, and only a fraction of analyte reacts therein [31]. A typical amperometric sensing system and its principle are shown in Fig. (**5**).

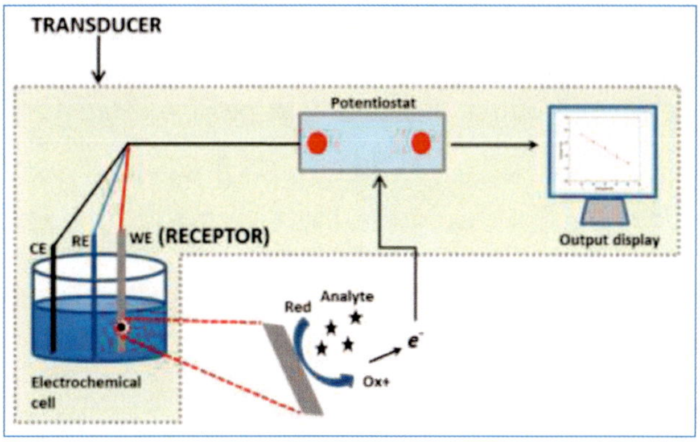

Fig. (5). Schematic representation of a typical amperometric sensing system [31].

Nanocatalyst in Electrochemical Sensor

Nanoparticles (NPs) of metals and metal oxides function as nanocatalysts due to their exceptionally high surface energy and large surface area to volume ratio [32]. In general, varieties of nanomaterials were applied in catalysis as a consequence of their superior magnetic, optical, thermal, and electrical properties. In addition, they are also used in sensing devices, in particular, electrochemical sensors. Different kinds of nanoparticles with different morphological features and compositions facilitate their application in different fields. Many types of nanoparticles such as metal nanoparticles, oxide nanoparticles, semiconductor nanoparticles, and even composite nanomaterials have played vital roles in varieties of electrochemical sensing electrodes, such as enzyme sensors, immunosensors, and DNA sensors. Among many nanomaterials, metal nanoparticles have exhibited tremendously significant catalytic attributes, which eventually resulted in their application to facilitate electrochemical reactions [33]. The fabricated/functionalized nano-electrodes with their better electrical conductivity, efficient mass transport, and surface adsorption capacity as well as super-sensitive signal features can be superior electrochemical sensors for heavy metal ions. Application of the functionalized nanomaterials electrode towards the electrochemical sensing of heavy metal ions is yet to be explored completely [34].

In most cases, the fabrication of electrochemical sensors requires surface functionalization of the nanomaterials used in the fabrication process. The process of functionalizing nanomaterials involves covalent and non-covalent approaches as per the requirement of various intermolecular interactions. The functionalized nanomaterials exhibit different attributes such as accelerated electron transfer reactions, efficient preselective transport, preferential aggregation, and bio-catalysis [18]. The selectivity, sensitivity, and efficiency of electrochemical sensors can be influenced by using chemical and biological surface modifiers. The electrochemical characteristics of the nanomaterial-based electrodes were believed to depend on the surface properties as the creation of functional groups at the surface can significantly enhance the rate of electron transfer [35].

The most familiar approach to incorporate a modifier into the surface of sensor nanomaterial is either by the formation of a nanofilm or by the doping of the nanoparticles. One of the fundamental requirements of the modifier is that it should have strong bonding features with the sensor and it should not dissolve in the surrounding medium. On the other hand, a modifier is expected to combine with the larger surface area of the electrode to enhance the conductivity, thereby an electron transfer process on the electrode surface and adsorption of metal ions will have to be greatly improved. The concentration of the analyte heavy metal ions is determined (Fig. **6**) according to the oxidation current produced during the electrochemical process [36].

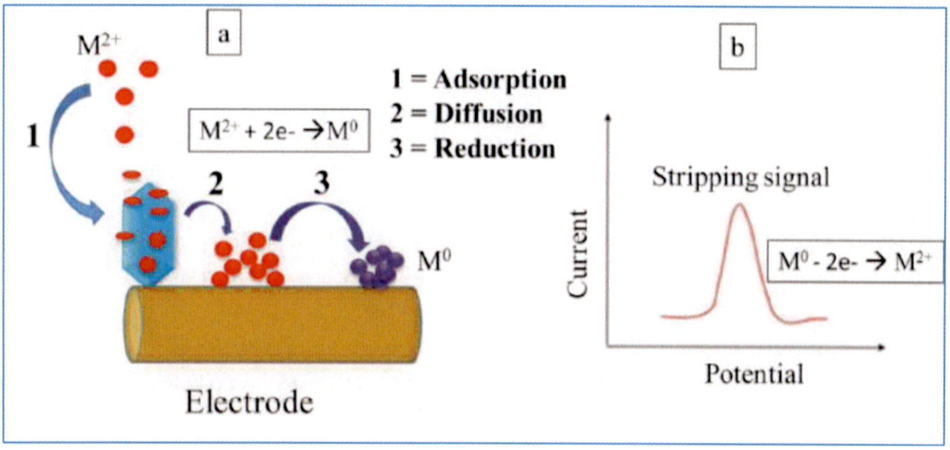

Fig. (6). The detection mechanism of heavy metal ions [36].

The functional organic compounds and noble and transition metal nanoparticles modifiers were successfully applied for the quantitative electrochemical determination of heavy metal ions. However, the sensitivity and selectivity of the organic functional group and nanoparticles depend upon the accumulation of heavy metal ions, which in turn is being influenced by the surface chemistry of these modifiers [37]. Incidentally, 3-D nanomaterials are anticipated to play a decent role in the latest electrochemical sensors to demonstrate quantum confinement effects. In addition to the organic molecules, metal and metal oxide nanoparticles have been considered as the most attractive materials due to their nontoxic and reliable nature, besides their superior mechanical, thermal, electrical, and chemical properties. The morphological features of the metal oxide nanoparticles are believed to influence the catalytic activity of sensor materials. Shivappa *et al.* reported that the nanorods and nanowires of CeO_2 exhibited superior sensing capacity and catalytic activity for the detection of heavy metal ions compared to the CeO_2 nanoparticles in the electrochemical sensors. The considerable long-term interaction of the attached moieties is a major factor in the efficiency of reliable sensors. Gong et al. reported the efficient sensing of Hg(II) based on a bimetallic Au-Pt nanoparticle/organic nanofibers. It was also concluded that the 3D nanofibers have provided a larger effective surface area and the Au-Pt nanoparticles served as the microelectrodes. However, developing a platform for sensitive detection of heavy metal ions without compromise of selectivity is always challenging. The electrochemical sensors-based nanomaterials catalyst, such as graphene applied as a carrier of electron transmission has an affinity to metal ions. The reduced graphene oxide as a carrier to compose with metal oxide would improve the adsorption property of heavy metal ions to enhance electrochemical performance [38]. To improve the efficiency of graphene in making better electrochemical sensors in terms of functionality and applicability, surface modification with suitable functional groups, molecules, biomolecules, or nanomaterials is necessary for the fabrication of different electrochemical sensors.

The electrochemical detection of heavy metal ions (HMIs) depends on the nature and type of nanoparticles involved. In recent years, the application of metal oxides nanoparticles to detect heavy metal ions has significantly enhanced scientific fundamentals and technological potentials. The metal and metal oxides such as gold, bismuth, cobalt oxide (Co_3O_4), silicon oxide (SiO_2), nickel oxide (NiO), and metal-supported graphene oxide/gold nanostructures were reported to be promising nanomaterials (Table **1**). Furthermore, they have found wide applications in electrochemical sensors, electrochromic devices, supercapacitors, and batteries. All these nanomaterials have been utilized for the detection of different heavy metal ions due to their electrochemical stability and adsorption capability [27].

Table 1. Nanomaterial-based electrochemical sensor platforms for the detection of heavy metal ions.

EC Sensing Platform	Technique	Heavy Metal Ions	Detection Range	Ref.
Fe_3O_4-NPs/TA/GCE	SWASV	Cd^{2+}, $Pb^{2+}$and Hg^{2+}	4.49-4.14 x10^2 μgL^{-1}	[39]
SnO_2 NPs/GCE	SWASV	Cd^{2+} and Pb^{2+}	-	[40]
CNT threads	OSWSV	Cu^{2+}, Cd^{2+}, $Pb^{2+}$and Zn^{2+}	-	[19]
3D MGF/Bi NPs	SWASV	Cd^{2+} and Pb^{2+}	1.00-1.20 x10^2 μgL^{-1}	[8]
Sn NPs-GSPEs	SWV	$Cd^{2+}$and Zn^{2+}	1.00-3.00 x10^1 μgL^{-1}	[23]
rGO/Au NPs	DPV	Pb^{2+}	1.67-1.67 x10^2 μgL^{-1}	[41]
GO/MWCNTs/Nafion/GCE	DPASV	Cd^{2+} and Pb^{2+}	5.00 101 -3.00 x10^1 μgL^{-1}	[42]
Au NP@TA-GCE	SWV	Hg^{2+}	2.01x10^{-5} -2.01 x10^1 μgL^{-1}	[25]
Co_3O_4 nanosheets/ITO	DPASV	Pb^{2+}	1.00-1.00 x10^2 μgL^{-1}	[27]

Au-NPs and their nanocomposites have exhibited superior potentiality in the sensor activities due to their ease and simplicity in the process of fabrication along with high chemical stability. TEM images of Fe_3O_4/TA (Tannic Acid) NPs revealed the presence of nearly spherical nanoparticles with uniform distribution, in the size range between 10 nm and 12 nm as shown in Fig. (**7**).

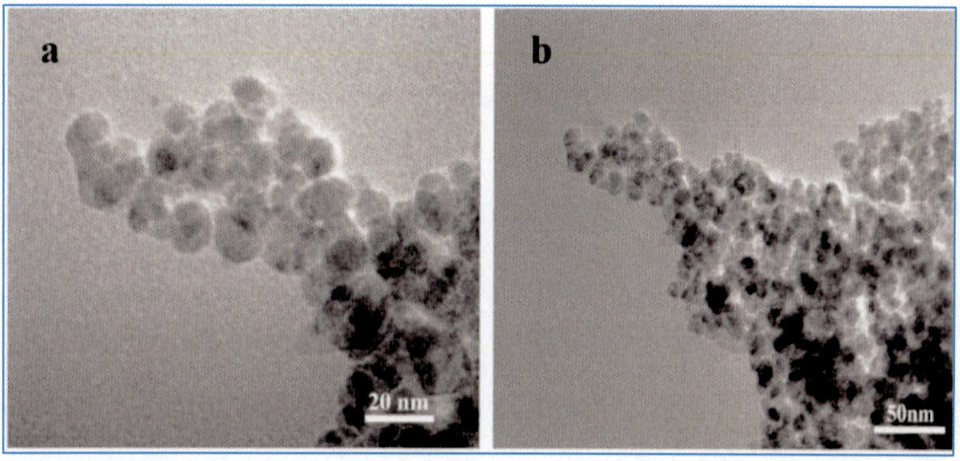

Fig. (7). TEM images of Fe_3O_4/TA nanomaterials (different magnification) [27].

The XRD pattern of Fe_3O_4/TA NPs (Fig. **8**) consists of several peaks which cores to the crystal planes (111), (220), (311), (511), (400) referred to the inverse spinel structure of Fe_3O_4 phase and the crystallite size of Fe_3O_4/TA nanoparticle is calculated using Scherrer's Equation [27].

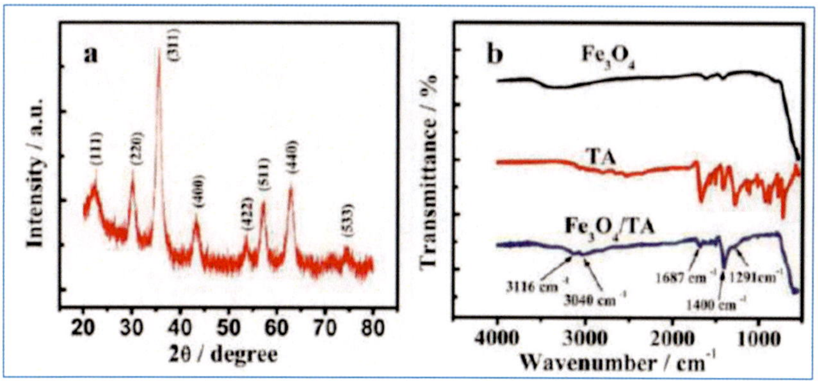

Fig. (8). (a) XRD pattern of Fe_3O_4/TA nanoparticles **(b)** The FTIR spectra of Fe_3O_4 nanoparticles (black), TA (red) and Fe_3O_4/TA nanoparticles (blue) [8, 39].

$$d = k\lambda/\beta\cos\theta$$

Where k is a constant (k = 0.9), λ is the wavelength of X-rays (1.54 Å), β is the full width at half maxima of diffraction peak line and θ is the half of the diffraction angle.

The metal oxides nanostructures have also been extensively applied after effective functionalization with the bare electrodes for the detection of HMIs. These characteristics originate from their strong adsorption capability due to their larger specific surface area and improved electron transfer kinetics. Other metal oxides such as MnO_2, ZnO, NiO, Fe_3O_4, ZrO_4, SnO_2, TiO_2, MgO, *etc*., have also been used as nanocatalysts in the sensing of HMIs [18].

In cyclic voltammetry, the bare electrode exhibits almost ideal anodic and cathodic peaks. Moreover, after modification by the nanomaterials, the anodic and cathodic peak currents were found to increase. This could be attributed to newly generated pathways for the electrons transfer by the functionalized nanoelectrodes. These new pathways enhance the electrochemical behavior of nanomaterial, resulting in the efficient ion transfer process on the electrode surface. A plot of peak currents (Ip) and the square root of scan rate ($v^{1/2}$) gives a linear relationship and the equation basically depends on the diffusion-controlled reactions, with negative intercept. The possibility of adsorption is also evident from the negative intercept value of the same. The diffusion coefficient was deduced by using the Randles-Sevcik equation.

$$Ip = 2.69\ 10^5 \text{ x } n^{3/2} AD^{1/2}V^{1/2} C_o$$

where,

Ip = Peak current (A),

n = Number of exchanged electrons,

T = temperature (K),

A = Surface area of the working electrode (cm^2),

D = Diffusion coefficient of the electroactive species (cm^2/s),

C_o = Bulk concentration of the electroactive species (mol/L) and

V = Scan rate (V/s).

The deduced diffusion coefficient value for BFS modified GCE was 7.19×10^{-4} cm^2/s.

The application of Fe_2O_3/graphene (G) electrode plated with Bi for electrochemical sensing of Zn(II), Cd(II), and Pb(II) has been reported. Fe_3O_4 nanoparticles decorated by terephthalic acid (TA) has been successfully applied to detect the Hg(II), Pb(II), and Cd(II) ions. The excellent current response with distinct peak potential was reported for the GCE/Fe_3O_4 decorated TA electrode. The current response can be attributed to the availability of TA on the surface of NPs, which facilitates the electron transfer from metal ions to the electrode surface, while the presence of the functional groups on TA provides an efficient route thatfor the electron transfer. As a result, the enhancement in the electron transfer kinetics of GCE/Fe_3O_4 electrode capped with TA yields good electrode material characteristics for the electrochemical sensing and detection of HMIs. A very important parameter for the stripping analysis is deposition potential. The decrease in peak currents with higher negative deposition potentials can be attributed to the hydrogen evolution at the CNT thread electrode. This influences the presence of an active electrode surface with mass transport of metal ions during metal deposition. The various plots of peak currents *versus* deposition potentials for all these metals ions, Cu^{2+}, Pb^{2+}, Cd^{2+}, and Zn^{2+} are shown in Fig. (**9**).

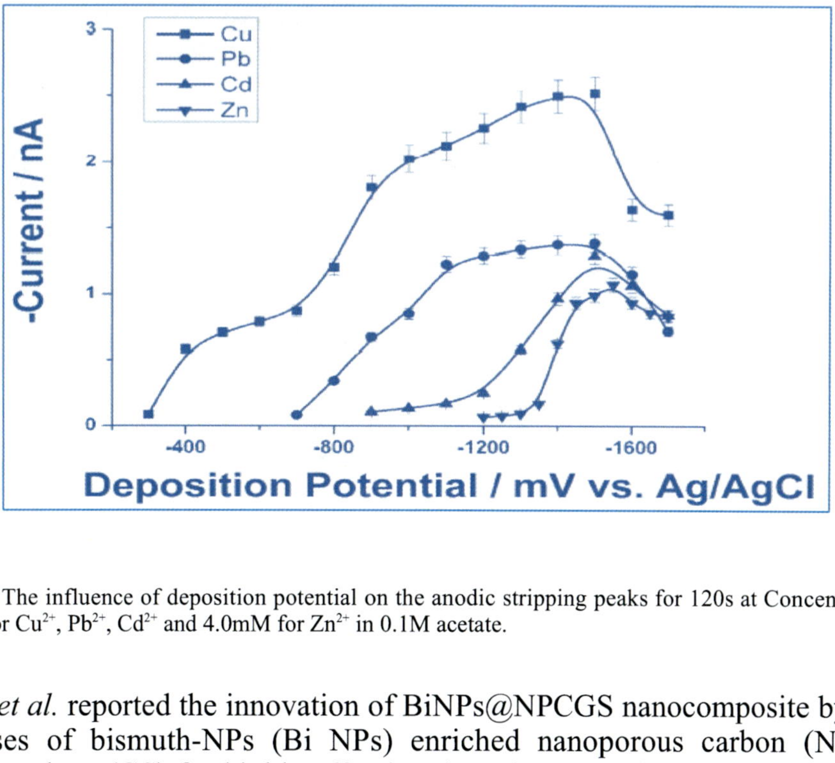

Fig. (9). The influence of deposition potential on the anodic stripping peaks for 120s at Concentrations of 3.0mM for Cu^{2+}, Pb^{2+}, Cd^{2+} and 4.0mM for Zn^{2+} in 0.1M acetate.

Lin C. *et al.* reported the innovation of BiNPs@NPCGS nanocomposite by in situ syntheses of bismuth-NPs (Bi NPs) enriched nanoporous carbon (NPC) on graphene sheet (GS) for highly effective detection of multiple heavy-metal ions. The Bi NPs@NPCGS/GCE exhibited comparatively better stripping peaks with greater peak currents than the bare GCE, BiNPs@RGO/GCE, BiNPs@NPC/GCE, and NPCGS/GCE, which substantiated better absorbability and conductivity of BiNPs@NPCGS towards the target heavy metal ions which is basically due to higher surface area to volume ratio. Ying H. *et al*, compared the cyclic voltammograms of bare GCE, $CoFe_2O_4$/GCE, and BiNPs@$CoFe_2O_4$/GCE, and, observed that the BiNPs@$CoFe_2O_4$/GCE exhibited the greatest peak current (Figs. **10** and **11**). The maximum response current at pH=5, the deposition potential (1.1V), and time (500s) were recorded for the modified electrode. The stripping voltammetric peak currents of Pb^{2+} and Cd^{2+} were found to reach a maximum of 1.1 V. While, the other decreasing potentials with competitive hydrogen evolution were found to increase and led to the stripping voltammetric peak current of Pb^{2+} and Cd^{2+} to continuously decrease [39].

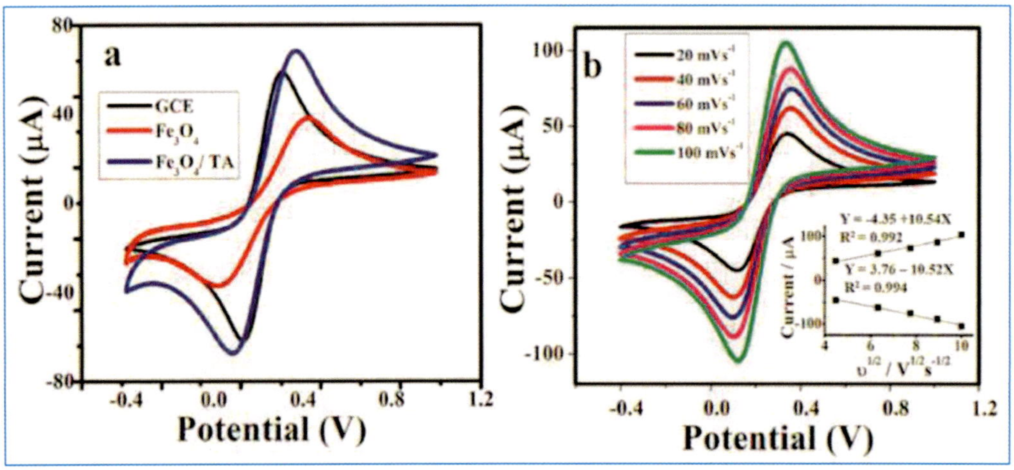

Fig. (10). The cyclic voltammogram response of **(a)** bare GCE, Fe_3O_4 NPs modified GCE and Fe_2O_4/TA NPs modified by GCE in the solution of 0.1MKCl and **(b)** Fe_3O_4/TA NPs modified GCE with scan rates 20, 40, 60, 80, 100mVs^{-1} [39].

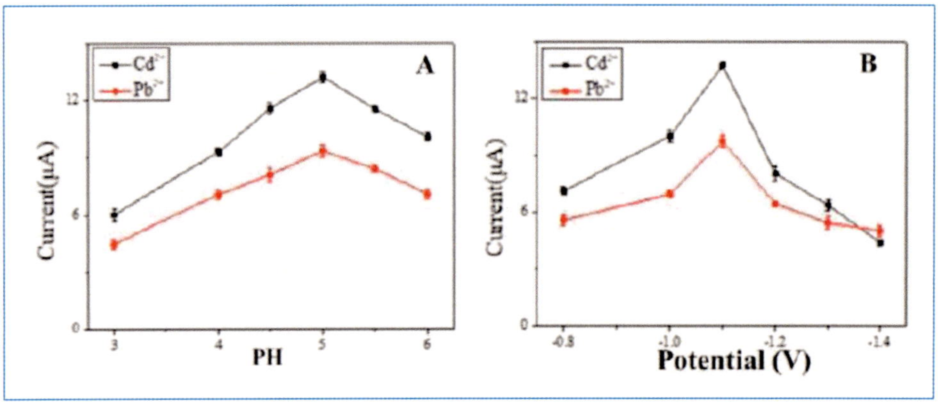

Fig. (11). Shows the SWASV response of 0.2 mM Pb^{2+} and 0.4 mM Cd^{2+} in acetate buffer at BiNPs@$CoFe_2O_4$/GCE [8].

At a lower pH value, a weak signal was generated, which could be attributed to the process of protonation on the graphene, whereas at a higher pH value, the hydrolysis occurred resulting in the decrease of signals. The effect of deposition showed potential on response signals. In addition, different deposition time was probed.

Li *et al.* designed a Cd^{2+} electrochemical sensing platform fabricated on Nafion-GN composite modified by mercury film electrode and experimented by differential pulse anodic stripping voltammetry. The GN-based sensing platform detected Cd^{2+} in a wide range of concentrations from 0.2 to 15 μgL^{-1}. The sensitivity of the technique is mostly based on the amount of substance transferred to the electrode surface during deposition. The mercury film electrode based on the Nafion-GN composite was applied for the determination of Zn^{2+}, Cd^{2+}, Pb^{2+}, and Cu^{2+} using the square-wave anodic stripping voltammetry. The square-wave anodic-stripping voltammetric (SWASV) responses were recorded effectively for the heavy metal ions at different electrodes [27].

CONCLUSION

A well-worn electrochemical sensing electrodes/nanoelectrodes were designed and developed using the supported/modified nanocomposites for the quantitative detection of heavy metal ions. The modified electrode nanocomposites involving various metal NPs (Au, Ag, Bi) and metal oxide NPs (Co_3O_4, MnO_2, ZnO, NiO, Fe_3O_4, ZrO_4, SnO_2, TiO_2, MgO) supported on graphene, graphene oxide, and CNTs have been designed and utilized as nanocatalysts for highly efficient sensing of multiple heavy-metal ions. Many heavy metal ions such as Hg^{2+}, Cu^{2+}, Pb^{2+}, Cd^{2+}, and Zn^{2+} have been successfully detected by the application of varieties of functionalized nanocatalysts. This chapter rationally explained the applicability of the nanocatalyst-modified electrode as a reliable sensor with futuristic applications.

CONSENT FOR PUBLICATION

Not applicable.

CONFLICT OF INTEREST

The author declares no conflict of interest, financial or otherwise.

ACKNOWLEDGEMENTS

Authors are grateful for the support rendered by Adama Science and Technology University, Ethiopia.

REFERENCES

[1] Ye, W.; Li, Y.; Wang, J.; Li, B.; Cui, Y.; Yang, Y.; Qian, G. Electrochemical Detection of Trace Heavy Metal Ions Using a Ln-MOF Modi Fi Ed Glass Carbon Electrode. *J. Solid State Chem.*, **2019**, *2020*(281)121032 [http://dx.doi.org/10.1016/j.jssc.2019.121032]

[2] Li, M.; Gou, H.; Al-ogaidi, I.; Wu, N. Nanostructured sensors for detection of heavy metals. *ACS Sustain. Chem.& Eng.*, **2013**, *1*(7), 713-723. [http://dx.doi.org/10.1021/sc400019a]

[3] Dai, X.; Wu, S.; Li, S. Progress on electrochemical sensors for the determination of heavy metal ions from contaminated water. *Progress on Electrochemical Sensors for the Determination of Heavy Metal Ions from Contaminated Water.*, **2018**, *3682*, 91-111. [http://dx.doi.org/10.1080/22243682.2018.1425904]

[4] Liu, F.; Zhang, Y.; Yin, W.; Hou, C.; Huo, D.; He, B. Sensors and Actuators B: Chemical A High – Selectivity Electrochemical Sensor for Ultra-Trace Lead (II) Detection Based on a Nanocomposite Consisting of Nitrogen-Doped Graphene / Gold Nanoparticles Functionalized with ETBD And. *Sens. Actuators B Chem.*, **2017**, *242*, 889-896. [http://dx.doi.org/10.1016/j.snb.2016.09.167]

[5] Yantasee, W.; Lin, Y.; Hongsirikarn, K.; Fryxell, G.E.; Addleman, R.; Timchalk, C. Review Electrochemical Sensors for the Detection of Lead and Other Toxic Heavy Metals. *Next Generation of Personal Exposure Biomonitors.*, **2007**, *115*(12), 1683-1690. [http://dx.doi.org/10.1289/ehp.10190] [PMID: 18087583]

[6] Bielská, L.; Lucia, Š.; Neuwirthová, N.; Cornelissen, G.; Hale, S.E. Science of the Total Environment Sorption. *Bioavailability and Ecotoxic Effects of Hydrophobic Organic Compounds in Biochar Amended Soils.*, **2018**, *624*, 78-86. [http://dx.doi.org/10.1016/j.scitotenv.2017.12.098] [PMID: 29247907]

[7] Shi, L.; Li, Y.; Rong, X.; Wang, Y.; Ding, S. Analytica Chimica Acta Facile Fabrication of a Novel 3D Graphene Framework / Bi Nanoparticle Fi Lm for Ultrasensitive Electrochemical Assays of Heavy Metal Ions. *Anal. Chim. Acta,* **2017**, *968*, 21-29. [http://dx.doi.org/10.1016/j.aca.2017.03.013] [PMID: 28395771]

[8] Huang, X.; Li, Z.; Yu, Z.; Deng, X.; Xin, Y. *Review Article Recent Advances in the Synthesis*; Properties, and Biological Applications of Platinum Nanoclusters, **2019**.

[9] Ussiri, D. A. N.; Johnson, C. E. Characterization of Organic Matter in a Northern Hardwood Forest Soil by 13 C NMR Spectroscopy and Chemical Methods. **2003**, *111*, 123-149.

[10] Bansod, B.; Kumar, T.; Thakur, R.; Rana, S.; Singh, I. Biosensors and Bioelectronics A Review on Various Electrochemical Techniques for Heavy Metal Ions Detection with Di Ff Erent Sensing Platforms. *Biosens. Bioelectron.*, **2017**, *94*(March), 443-455. [http://dx.doi.org/10.1016/j.bios.2017.03.031] [PMID: 28340464]

[11] Armienta, A.; Argyraki, A.; Durães, N. *Inorganic Pollutants in Soils*; Elsevier Inc., **2018**. [http://dx.doi.org/10.1016/B978-0-12-849873-6.00006-6]

[12] Zhang, T.; Jin, H.; Fang, Y.; Guan, J.; Ma, S.; Pan, Y.; Zhang, M.; Zhu, H.; Liu, X.D.; Du, M.L. Detection of Trace Cd^{2+}, Pb^{2+} and Cu^{2+} Ions *via* Porous Activated Carbon Supported Palladium Nanoparticles Modi Fi Ed Electrodes Using SWASV. *Mater. Chem. Phys.*, **2018**, *2019*(225), 433-442. [http://dx.doi.org/10.1016/j.matchemphys.2019.01.010]

[13] Vimala, A.; Vedhi, C. *Modified Electrode Used For Detection of Heavy Metals*; No. April, **2018**.

[14] Nguyen, T. L.; Cao, V. H.; Hai, T.; Pham, Y.; Le, T. G. Platinum Nanoflower-Modified Electrode as a Sensitive Sensor for Simultaneous Detection of Lead and Cadmium at Trace Levels. **2019**. [http://dx.doi.org/10.1155/2019/6235479]

[15] March, G.; Dung, T.; Piro, B. Modified Electrodes Used for Electrochemical Detection of Metal Ions in Environmental Analysis. **2015**, 241-275.
[http://dx.doi.org/10.3390/bios5020241]

[16] Zhao, D.; Guo, X.; Wang, T.; Alvarez, N.; Shanov, V. N. Simultaneous detection of heavy metals by anodic stripping voltammetry using carbon nanotube thread. **2014**, 488-496.
[http://dx.doi.org/10.1002/elan.201300511]

[17] Aragay, G.; Merkoc, A. Nanomaterials application in electrochemical detection of heavy metals. *Electrochim. Acta,* **2012**, *84*, 49-61.
[http://dx.doi.org/10.1016/j.electacta.2012.04.044]

[18] Bakker, E.; Chumbimuni-torres, K.Y. Modern directions for potentiometric sensors. **2008**.

[19] Rashmi, B.N.; Harlapur, S.F.; Avinash, B.; Ravikumar, C.R.; Nagaswarupa, H.P.; Anil Kumar, M.R.; Gurushantha, K.; Santosh, M.S. Facile green synthesis of silver oxide nanoparticles and their electrochemical, photocatalytic and biological studies. *Inorg. Chem. Commun.,* **2019**, *2020*(111)107580
[http://dx.doi.org/10.1016/j.inoche.2019.107580]

[20] Cui, L.; Wu, J.; Ju, H. Electrochemical sensing of heavy metal ions with inorganic, organic and bio-materials. *Biosens. Bioelectron.,* **2015**, *63*, 276-286.
[http://dx.doi.org/10.1016/j.bios.2014.07.052] [PMID: 25108108]

[21] Wang, J. *Modified Electrodes for Electrochemical Sensors.,* **1990**, *3*, 255-259.

[22] Simões, F. R.; Xavier, M. G. Electrochemical Sensor Electrochemical Techniques for Envi-Ronmental Analysis **2017**.

[23] Bakker, E.; Telting-diaz, M. *Electrochemical Sensors.,* **2002**, *74*(12), 2781-2800.

[24] Qu, C.; Chen, W.; Hu, X.; Cai, P.; Chen, C.; Yu, X.Y.; Huang, Q. Heavy metal behaviour at mineral-organo interfaces: Mechanisms, modelling and influence factors. *Modelling and in Fl Uence Factors.,* **2019**, *131*(1)104995
[http://dx.doi.org/10.1016/j.envint.2019.104995] [PMID: 31326822]

[25] Trachioti, M.G.; Hrbac, J.; Prodromidis, M.I. Determination of Cd and Zn with " Green " screen - printed electrodes modified with instantly prepared sparked tin nanoparticles. *Sens. Actuators B Chem.,* **2017**.
[http://dx.doi.org/10.1016/j.snb.2017.10.039]

[26] Simpson, I.P.C.; Mar, D.; Us, C.A.; Petisce, J.R.; Diego, S.; Us, C.A.; Carr-brendel, V.; Brauker, J.H.; Diego, S.; Knobbe, F. (12) United States Patent (45) Date of Patent. **2006**, *2*(12).

[27] Shang, J.; Zhao, M.; Qu, H.; Li, H.; Gao, R.; Chen, S. New Application of P-n Junction in Electrochemical Detection: The Detection of Heavy Metal Ions. *J. Electroanal. Chem. (Lausanne),* **2019**, *855*(August)113624
[http://dx.doi.org/10.1016/j.jelechem.2019.113624]

[28] Chang, J.; Zhou, G.; Christensen, E.R.; Heideman, R.; Chen, J. Graphene-based sensors for detection of heavy metals in water: a review. *Anal. Bioanal. Chem.,* **2014**, *406*(16), 3957-3975.
[http://dx.doi.org/10.1007/s00216-014-7804-x] [PMID: 24740529]

[29] Cui, J.; Xu, S.; Wang, L. Monolayer MoS_2 Decorated Cu_7S_4-Au Nanocatalysts for Sensitive and Selective Detection of Mercury (II). **2017**, *60*(4), 352-360.
[http://dx.doi.org/10.1007/s40843-017-9019-4]

[30] Long, Y.; Mao, X.; Nan, S.; Fang, H.; Yan, H. An exploration of the relationship between adsorption and bioavailability of pesticides in soil to earthworm. **2006**, *141*, 428-433.
[http://dx.doi.org/10.1016/j.envpol.2005.08.058]

[31] Privett, B. J.; Shin, J. H.; Schoenfisch, M. H. Electrochemical Sensors. **2010**, 4723-4741.
[http://dx.doi.org/10.1021/ac101075n]

[32] Tchounwou, P. B.; Yedjou, C. G.; Patlolla, A. K.; Sutton, D. J. Heavy metals toxicity and the environment. **2014**, 4723-4741.
[http://dx.doi.org/10.1007/978-3-7643-8340-4]

[33] Yang, M.; Jiang, T.; Guo, Z.; Liu, J.; Sun, Y.; Chen, X. Sensors and actuators b: chemical sensitivity and selectivity sensing cadmium (ii) using amination functionalized porous SnO_2 nanowire bundles-room temperature ionic liquid nanocomposite: combined efficient cation capture with control experimental Co. *Sens. Actuators B Chem.,* **2017**, *240*, 887-894.
[http://dx.doi.org/10.1016/j.snb.2016.09.060]

[34] Deshmukh, S.; Kandasamy, G.; Kant, R.; Bhattacharya, G.; Banerjee, D.; Maity, D.; Deshusses, M.A.; Sinha, S. Terephthalic acid capped iron oxide nanoparticles for sensitive electrochemical detection of heavy metal ions in water. *J. Electroanal. Chem. (Lausanne),* **2017**, *788*, 91-98.
[http://dx.doi.org/10.1016/j.jelechem.2017.01.064]

[35] Jin, Z.; Yang, M.; Chen, S.; Liu, J.; Li, Q.; Huang, X. Tin oxide crystals exposed by low-energy {110} facets for enhanced electrochemical heavy metal ions sensing: X-Ray absorption fine structure experimental combined with density-functional theory evidence. **2017**.
[http://dx.doi.org/10.1021/acs.analchem.6b04977]

[36] Zhu, Y.; Pan, D.; Hu, X.; Han, H.; Lin, M.; Wang, C. Sensors and actuators b: chemical an electrochemical sensor based on reduced graphene oxide / gold nanoparticles modified electrode for determination of iron in coastal waters. *Sens. Actuators B Chem.,* **2017**, *243*, 1-7.
[http://dx.doi.org/10.1016/j.snb.2016.11.108]

[37] Huang, H.; Chen, T.; Liu, X.; Ma, H. Ultrasensitive and simultaneous detection of heavy metal ions based on three-dimensional graphene-carbon nanotubes hybrid electrode materials. *Anal. Chim. Acta,* **2014**, *852*, 45-54.
[http://dx.doi.org/10.1016/j.aca.2014.09.010] [PMID: 25441878]

[38] Zhou, S.; Wang, J.; Gan, L.; Han, X.; Fan, H.; Mei, L.; Huang, J.; Liu, Y. Individual and simultaneous electrochemical detection toward heavy metal ions based on l-cysteine modi fi ed mesoporous $MnFe_2O_4$ Nanocrystal Clusters. *J. Alloys Compd.,* **2017**, *721*, 492-500.
[http://dx.doi.org/10.1016/j.jallcom.2017.05.321]

[39] Liao, J.; Tao, Z.; Lin, S. Theoretical and experimental insights into the electrochemical heavy metal ion sensing with nonconductive nanomaterials. *Curr. Opin. Electrochem.,* **2019**, *17*, 1-6.
[http://dx.doi.org/10.1016/j.coelec.2019.04.006]

[40] Mourya, A.; Mazumdar, B.; Sinha, S.K. Determination and quantification of heavy metal ion by electrochemical method. *J. Environ. Chem. Eng.,* **2019**, *7*(6)103459
[http://dx.doi.org/10.1016/j.jece.2019.103459]

[41] Jin, W.; Maduraiveeran, G. Nanomaterial-based environmental sensing platforms using state-of-the-art electroanalytical strategies. *J. Analy. Sci. Tech.,* **2018**, *9*(18).
[http://dx.doi.org/10.1186/s40543-018-0150-4]

[42] Senthilkumar, N.; Nandhakumar, E.; Priya, P.; Soni, D.; Vimalan, M.; Vetha Potheher, I. Synthesis of ZnO Nanoparticles Using Leaf Extract of: *Tectona Grandis* (L.) and Their Anti-Bacterial, Anti-Arthritic, Anti-Oxidant and *in vitro* Cytotoxicity Activities. *New J. Chem.,* **2017**, *41*(18), 10347-10356.
[http://dx.doi.org/10.1039/C7NJ02664A]

CHAPTER 12

Fundamentals and Advances in Transition Metal Oxide-Based Nano-Electrocatalyst for OER

Athira Krishnan[1,*], **U. S. Krishnanunni Namboothiri**[1], **R. Ajay**[1] and **J. Anakha**[1]

[1] *Department of Chemistry, Amrita Vishwa Vidyapeetham, Amritapuri, Kerala-690 525, India*

Abstract: Oxygen evolution reaction (OER) is the anodic reaction in the overall water splitting process. The four electron-proton reaction is thermodynamically non-spontaneous and kinetically slow. Hence, the need for an electrocatalyst is urgently required to achieve faster kinetics for OER and to enhance the rate of overall water-splitting reaction. The chapter discusses the fundamentals; which include mechanisms of OER, the characteristic features as well as kinetic performance parameters of OER electrocatalyst and advances in nano-electrocatalysis; which includes recent developments in TMO based electrocatalyst, strategies for improving the performance of an electrocatalyst; catalyst other than TMOs. The present discussion offers profound insight and interest on catalysts in oxygen evolution reactions.

Keywords: Electrocatalyst, Nano-catalyst, Overpotential, Oxygen evolution reaction, Transition metal oxides, Water splitting process.

INTRODUCTION

The beauty of the earth is kept alive only through sustainable development. So that, it is desirable to utilize energy properly. As the energy demand is increasing day by day, we must search for alternative renewable energy resources rather than depending on limited resources like fossil fuels. These widely used fuels such as coal, petroleum, natural gas, *etc*., are not only limited, but they aren't eco-friendly also. Around these years, many ideas have been proposed for achieving the simplest, reliable, and efficient energy conversion systems which include fuel cells, alkaline water electrolysis, metal-air batteries, *etc*. In all these systems, oxygen evolution reaction (OER) plays a major role along with oxygen reduction reaction (ORR) or hydrogen evolution reaction (HER). OER is simply an electrochemical process in which molecular oxygen is evolved. The electrolysis of water is a highly significant reaction as a means of clean hydrogen production.

* **Corresponding author A. Krishnan:** Department of Chemistry, Amrita Vishwa Vidyapeetham, Amritapuri, Kerala-690 525, India; Tel: 9142035497; E-mail: athikrishnan91@gmail.com

These reactions occur in a two-electrode system of which oxygen reduction reaction (ORR) or hydrogen evolution reaction (HER) takes place at the cathode, whereas oxygen evolution reaction (OER) takes place at the anode. OER is pH-dependent and occurs *via* different routes in acidic/basic/neutral environments.

The equilibrium half-cell potentials at standard conditions for OER are shown as follows [1];

$$4OH^- \leftrightarrow 2\,H_2O_{(l)} + O_{2(g)} + 4e^- \qquad E° = 0.404\text{ V (alkaline solution)} \quad \textbf{(1)}$$

$$2H_2O_{(l)} \leftrightarrow 4H^+ + O_{2(g)} + 4e^- \qquad E° = 1.230\text{ V (acidic/neutral solution)} \quad \textbf{(2)}$$

The equilibrium half-cell potentials at standard conditions for HER and ORR are [2, 3];

$$\text{HER: } 2H^+_{(aq)} + 2e \leftrightarrow H_{2(g)} \qquad E° = 0.000\text{ V } (vs\text{ SHE}) \quad \textbf{(3)}$$

$$\text{OER: } O_{2(g)} + H_2O_{(l)} + 2e \leftrightarrow HO_2^- + OH^- \qquad E° = 0.695\text{ V } (vs\text{ SHE}) \quad \textbf{(4)}$$

From this, it is evident that OER involves the transfer of four electrons as well as four protons [4]. Whereas at the cathode, HER is a two-electron transfer reaction *i.e.* evolution of a single oxygen molecule involves several single-electron/proton transfer steps. Both the reactions are slow in kinetics, especially OER as it involves four electron-proton coupled transfers and also the higher overpotential requirement to overcome the energy barrier associated with each step. While considering the practical application of this reaction, this overpotential has resulted in a huge loss of energy and hence a catalyst is urgently required to lower the activation barrier for the oxygen evolution reaction.

An ideal OER catalyst includes low overpotential, high conductivity, large surface area, active surface sites, high stability, and large earth-earth abundance metal-based catalysts like RuO_2 and IrO_2 show excellent electrocatalytic activity on OER. They are known as benchmark electrocatalysts for OER and are applicable for a wide range of pH conditions [5]. However, being costly and meagerly available, we can't use these catalysts on a large scale. Recently, numerous alternatives have been discovered. Oxides and non-oxides of other transition metals, organometallics, non-metal compounds like carbon nanotube, *etc.* show great results on electrocatalysis. Among these, transition metal oxide (TMO) based electrocatalysts have a supreme position in OER catalysis.

There are wide varieties of transition metals and their oxides. Besides, they are abundant to mostly first-row transition elements [5]. Flexibility towards alloying, doping, and nanoparticle synthesis makes them very special. Certain oxides of iron, nickel, and cobalt have been reported to act as efficient OER catalysts in alkaline media. While corrosion of these materials leaves a negative impact. This issue is found to be rectified by alloying those metals with other transition metals. For example, iron can be alloyed with Cu, Mo, Ni, Zn and/or Co, which enhances the corrosion resistivity of iron-based catalysts [5]. While discussing the application of nano-science in electrocatalyst, nanoparticles of lithium-induced $NiFeO_x$ are a great example [6]. This nano-electrocatalyst has shown higher activity as well as stability than the benchmark electrocatalysts in alkaline medium. Nano-electrocatalysts are very much effective than the normal catalysts typically due to their small size (10-80 nm), achieving a very large catalytic surface area [7]. Also, they show exceptionally different properties which they did not possess at their macroscopic stages. Considering all these, we can expect a wide range of applications of these transition metal-oxide based nano-electrocatalyst in OER. The chapter mainly discusses the mechanism of TMO based OER, features of a nano-electrocatalyst for the reaction, kinetic parameters used to express the catalytic efficiency, developments in TMO based catalytic system and catalysts other than TMOs, which are significant for OER.

FUNDAMENTALS OF OER

Thermodynamically, OER is an uphill reaction with slow kinetics and hence hard to achieve. Theoretically, evolution happens at 1.23 V. But practically occurs at a higher potential. The difference between theoretical and actual potential values is called overpotential [8]. OER is a complex four electron/four proton transfer reaction, that requires higher overpotential and hence acts as the bottleneck of overall water electrolysis. TMO's constitute the most active, stable and low cost electrocatalysts for enhancing the kinetics of OER from different chemical environments. They are efficient and hence widely used in alkaline water electrolytes. Among the different metal hydroxides, the highly conductive KOH solution is the most common electrolyte in use. Acid electrolytes, due to the highest molar conductivity of protons, could minimize voltage loss at high current density and also offer lower resistance. Mainly, noble metals such as Ir, Ru, and their oxides are well known for their catalytic activity for application in acid electrolytes. Compared to acid and alkaline electrolytes, OER reaction follows slower kinetics in neutral environment due to the low conductivity and the lower concentration of proton and hydroxyl ions.

Electrocatalytic OER reaction is pH-dependent. The starting material in an acidic medium is water. Whereas, OER from alkaline as well as neutral medium initiates with hydroxide ions. The mechanisms of OER from acidic, basic as well as. The mechanisms of OER from acidic, basic as well as the neutral environments are evidenced from many research publications. The general mechanism of OER from the different chemical environments is given in Fig. (**1**). The mechanism explores that the intermediate is common; MO, MOH, and MOOH, during OER from acidic/basic/neutral environment, where M, is the active catalytic site. By theory, Gibb's free energy change associated with each step is 1.23 eV.

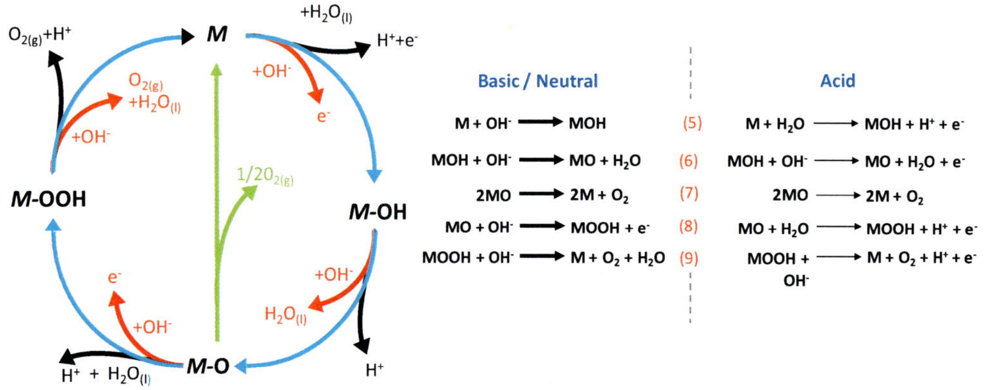

Fig. (1). Mechanism of OER from neutral/ basic/acidic medium [8].

But in actual practice, the reaction with the highest ΔG value limits the rate of the reaction. ΔG_{MOH}- ΔG_{MOOH}, is ~ 3.2 eV and is independent of the nature of metal and metal oxide [9]. Hence,

$$\eta = \frac{\Delta G_{MOOH} - \Delta G_{MOH}}{2e^-} - 1.23\ \text{V} = 0.37\ \text{V} \quad \textbf{(10)}$$

Equation (8) is highly significant in predicting the activity of the OER catalyst that follows the mechanism discussed. ΔG_{MO}- ΔG_{MOH} could be used as a descriptor to represent the trend in the activity of catalysts towards OER. The volcano type plot obtained is shown in Fig. (**2a**). The plots indicate that surfaces that bind oxygen too strongly and too loosely exhibit lesser activity towards OER. If the surface binds oxygen tightly, MOOH formation limits the rate of the reaction. However, if the binding is weak, then the rate of OER is limited by MOH formation. That is, the reaction happens at a considerable rate only if the interaction between the intermediate and the metal surface is just right, that is not too strong or too weak, as proposed by Sabatier principle. As discussed, MOH formed in the primary process changed to MO. There are two pathways observed for the generation of O_2 from MO. One is a green route in which 2MO combined to produce O_2. The second route is where MO changes to MOOH and then it decomposes to give molecular oxygen [10].

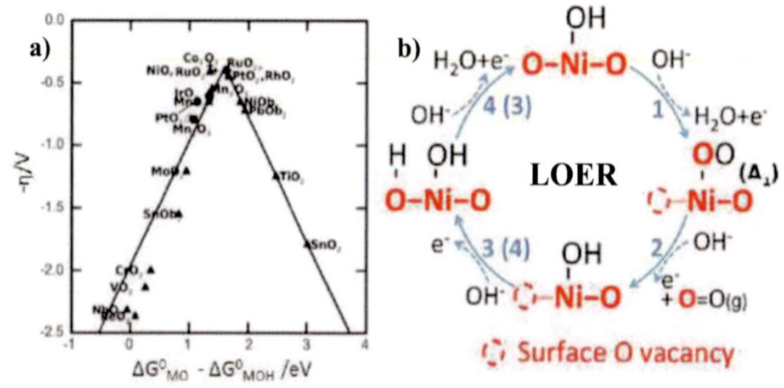

Fig. (2). (a) Volcano plot showing activity of transition metal oxide [2] **(b)** LOER mechanism of OER [9].

With the advancements in OER electrocatalysts, the conventional adsorbate evolution mechanism cannot offer a satisfactory explanation for the activity of many of the catalytic systems. In certain cases, materials cannot obey Sabatier's principle and also some system reveals pH-dependent activity. On account of the need to explain the activity of such systems, more mechanisms are put forward and the one in which lattice oxygen of catalyst is directly involved in the evolution of oxygen receives greater attention (LOER mechanism). The mechanism involving lattice oxygen was first reported in 1976 [11]. Later on, in 1987, M.W. Mehrens and J. Heitbaum, explained the activity of Ru/RuO_2 electrode by LOER, with the evidence from isotopic labeling [12]. Recently many research reports are available on lattice oxygen-interven OER [12 - 20]. Although, LOER mechanism is applicable to OER from both alkaline and acidic medium, the mechanism is more established in alkaline medium due to the lack of stability of most of the LOER active catalyst in acidic medium [21, 22]. In TMO catalyzed OER, sometimes, we observed a competition between conventional mechanism and LOER. It has been reported that highly crystalline TMO's with higher oxygen vacancy concentration, has a preference of LOER mechanism. On contrast pyrochlore, structured metal oxides with high crystallinity does not involve lattice oxygen in OER performance, while doped-pyrochlore crystal liberates oxygen *via* LOER, due to enhanced concentration of oxygen vacancy [23]. Fig. (**2b**) depicted the lattice oxygen-mediated O_2 evolution reaction.

KINETICS PARAMETERS OF OER

Thermodynamically, OER is a non-spontaneous reaction with slow kinetics. High performance electrocatalysts developments are actively progressing for controlling the kinetics of rate determining step (RDS) of OER reaction. Highly active electrocatalyst is necessary to upgrade the reaction rate. The parameters that express the efficiency of an electrocatalyst are as follows:

Overpotential(η)

It is the difference between the theoretical reduction potential and actual voltage needed [10]. Overpotential is one of the most widely used and most desired performance parameter of an electrocatalyst. In ideal state, the applied voltage (E) should be equal to the potential at equilibrium (E_{eq}), whereas in actual OER, equilibrium potential and the applied potential are different, termed as overpotential ($\eta = E_{eq} - E$). Lower the catalytic overpotential towards OER, and better will be its electrocatalytic activity towards the reaction. Hence, a low overpotential is usually an indicator of efficiency of a catalyst.

Exchange Current Density (J°)

When electrode reaction reaches equilibrium, the rates of anodic and cathodic reaction are the same. Cathodic current and anodic current balance each other with Net current zero. At equilibrium, $j = j_a = j_c$, where, the overall current (j) is the sum of anodic (j_a) and cathodic (j_c) currents. Exchange current density is given by, j° = j/A; A, the area of the electrode. The exchange current densities consider the intrinsic rate of electron transfer between analyte and the electrode. It provides insight into the structure and bonding in the analyte and the electrode. It is used to narrate the ease with which an electrode reaction can happen. It depends critically on the nature of surface roughness of electrode [8].

Tafel Slope (b)

The current density(j°) and the applied overpotential (η) can be described by Butler-Volmer equation, which relates the rate of an electrochemical reaction to overpotential [10].

$$j = j^{\circ}\{\exp\left(\frac{\alpha_a nFE}{RT}\right) + exp\frac{\alpha_a nFE}{RT}\} \quad \textbf{(11)}$$

Where, α_a is the transfer coefficient, F is Faraday's constant, T is absolute temperature. The overall current is mainly due to the anodic end under high anodic potential conditions while the contribution from the cathodic part is trivial, and the equation can be given as,

$$j \approx j^{\circ}exp\frac{\alpha_a nF\eta}{RT} \quad \textbf{(12)}$$

This is known as Tafel equation. Tafel equation is an approximation of the Butler-Volmer equation at large over potential. Tafel slope for a single electron reaction is given by;

$$b=\frac{\partial\eta}{\partial\log(\iota)}=\frac{2.303}{\alpha F} \qquad \textbf{(13)}$$

The Tafel slope reveals how efficiently an electrode can produce current in response to change in applied potential. Lesser the Tafel slope, lesser overpotential is required to get high current density. Hence, the rate of OER will be higher [8].

Charge Transfer Resistance (R_{ct})

It is the resistance offered to the process of electron transfer at the electrode/electrolyte interface. The resistance to charge transfer for oxygen evolution reaction is small at lower over potential but increases with increasing overpotential [24]. The resistance for both the production of intermediates and charge transfer decreases the rate of the reaction. Lower the R_{ct} value, the higher will be the charge transfer reaction.

Double Layer Capacitance (C_{dl})

It is the parameter for determining electrochemically active surface area. The approach of electrocatalysis is to determine the Helmholtz double layer capacitance and to deduce the electrochemical active surface area. The double layer capacitance can be attained by two different methods; First, using differential capacitance measurement and second, impedance spectroscopy. C_{dl} can be used to calculate the electrochemically active surface area (ESCA) by the equation [25],

$$\text{ESCA}=\frac{C_{dl}}{C_{s}} \qquad \textbf{(14)}$$

Where, C_{dl} is double layer capacitance, and C_s is the specific conductance.

Turn Over Frequency (TOF)

It is the number of oxygen molecules evolving from a catalyst site per unit time. TOF is obtained by examining the total metal loading of the catalysts. It is used to compare the efficiency of active sites between different catalysts and quantifies the specific activity of a catalytic center for a specific reaction under defined reaction conditions by the number of catalytic cycles occurs at the center per unit time.

Geometric Current Density (j_g)

j_g depends literally on catalyst loading and electronic configuration. When the loading increases, activity increases.

Specific Current Density (j_s)

It expresses the current density at specific overpotential normalized by the active surface area of catalyst. It is obtained by dividing the current by the electrochemical surface area. It is another metric for representing intrinsic activity of an electrocatalyst.

Faraday Efficiency

It is described as the efficiency with which electrons are transferred in a system. The total energy efficiency is the product of Faradic efficiency and voltage efficiency. It is measured through bulk electrolysis where a known quantity of reagent is stoichiometrically converted to product by the current passed. In OER, some electrons are rerouted to the formation of hydrogen peroxide, and this fraction of electrons so diverted represents a faradaic loss and vary in dissimilar systems. It is defined as the ratio of the measured amount of oxygen produced to the theoretical amount of oxygen produced. According to Faraday's law, it can be calculated as:

$$n = \frac{l_{el}}{(l_{el} + l_{loss})} \tag{15}$$

Where, l_{el} is the current that causes water splitting; $(l_{el}+l_{loss})$ is the sum of the current in the water.

FEATURES OF ELECTROCATALYSTS FOR OER

OER is a four electron-proton coupled reaction in which molecular oxygen is generated through four different steps. As oxygen evolution is associated with high overpotential, it is essential to develop a highly active catalyst to overcome the activation barrier and to make the process more productive. They can be either in the form of electrodes or may be coated on the surface of the electrode. It assists the transfer of electrons across the electrode-electrolyte interface. Electrocatalysts act by lowering the excess energy consumed for overcoming the activation barrier associated with a redox reaction without a change in its reaction equilibrium. The major features that dictate the activity of electrocatalyst are electrochemical surface area, electronic structure, the geometry of the catalyst, chemical, and mechanical stability.

Electrochemical Surface Area

Electrochemical surface area refers to the area of electroactive material that is in contact with the reacting species. It constitutes the area of the electrode that is utilized for charge transfer, which is favorable to the electrolytes. OER catalyst depends on the number of active sites on the catalyst surface. The more the exposed surface area, the more will be the active sites available for the electrochemical reaction. This leads to the lowering of activation energy needed for the reaction and causes an overall increase in reaction rate. It is believed that the electrocatalysts with increased electrochemically active surface area require lower overpotential for OER.

Electronic Property

OER reaction proceeds through a series of intermediates such as M-OH, M-O, M-OOH, M-OO, all possessing M-O bonds. The binding strength of metal with oxygen in each intermediate depends on the e_g orbital occupancy. The e_g orbitals will have strong spatial overlap with those of adjoining O 2p orbitals, forming σ-bonding and σ*- antibonding states with partial metal and oxygen character. Stronger the binding capacity, lower is its e_g orbital occupancy [26]. The metal cation-based electrocatalysts exhibit a higher electrocatalytic activity when the e_g orbital occupancy of surface metal ions is 1, which increases the covalency of M-O bonds and is regarded as the optimal electronic configuration for OER activity. The e_g orbital occupancy of active metal cation in the octahedral site is determined by oxidation state and spin state [27]. The number of d electrons determined by the oxidation state of the metal cation, for example, Mn^{3+} ($3d^4$) and Fe^{4+} ($3d^4$) afford $t_{2g}^3e_g^1$ with a high spin state. Similarly, Ni^{3+} ($3d^7$) afford $t_{2g}^6e_g^1$ configuration with a low spin state while Ni^{3+} ($3d^8$) affords $t_{2g}^6e_g^2$. In spin state, the low spin state Co^{3+} possesses the electronic configuration of (LS $t_{2g}^6e_g^0$) changing its electronic configuration to an intermediate spin state ($t_{2g}^5e_g^1$), and has been expected to significantly improve the OER activity.

Chemical and Mechanical Stability

Stability in acidic as well as an alkaline medium is a highly desired characteristic of OER catalyst. As the conductivity of the acidic and alkaline solution is high, the kinetics of OER is greater in acidic/alkaline solution compared to the neutral medium. Hence industries prefer OER from an acidic/alkaline medium rather than a neutral environment. Alkaline solutions are more often used in industrial processes. The transition metal-based catalysts are stable in alkaline solutions and display good conductivity.

Geometry

The electrocatalytic activity of catalyst with rutile, perovskite, spinel, layer structure, and rock salt structure could contribute towards oxygen evolution reaction. Among these structures, rutile is thermodynamically most stable. Rutile type RuO_2 exhibits excellent OER catalytic activity in both acidic and alkaline electrolytes.

TRANSITION METAL OXIDES IN OER

Transition metal oxides are very much favored as catalysts in OER due to their half-filled d orbitals that give them the ability to exist in different oxidation states. They are very good choice due to their low cost, abundant sources, and high resistance to corrosion [1]. Beer was the first scientist to come up with conductive oxide electrodes in oxygen electrocatalysis. Trasatti collected ideas from Tseung and Jasem and put forward his work on oxygen evolution on spinel, rutile, and perovskite oxides during 1980-1984 [28]. OER is a kinetically sluggish process. Usage of catalysts like the transition metal oxides, hydroxides/ layered double hydroxides, and also perovskite oxides/hydroxides have been found to enhance the OER catalytic activity [29].

The most widely used transition metal oxides in OER catalysis are the abundant transition metal oxides of the type FeO_x, CoO_x, $NiO_{x,}$ and MnO_x. The noble metal Ir and Ru oxides, due to their scarcity and high cost, have only limited use in large-scale applications. CoO_x is vastly used in OER due to the multiple oxidation states of the cobalt metal ion, $Co^{2+/3+/4+}$. The phase transformation of cobalt-based substances from oxides to hydroxides or oxyhydroxides is of great importance. Mesoporous Co_3O_4 nanowires have a higher OER activity than the original Co_3O_4, as, mesoporous structures result in a higher surface area and hence a greater number of active sites. Nanowire forms improve charge transportation. The oxygen vacancies help in improving the electrical conductivity. NiO is a good choice for OER as they are resistant to corrosion in alkaline media. Furthermore, their performance can be made excellent by modulating their surface area, microstructure, and particle size. The use of nickel in higher oxidation states, like Ni^{3+}, gave optimal OER results. Recently, the nickel-based conjugates and nickel oxide nanostructures are good OER catalysts due to their quick response, good corrosion resistance, plentiful sources, low cost, high surface kinetic reaction, and good stability [30]. It has been found that the incorporation of other metals like Fe and Co, improves the OER activity of NiO by enhancing the electric conductivity. Manganese oxide is another attractive choice for oxygen evolution. In alkaline solutions, they exhibit the good activity of OER. This is because of their nanostructured form that speeds up the occurrence of active sites at particular potentials [1]. Mixed metal oxides of type AB_xO_y are important electrocatalysts for OER. The activity of metal oxides is enhanced by decreasing their particle size to the nanoscale. The good electrical conductivity and electrochemical activity of $NiCo_2O_4$ is of high use in OER electrocatalysis. Mixed Ni-Fe catalysts are also used abundantly in OER in alkaline solutions.

Table 1. Catalytic parameters for metal oxide electrocatalysts for OER.

Catalysts	Electrolyte	Overpotential (η=10 mA/cm²)	Ref.
CoO_x/Au	0.1M KOH	351	[2]
$ZnCo_2O_4$	1M KOH	390	[31]
MnO_2	0.1M KOH	430	[32]
MnO_2-CoP_3/Ti	1M KOH	288	[33]
CoO_x NPs/BNG	0.1M KOH	295	[34]
CoO_x (N-doped graphene)	0.1M KOH	370	[35]
Fe_20_3/CNT_{50}	0.1M KOH	383	[36]
Fe-NiO	1M KOH	206	[37]
FeCoMo	1M KOH	277	[38]
Co_30_4	0.1M KOH	340	[39]
$CoFeO_x$	1M KOH	240	[40]
$Zn_xCo_{3-x}O_4$	1M KOH	320	[2]

It is observed that certain catalysts offer remarkable activity because of their active structure [1]. Such structures include perovskite, spinel, ferrite, and pyrochlore. Perovskite oxides/hydroxides of type ABO_x (where A is the alkaline-earth or rare-earth metal and B is the transition metal) finds good as catalysts in OER because of the flexibleness in their structure and composition. Iron-based perovskite oxides, for instance, the cubic $BaFeO_3$-δ perovskite oxide, are ideal to study the OER activity of the Fe site as the perovskite lattice can make the potentially active group Fe^{4+} stable in the OER process [41]. Spinel transition metal oxides of the type $A_xB_{3-x}0_4$ (where A, B = Ni, Zn, Mn, Co, *etc.*) play a great role as catalysts in the OER [41]. Spinel-based transition metal oxides have been found to exhibit good catalytic activity due to their tunable composition, flexibility in their structure, and morphology [42]. They are also a source of surface redox centers that help in oxygen adsorption and stimulation. Duan *et al.* had found that the OER catalytic performance for $MnCo_2O_4$ was found to be enhanced when an amalgamation of spinel compounds and N-doped carbon was used. Nickel ferrite spinel-based catalysts have also been found to be efficient for OER. The partial replacement of Fe by Ni in Fe_3O_4 spinel structure increased the

OER performance of the transition metal oxide to about 1.8 times more than that of Fe_3O_4 [43]. Metal oxides with pyrochlore structure of the type $A_2B_2O_{7-y}$ (where A = Pb, Bi and B = Ru, Os or Ir) was found as good catalysts for the oxygen evolution reaction. The lead and bismuth ruthenate pyrochlores exhibited a good activity compared to the IrO_2 catalyst. Table 1 recorded the established TMO-based electrocatalysts for OER [44].

STRATEGIES FOR ENHANCING ACTIVITY OF OER ELECTROCATALYSTS

The catalytic activity of the transition metal oxides can be enhanced by certain strategies that are discussed below.

Surface Vacancy/Defect

In a transition metal oxide catalyst, the reaction rate is affected by only small fraction of the total active sites in it. The introduction of artificial vacancy/defects is a promising strategy to enhance the performance of transition metal oxides in OER electrocatalysis. This helps in the creation of more active sites, optimizes the surface-adsorption ability for reaction intermediates, and at the same time enhances the surface hydrophilic property, thereby promoting electrocatalytic activity [45].

Heteroatom Doping

Introducing heteroatoms destroys the recurrence of the lattice. This leads to a change in the electronic structure. This can help in improving the adsorption abilities of the intermediates formed in the reaction, thus increasing the catalytic activity. Doping of ferrite with Co, Cu, Ni, Mn, and nitrogen doping, are some ways of enhancing the OER activity of metal oxides. The electronic conductivity increases and the adsorption strength of OH^- increases after nitrogen doping, and it accelerates the reaction kinetics. The defect densities can tune electronic properties and even provide active sites [45].

Reducing Electrical Resistance by Conductive Supports

Often the nano-size of the OER catalysts gives only limited contact between the catalyst and the electrode. This leads to a huge loss of energy and ultimately suppresses the OER activity. The electronic conductivity and dispersion of metal oxides can be enhanced by employing carbon substrates of high surface area like graphene and carbon black [2].

Synergistic Effects

Incorporating multi-metal oxides generates a synergistic effect, giving an increased activity to the catalyst [2]. For example, in the systematic study on OER activity of $AA'_3B_4O_{12}$ -type quadruple perovskite oxides, in which multiple transition metal ions are located at distinct crystallographic sites, the quadruple perovskite framework provides suitable synergistic interactions to increase OER catalytic activity. These findings show a simple way to enhance the OER activity of transition metal oxides consisting of multiple transition metal elements. $Fe_2O_3/LaNiO_3$ revealed an increased catalytic activity than their single components due to the synergistic interaction between the two metal oxides with the particular forms and higher number of active sites.

Cation Vacancy

Efficiently adjusting the activity of reaction sites by vacancy-induced e_g electron configuration regulation of the active atoms is a good strategy in improving catalytic activity. The manufacturing of $CoSe_2$ ultrathin nanosheets with cation vacancy by a liquid exfoliation method for OER is based on this strategy [45].

Anion Vacancy

Introduction of anion vacancies in transition metal oxide catalysts is a good way to control electrocatalytic activity. For example, MoS_2 nanosheets with S-vacancy can be produced by Zn reducing method [45].

Increasing Active Sites by Extending Catalytic Active Surface

Many of the research works give importance to reducing the size of transition metal oxides into nano range, to achieve increased surface area to volume ratio to enhance the catalytic activity. The nano-sized transition metal oxide catalysts possess improved performance activity than their vast complements [2]. Researchers have investigated a variety of 2-D ultrathin nanostructured materials to use as catalysts for OER.

High Surface Area Catalysts

The high surface areas of porous metal oxides are of high importance in OER for good catalytic activity. For example, Yang *et al.* synthesized mesoporous Co_3O_4 with the help of porous silica through a method that involves nano-casting [4].

OTHER CATALYSTS IN OER

Transition metal-oxide-based catalysts have a supreme position in OER catalysis, even though they have limitations like low selectivity, impurity interference, unclear active sites, etc [8]. The section introduces catalysts other than TMOs, which are found significant in OER. Primarily, we can consider metal (including transition metals) based catalysts other than the oxides which include chalcogenides (sulfide, selenides or telluride), pnictides (nitrides and phosphides) [10]. hydroxides, oxy-hydroxides, and organometallics. These metal-chalcogenides have faster charge transfer abilities and also show anti-corrosive resistance in both acid and alkaline mediums. Whereas, pnictides shows inertness even in harsh electrolytes. A few transition metal hydroxides and oxy-hydroxides have shown more stability in OERs compared to their oxides. As an example, $Ni(OH)_2$/NiOOH is much more stable and as efficient as its respective oxide NiO [46]. The layered metal hydroxides (LDH) contain cationic brucite-like layers divided by interlayer anions that are convertible. Their open forms allow the speedy diffusion process and a fast electron transfer. Hence, the active sites are available and result in good electrocatalytic activity [1]. The first row transition metals based on layered double hydroxides are now of utmost importance in the field of OER catalysis due to their layered structure and the adaptability of including mixed-valence transition metal ions into the LDH forms in various compositions [47]. Coming to organometallics, they are not widely used due to their meager stability and low conductivity.

When we look into non-metal-based OER catalysts, we observe a wide variety of compounds that we can use depending on various applications. The most important is the carbon-based electrocatalysts. High activity with more active sites and stable lifetime makes them special among OER catalysts [8]. Compounds such as carbon black, graphene, carbon nanotube (CNTs) are the most discussed carbon-based electrocatalysts. Doping these compounds with N, B, P, S *etc.* enhances the conductivity as well as inertness in harsh electrolytic conditions. Similarly, (h-BN) hexagonal boron nitride which resembles graphene is also an excellent OER catalyst. Not only h-BN, but there is also L-BN which is laser modified boron nitride, doped with C, O [48]. L-BN has higher corrosion resistance & electrical conductivity. Besides carbon, black phosphorus materials have been used in OER. They are anisotropic layered structures. By adjusting the thickness of these layers, we can control its electrical conductivity, which is a special feature of black phosphorus material.

CONCLUSION

Transition metal oxides are promising materials used as an efficient catalyst in OER from acidic/alkaline/neutral medium. High conductivity, large surface area, presence of active surface sites, high stability, and large earth-abundance has made TMOs attractive in OER. Mixed metal oxides, composites, or alloys are more efficient and widely employed rather than their metal oxides. Compounds with geometries such as rutile, perovskite, spinel, layer structure, and rock salt structure receive prominence in catalytic chemistry. Properly surface-activated TMOs are undoubtedly the most promising candidates for OER.

CONSENT FOR PUBLICATION

Not applicable.

CONFLICTS OF INTEREST

The author declares no conflict of interest, financial or otherwise.

ACKNOWLEDGEMENTS

The authors acknowledge Dr. S.M.A. Shibli, Professor, Department of Chemistry, the University of Kerala for proper guidance and support.

REFERENCES

[1] Tahir, M.; Pan, L.; Idrees, F.; Zhang, X.; Wang, L.; Zou, J-J.; Wang, Z.L. Electrocatalytic oxygen evolution reaction for energy conversion and storage: A comprehensive review. *Nano Energy,* **2017**, *37*, 136-157. [http://dx.doi.org/10.1016/j.nanoen.2017.05.022]

[2] Zhang, L.; Fan, Q.; Li, K.; Zhang, S.; Ma, X. First-row transition metal oxide oxygen evolution electrocatalysts: regulation strategies and mechanistic understandings. *Sustain. Energy Fuels,* **2020**, *4*(11), 5417-5432. [http://dx.doi.org/10.1039/D0SE01087A]

[3] Osgood, H.; Devaguptapu, S.V.; Xu, H.; Cho, J.; Wu, G. Transition metal (Fe, Co, Ni, and Mn) oxides for oxygen reduction and evolution bifunctional catalysts in alkaline media. *Nano Today,* **2016**, *11*(5), 601-625. [http://dx.doi.org/10.1016/j.nantod.2016.09.001]

[4] Song, F.; Bai, L.; Moysiadou, A.; Lee, S.; Hu, C.; Liardet, L.; Hu, X. Transition Metal Oxides as Electrocatalysts for the Oxygen Evolution Reaction in Alkaline Solutions: An Application-Inspired Renaissance. *J. Am. Chem. Soc.,* **2018**, *140*(25), 7748-7759. [http://dx.doi.org/10.1021/jacs.8b04546] [PMID: 29788720]

[5] Ishaque, M.; Shah, A.; Iftikhar, F.J.; Akbar, M. Development of transition metal based electrolyzer for efficient oxygen evolution reaction. *J. Renew. Sustain. Energy,* **2020**, *12*(2), 024102. [http://dx.doi.org/10.1063/1.5123234]

[6] Cheng, Y.; Jiang, S.P. Advances in electrocatalysts for oxygen evolution reaction of water electrolysis-from metal oxides to carbon nanotubes. *Prog. Nat. Sci.,* **2015**, *25*(6), 545-553. [http://dx.doi.org/10.1016/j.pnsc.2015.11.008]

[7] Chaturvedi, S.; Dave, P.N.; Shah, N.K. Applications of nano-catalyst in new era. *J. Saudi Chem. Soc.,* **2012**, *16*(3), 307-325. [http://dx.doi.org/10.1016/j.jscs.2011.01.015]

[8] Yuan, N.; Jiang, Q.; Li, J.; Tang, J. A review on non-noble metal based electrocatalysis for the oxygen evolution reaction. *Arab. J. Chem.,* **2020**, *13*(2), 4294-4309. [http://dx.doi.org/10.1016/j.arabjc.2019.08.006]

[9] Rong, X.; Parolin, J.; Kolpak, A.M. A fundamental relationship between reaction mechanism and stability in metal oxide catalysts for oxygen evolution. *ACS Catal.,* **2016**, *6*(2), 1153-1158. [http://dx.doi.org/10.1021/acscatal.5b02432]

[10] Suen, N-T.; Hung, S-F.; Quan, Q.; Zhang, N.; Xu, Y-J.; Chen, H.M. Electrocatalysis for the oxygen evolution reaction: recent development and future perspectives. *Chem. Soc. Rev.,* **2017**, *46*(2), 337-365. [http://dx.doi.org/10.1039/C6CS00328A] [PMID: 28083578]

[11] Damjanovic, A.; Jovanovic, B. Anodic oxide films as barriers to charge transfer in O_2 evolution at Pt in acid solutions. *J. Electrochem. Soc.,* **1976**, *123*(3), 374-381. [http://dx.doi.org/10.1149/1.2132828]

[12] Wohlfahrt-Mehrens, M.; Heitbaum, J. Oxygen evolution on Ru and RuO_2 electrodes studied using isotope labelling and on-line mass spectrometry. *J. Electroanal. Chem. Interfacial Electrochem.,* **1987**, *237*(2), 251-260. [http://dx.doi.org/10.1016/0022-0728(87)85237-3]

[13] Bockris, J.O.; Otagawa, T. The electrocatalysis of oxygen evolution on perovskites. *J. Electrochem. Soc.,* **1984**, *131*(2), 290-302. [http://dx.doi.org/10.1149/1.2115565]

[14] Mefford, J.T.; Rong, X.; Abakumov, A.M.; Hardin, W.G.; Dai, S.; Kolpak, A.M.; Johnston, K.P.; Stevenson, K.J. Water electrolysis on La(1-x)Sr(x)CoO(3-δ) perovskite electrocatalysts. *Nat. Commun.,* **2016**, *7*(1), 11053. [http://dx.doi.org/10.1038/ncomms11053] [PMID: 27006166]

[15] Bockris, J.O.; Otagawa, T. Mechanism of oxygen evolution on perovskites. *J. Phys. Chem.,* **1983**, *87*(15), 2960-2971. [http://dx.doi.org/10.1021/j100238a048]

[16] Trześniewski, B.J.; Diaz-Morales, O.; Vermaas, D.A.; Longo, A.; Bras, W.; Koper, M.T.; Smith, W.A. *In Situ* Observation of Active Oxygen Species in Fe-Containing Ni-Based Oxygen Evolution Catalysts: The Effect of pH on Electrochemical Activity. *J. Am. Chem. Soc.,* **2015**, *137*(48), 15112-15121. [http://dx.doi.org/10.1021/jacs.5b06814] [PMID: 26544169]

[17] Fabbri, E.; Schmidt, T.J. Oxygen Evolution Reaction—The Enigma in Water Electrolysis. *ACS Catal.,* **2018**, *8*(10), 9765-9774. [http://dx.doi.org/10.1021/acscatal.8b02712]

[18] Krishnan, A.; Shibli, S.M.A. Electroactive P-Ani/core-shell/TiO_2/TiO_2-WO_3 employed surface engineering of Ni-P electrodes for alkaline hydrogen evolution reaction. *J. Ind. Eng. Chem.,* **2020**, *87*(133), 198-212. [http://dx.doi.org/10.1016/j.jiec.2020.04.002]

[19] Hardin, W.G.; Mefford, J.T.; Slanac, D.A.; Patel, B.B.; Wang, X.; Dai, S.; Zhao, X.; Ruoff, R.S.; Johnston, K.P.; Stevenson, K.J. Tuning the electrocatalytic activity of perovskites through active site variation and support interactions. *Chem. Mater.,* **2014**, *26*(11), 3368-3376. [http://dx.doi.org/10.1021/cm403785q]

[20] Yoo, J.S.; Rong, X.; Liu, Y.; Kolpak, A.M. Role of lattice oxygen participation in understanding trends in the oxygen evolution reaction on perovskites. *ACS Catal.,* **2018**, *8*(5), 4628-4636.

[http://dx.doi.org/10.1021/acscatal.8b00612]

[21] Grimaud, A.; Diaz-Morales, O.; Han, B.; Hong, W.T.; Lee, Y.L.; Giordano, L.; Stoerzinger, K.A.; Koper, M.T.M.; Shao-Horn, Y. Activating lattice oxygen redox reactions in metal oxides to catalyse oxygen evolution. *Nat. Chem.,* **2017**, *9*(5), 457-465. [http://dx.doi.org/10.1038/nchem.2695] [PMID: 28430191]

[22] Binninger, T.; Mohamed, R.; Waltar, K.; Fabbri, E.; Levecque, P.; Kötz, R.; Schmidt, T.J. Thermodynamic explanation of the universal correlation between oxygen evolution activity and corrosion of oxide catalysts. *Sci. Rep.,* **2015**, *5*(1), 12167. [http://dx.doi.org/10.1038/srep12167] [PMID: 26178185]

[23] Fabbri, E.; Nachtegaal, M.; Binninger, T.; Cheng, X.; Kim, B.J.; Durst, J.; Bozza, F.; Graule, T.; Schäublin, R.; Wiles, L.; Pertoso, M.; Danilovic, N.; Ayers, K.E.; Schmidt, T.J. Dynamic surface self-reconstruction is the key of highly active perovskite nano-electrocatalysts for water splitting. *Nat. Mater.,* **2017**, *16*(9), 925-931. [http://dx.doi.org/10.1038/nmat4938] [PMID: 28714982]

[24] Kalimuthu, V.S.; Attias, R.; Tsur, Y. Electrochemical impedance spectra of RuO_2 during oxygen evolution reaction studied by the distribution function of relaxation times. *Electrochem. Commun.,* **2020**, *110*, 106641. [http://dx.doi.org/10.1016/j.elecom.2019.106641]

[25] Connor, P.; Schuch, J.; Kaiser, B.; Jaegermann, W. The determination of electrochemical active surface area and specific capacity revisited for the system MnOx as an oxygen evolution catalyst. *Z. Phys. Chem.,* **2020**, *234*(5), 979-994. [http://dx.doi.org/10.1515/zpch-2019-1514]

[26] Li, X.; Cheng, Z.; Wang, X. *Understanding the mechanism of oxygen evolution reaction (OER) with the consideration of spin,* **2020**.

[27] Xue, Z.; Li, Y.; Zhang, Y.; Geng, W.; Jia, B.; Tang, J.; Bao, S.; Wang, H-P.; Fan, Y.; Wei, Z.; Zhang, Z.; Ke, Z.; Li, G.; Su, C-Y. Modulating electronic structure of metal-organic framework for efficient electrocatalytic oxygen evolution. *Adv. Energy Mater.,* **2018**, *8*(29), 1801564. [http://dx.doi.org/10.1002/aenm.201801564]

[28] Hong, W.T.; Risch, M.; Stoerzinger, K.A.; Grimaud, A.; Suntivich, J.; Shao-Horn, Y. Toward the rational design of non-precious transition metal oxides for oxygen electrocatalysis. *Energy Environ. Sci.,* **2015**, *8*(5), 1404-1427. [http://dx.doi.org/10.1039/C4EE03869J]

[29] Cai, Z.; Yao, Q.; Chen, X.; Wang, X. Nanomaterials with different dimensions for electrocatalysis. In: *Novel Nanomaterials for Biomedical, Environmental and Energy Applications*; Elsevier, **2019**; pp. 435-464. [http://dx.doi.org/10.1016/B978-0-12-814497-8.00014-X]

[30] Sekar, S.; Kim, D.Y.; Lee, S. Excellent oxygen evolution reaction of activated carbon-anchored nio nanotablets prepared by green routes. *Nanomaterials (Basel),* **2020**, *10*(7), 1382. [http://dx.doi.org/10.3390/nano10071382] [PMID: 32679812]

[31] Kim, T.W.; Woo, M.A.; Regis, M.; Choi, K-S. Electrochemical synthesis of spinel type $ZnCO_2O_4$ electrodes for use as oxygen evolution reaction catalysts. *J. Phys. Chem. Lett.,* **2014**, *5*(13), 2370-2374. [http://dx.doi.org/10.1021/jz501077u] [PMID: 26279561]

[32] Meng, Y.; Song, W.; Huang, H.; Ren, Z.; Chen, S-Y.; Suib, S.L. Structure-property relationship of bifunctional MnO_2 nanostructures: highly efficient, ultra-stable electrochemical water oxidation and oxygen reduction reaction catalysts identified in alkaline media. *J. Am. Chem. Soc.,* **2014**, *136*(32), 11452-11464. [http://dx.doi.org/10.1021/ja505186m] [PMID: 25058174]

[33] Xiong, X.; Ji, Y.; Xie, M.; You, C.; Yang, L.; Liu, Z.; Asiri, A.M.; Sun, X. MnO_2-CoP_3 nanowires

array: An efficient electrocatalyst for alkaline oxygen evolution reaction with enhanced activity. *Electrochem. Commun.,* **2018**, *86*, 161-165.
[http://dx.doi.org/10.1016/j.elecom.2017.12.008]

[34] Tong, Y.; Chen, P.; Zhou, T.; Xu, K.; Chu, W.; Wu, C.; Xie, Y. A bifunctional hybrid electrocatalyst for oxygen reduction and evolution: cobalt oxide nanoparticles strongly coupled to b,n-decorated graphene. *Angew. Chem. Int. Ed. Engl.,* **2017**, *56*(25), 7121-7125.
[http://dx.doi.org/10.1002/anie.201702430] [PMID: 28523861]

[35] Zhou, T.; Xu, W.; Zhang, N.; Du, Z.; Zhong, C.; Yan, W.; Ju, H.; Chu, W.; Jiang, H.; Wu, C.; Xie, Y. Ultrathin cobalt oxide layers as electrocatalysts for high-performance flexible Zn-air batteries. *Adv. Mater.,* **2019**, *31*(15), e1807468.
[http://dx.doi.org/10.1002/adma.201807468] [PMID: 30785222]

[36] Bandal, H.A.; Jadhav, A.R.; Chaugule, A.A.; Chung, W-J.; Kim, H. Fe_2O_3 hollow nanorods/CNT composites as an efficient electrocatalyst for oxygen evolution reaction. *Electrochim. Acta,* **2016**, *222*, 1316-1325.
[http://dx.doi.org/10.1016/j.electacta.2016.11.107]

[37] Wu, Z.; Zou, Z.; Huang, J.; Gao, F. Fe-doped NiO mesoporous nanosheets array for highly efficient overall water splitting. *J. Catal.,* **2018**, *358*, 243-252.
[http://dx.doi.org/10.1016/j.jcat.2017.12.020]

[38] Fei Liu, P.; Yang, S.; Rong Zheng, L.; Zhang, B.; Gui Yang, H. Mo 6^+ activated multimetal oxygen-evolving catalysts. *Chem. Sci. (Camb.),* **2017**, *8*(5), 3484-3488.
[http://dx.doi.org/10.1039/C6SC04819F]

[39] Menezes, P.W.; Indra, A.; González-Flores, D.; Sahraie, N.R.; Zaharieva, I.; Schwarze, M.; Strasser, P.; Dau, H.; Driess, M. High-performance oxygen redox catalysis with multifunctional cobalt oxide nanochains: morphology-dependent activity. *ACS Catal.,* **2015**, *5*(4), 2017-2027.
[http://dx.doi.org/10.1021/cs501724v]

[40] Liardet, L. Synthesis and Application of Transition Metal Oxides as Oxygen Evolution Catalysts. *Infoscience,* **2019**.http://infoscience.epfl.ch/record/268825

[41] Zhu, K.; Liu, H.; Li, X.; Li, Q.; Wang, J.; Zhu, X.; Yang, W. Oxygen evolution reaction over Fe site of BaZr x Fe 1-x O 3-δ perovskite oxides. *Electrochim. Acta,* **2017**, *241*, 433-439.
[http://dx.doi.org/10.1016/j.electacta.2017.04.167]

[42] Park, M-S.; Kim, J.; Kim, K.J.; Lee, J-W.; Kim, J.H.; Yamauchi, Y. Porous nanoarchitectures of spinel-type transition metal oxides for electrochemical energy storage systems. *Phys. Chem. Chem. Phys.,* **2015**, *17*(46), 30963-30977.
[http://dx.doi.org/10.1039/C5CP05936D] [PMID: 26549729]

[43] Janani, G.; Chae, Y.; Surendran, S.; Sim, Y.; Park, W.; Kim, J.K.; Sim, U. Rational design of spinel oxide nanocomposites with tailored electrochemical oxygen evolution and reduction reactions for ZincAir batteries. *Appl. Sci. (Basel),* **2020**, *10*(9), 9.
[http://dx.doi.org/10.3390/app10093165]

[44] Parrondo, J.; George, M.; Capuano, C.; Ayers, K.E.; Ramani, V. Pyrochlore electrocatalysts for efficient alkaline water electrolysis. *J. Mater. Chem. A Mater. Energy Sustain.,* **2015**, *3*(20), 10819-10828.
[http://dx.doi.org/10.1039/C5TA01771H]

[45] Huang, J.; Jiang, Y.; An, T.; Cao, M. Increasing the active sites and intrinsic activity of transition metal chalcogenide electrocatalysts for enhanced water splitting. *J. Mater. Chem. A Mater. Energy Sustain.,* **2020**, *8*(48), 25465-25498.
[http://dx.doi.org/10.1039/D0TA08802A]

[46] Burke, M.S.; Enman, L.J.; Batchellor, A.S.; Zou, S.; Boettcher, S.W. Oxygen evolution reaction electrocatalysis on transition metal oxides and (OXY)hydroxides: activity trends and design principles. *Chem. Mater.,* **2015**, *27*(22), 7549-7558.

[http://dx.doi.org/10.1021/acs.chemmater.5b03148]

[47] Long, X.; Wang, Z.; Xiao, S.; An, Y.; Yang, S. Transition metal based layered double hydroxides tailored for energy conversion and storage. *Mater. Today,* **2016**, *19*(4), 213-226. [http://dx.doi.org/10.1016/j.mattod.2015.10.006]

[48] Liu, H.; Zhang, X-H.; Li, Y-X.; Li, X.; Dong, C-K.; Wu, D-Y.; Tang, C-C.; Chou, S-L.; Fang, F.; Du, X-W. Conductive boron nitride as promising catalyst support for the oxygen evolution reaction. *Adv. Energy Mater.,* **2020**, *10*(25), 1902521. [http://dx.doi.org/10.1002/aenm.201902521]

CHAPTER 13

Need and Scope of Nanocatalysis for Synthesis and Sensing

H.C. Ananda Murthy[1,*], **C.R. Ravikumar**[2], **Tegene Desalegn**[1] and **Aschalew Tadesse**[1]

[1] *Department of Applied Chemistry, School of Applied Natural Science, Adama Science and Technology University, P O Box 1888, Adama, Ethiopia*

[2] *Research Center, Department of Science, East West Institute of Technology, VTU, Bengaluru560091, India*

Abstract: The process of functionalization of nanomaterials with functional groups, molecules, or biomolecules significantly influences the performance of nanomaterials and nanoparticles and enables them to act efficiently in the design and fabrication of nanocatalysts and electrochemical sensors which can be applied in different fields for the detection of analytes such as biomarkers, contaminants, preservatives *etc.* Nanocatalysis is indispensable in the real sense for its benefits that come from both the homogenous and heterogeneous catalysts, such as greater efficiency, stability, an recyclability. Due to the unique properties of nanomaterial-based catalysts, nanocatalysts exhibit extraordinary potential applications in various sectors, like cosmetics, energy, power, petroleum, drug delivery, fuels, biomedicine, and herbal drug synthesis. This book chapter discusses the need for nanocatalysis and its scope, focusing on synthesis and sensing properties of varieties of nano-catalytic materials for multifunctional applications.

Keywords: Nanocatalysis, Nano-catalytic materials, Nanoscience, Scope, Synthesis.

INTRODUCTION

Nanoscience is considered the central science among the natural science disciplines in the 21st century. Recently, several publications have been released on the areas of nanotechnology; this may show the presence of great research interest in the field. Besides, nanomaterials have tremendous potential in several

* **Corresponding Author H.C. Ananda Murthy:** Department of Applied Chemistry, School of Applied Natural Science, Adama Science and Technology University, P.O. Box 1888, Adama, Ethiopia; Tel: +251 988683640; Fax: +251-221-100038; E-mail: anandkps350@gmail.com

smart superhydrophobic applications. Smart food packaging and bacteria detection, drug delivery at a specific area of the human body, and blood purification, as an antibacterial agent and superhydrophobic properties in the textiles industries, and as sterilization and light-transmission controlling glasses [1].

At the pinnacle of the 19th century, besides the expansion of nanoscience, nanocatalysis has emerged as one of the dominant topics at the border between homogeneous and heterogeneous catalysis. This is also reported to be a novel solution that can answer the active conditions for catalyst development (Fig. **1**). The nanocatalysts have valuable potential as homogenous and heterogeneous catalysts. The efficiency, stability, selectivity, and simple regeneration/recycling properties are the countable merits of the nanocatalysts. The nanocatalyst also gives improved activity different from regular, non-nano materials; this is due to the nano dimensional properties [2].

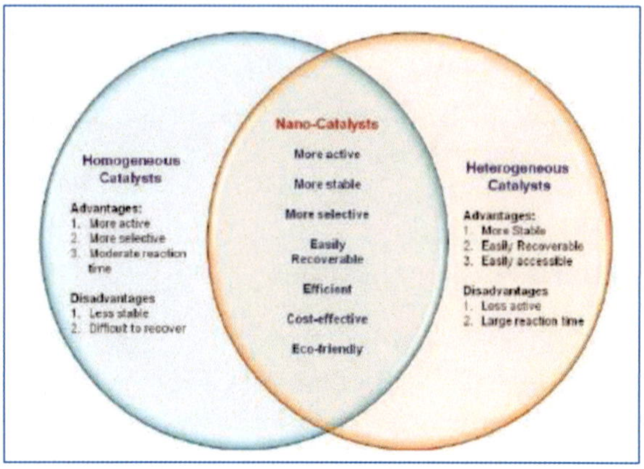

Fig. (1). A schematic showing the differences among the homogeneous, heterogeneous, and nano-catalysts [4].

Catalysis has a crucial purpose for the advancement of the chemical industry. The efficient use of natural resources and diminishing the contamination in the advancement of industrial technology is the countable use of catalysis. Since the last century, catalysis has gained the base of the mass making of the chemicals and varieties of petrochemical products. Catalysis has displayed some new difficulties and occasions as a consequence of the advancement in the field of nanotechnology. Recently, some achievements have been accomplished as per the earlier reports concerning metal-based nanomaterials as heterogeneous catalysts.

Catalysts have crucial catalytic properties and selectivity for particular reactions. The coupling, redox, and electrochemical reactions are countable catalytic chemical reactions that occur by nanocatalysts [3].

NANOCATALYSIS

Nanoparticles as catalysts are one of the principal nanomaterials. In the last decades, several nanosized materials such as titanium dioxide, zinc oxide, iron oxides, aluminum, and silica have been employed as a catalyst. The large surface active sites of nanoparticles have an honest and helpful consequence on the reaction rate. The geometry/structure and morphology-dependent properties of materials at their nanoscale crystallite size can affect the reactant progress of a material. Therefore, optimizing the structure, shape and size of the nanocatalyst during the synthesis should also be taken into consideration [4].

The concept of applying nanoparticles (NPs) for varieties of applications has been found in history for many centuries. The Lycurgus Cup, basically a dichroic glass of gold and silver NPs, exhibited opaque green when viewed from the outside and glowing red when seen from the inside. Similarly, the Damascus saber blades, used during the period between 300 BCE and 1700 CE, were found to exhibit excellent strength, toughness, and resistance to shattering, as they contained carbon nanotubes and cementite nanowires. Iron NPs were applied as a nanocatalyst for a very old Haber-Bosch process for the synthesis of ammonia. At the beginning of the 20th century, Ostwald experimented to increase the surface area of a catalyst for surface-dependent catalysis. Nowadays, a large number of nanomaterials have been utilized in the industries for varieties of synthetic processes of broad commercial applications [5].

The nanoparticles of metal or metal oxides or mixed oxides have a proven record of serving either as stable heterogeneous catalysts or as support materials for various industrially important processes. Similarly, nanostructured catalysts can also serve as heterogeneous catalysts; in which case, separation of catalysts from the reaction mixtures is much easier. In addition, highly active and stable nanocatalysts can be fabricated and functionalized in various ways in order to control the sizes, shapes, and morphologies of nanomaterials effectively [5]. The homogeneous catalysts were employed to a set of limited industrial processes. But, heterogeneous catalysts were found to be employed extensively in the industry due to their advanced features such as ease of separation and recovery as well as their capacity to withstand severe reaction conditions. These heterogeneous catalysts exhibit lower efficiency than their homogeneous counterparts, possibly due to limited diffusion and decreased active sites [6]. The application of NPs has gained significant momentum in the recent past, especially in global environmental problems, which could be attributed to their enhanced catalysis via increased surface area to volume (SAR) ratio [7].

Nanocatalysis has been considered to be the most promising field of research and innovation, where NPs play a significant role as nanocatalysts with superior properties with an eco-friendly nature. A large number of organic and inorganic nanocatalyst stabilizers have also been applied, but they were found to damage the catalytic activity. Among many heterogeneous catalyst supports, carbon nanomaterials such as graphene, reduced graphene oxides, and carbon nanotubes have proven record of being synergistically active [8].

General Properties of Nanocatalyst

Compared to bulk materials, NPs were found to exhibit substantially higher SAR and combined effects of electronic confinement to a small region resulting in the possibility of tuning the catalytic process. The optimization during adsorption and catalytic reactions can be executed by suitably controlling the morphology of the catalyst NPs [6].

The utilization of metal NPs, metal oxide NPs and supported nanocomposites in catalysis/nanocatalysis is extremely beneficial they mimic metal surface activation and thus impart selectivity and sensitivity to heterogeneous catalysis. But it is expected that the NPs must possess the following properties to be efficient and effective:

i. Nanosize
ii. Surface composition
iii. Suitable particle morphology and
iv. Recyclability [2].

Morphology and Surface Features of Nanocatalysts

The morphological features of the nanocatalysts have been proved to play a significant role in influencing the mechanistic aspects of catalysis. The varieties of morphological forms of the nanocatalyst materials include nanoparticles, nanowires, nanoflakes, nanoflowers, nanocubes, nanorods, *etc.* Each morphological form is unique with a set of physicochemical properties. The size and shapes of nanomaterials can be tuned by the application of stabilizing agents and/ capping agents (phytochemicals, surfactants, polymers) [2]. The properties of the nanomaterials can be altered in a controlled way, to get special Physico-chemical properties with reduced size of the nanostructure [9]. The degradation of a very large number of hazardous pollutants has been achieved by using several metal oxide NPs which could be attributed to enhanced SAR and catalytic properties [10].

Synthesis of Nanocatalysts

The conventional synthesis approaches for the nano-catalytic materials failed to follow a precise control to influence the size and shape. The activity, selectivity, durabilityand recyclability can be influenced by significantly altering the morphological features of nanomaterials. The morphology and surface features of nanocatalysts get significantly affected by the process of functionalization with the regulation of the reaction conditions, such as reaction time, temperature, pressure, and reactant concentrations [1].

In the 1850s, Faraday prepared metal NPs by the reduction of metal ions, which is still being considered as a common and the simplest way of the synthesis of NPs. In the 1920s metal NPs were applied to catalyze chemical processes. A few decades later in the 1950s metal NPs with reduced size were applied in the refinery industry. Thus, a reduction in size occurred from 100 nm to less than 1 nm. The mechanism of top-down and bottom-up methods is presented in Fig. (**2**). In addition, several methods were developed such as mechanical grinding, electrochemical, green synthesis, microemulsion, solvothermal processing of precursor solutions *etc.* [1]. The most common top-down and bottom-up methods for the synthesis of nanocatalysts have also been depicted in Fig. (**3**).

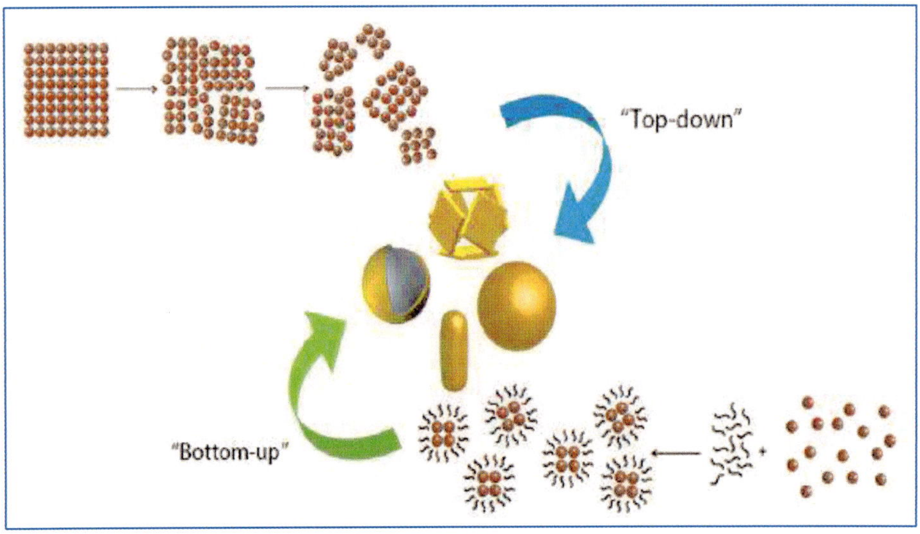

Fig. (2). The top-down and bottom-up mechanism for producing controlled nanomaterials [11].

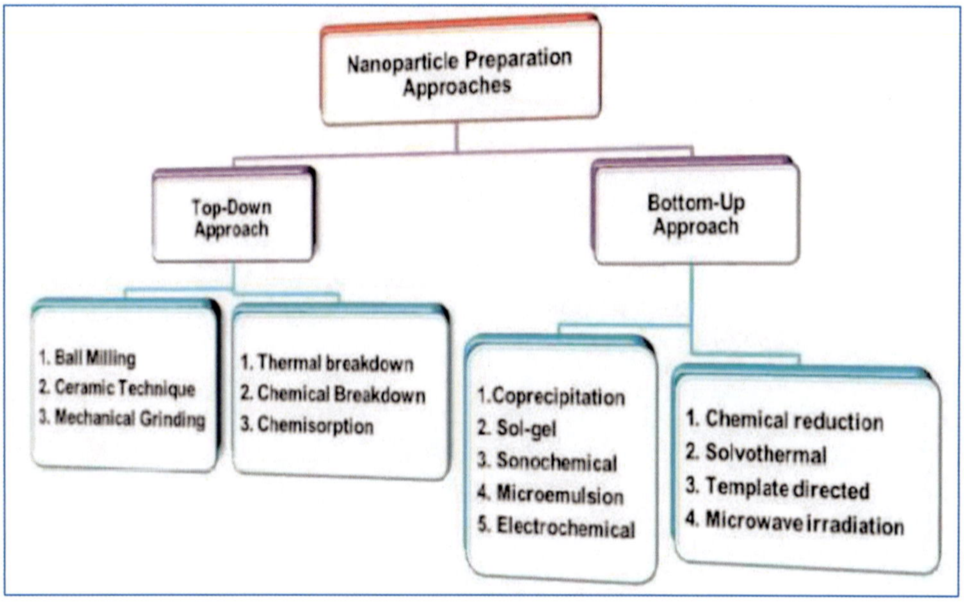

Fig. (3). Summary of nanoparticle preparation methods [4].

Synthesis of Nanoparticles

The synthesis of stable, efficient, and multifunctional NPs in the size range of 1-100 nm is the primary objective of the research in nanoscience. The varieties of synthetic approaches for the nanoparticles are depicted in Fig. (**3**). These major synthetic processes are classified under top-down and bottom-up categories (Table **1**) [4].

Table 1. The overview of the top-down and bottom-up approaches used for the synthesis of nanoparticles [11].

Top-down Methods	Bottom-up Methods
Mechanical Grinding	Green
Metal Vapour	Sol-gel
Thermal breakdown	Microemulsion
Spontaneous chemisorption	Co-precipitation
Chemical breakdown	Sonochemical
	Electrochemical
	Microwave Irradiation
	Template Directed
	Chemical Reduction

Assembling Strategies

A very well controlled synthesis of nanomaterials for catalysis and the characterization of nanocatalysts should support to correlation the structural properties, with their physical and chemical properties. For decorated nanocatalysts, the synthesis should integrate strategies to control, active structure, decorated structure, and their assembly. This also throws a significant challenge in many functionalization and assembly approaches for the fabrication of complex and robust nanostructured catalysts. Varieties of strategies (Fig. **4**) have been evaluated to arrive at an optimal control of active phase location in a supported/fabricated nanocatalyst [2].

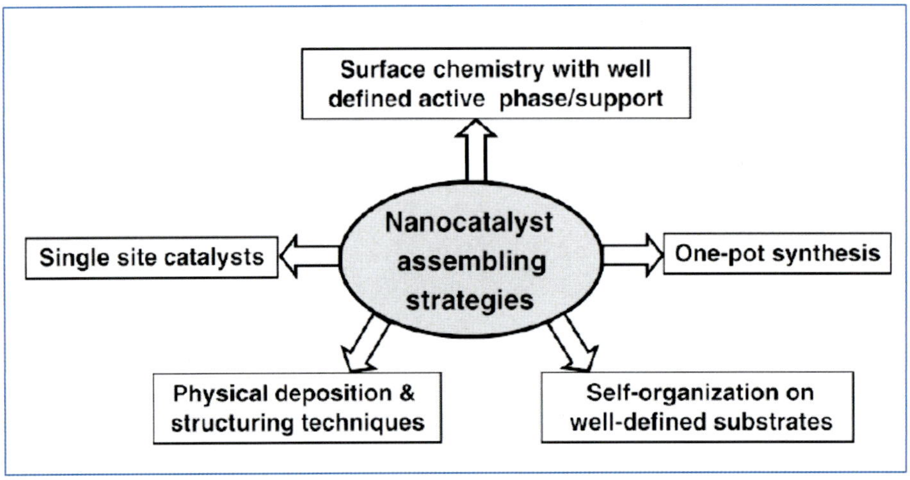

Fig. (4). A schematic of assembling strategies for nanocatalysts [2].

The nanocatalyst can be categorized into two types; unsupported nanocatalysts and supported nanocatalysts. An unsupported nanocatalyst contains only the active material, such as metal NPs, metal oxide NPs, metal compounds and metal complexes that are used in the desired chemical transformation. On the other hand, a supported nanocatalyst is one in which case, an active material is based on another support material such as graphene, reduced graphene oxide, carbon nanotubes *etc*. The support materials offer solid support for the nanocatalyst without affecting the activity of NPs.

Many researchers have reported the synthesis of the supported nanocatalysts. For example, Cu NPs were decorated on the molecular sieves of manganese oxide; ruthenium(II) complexes were also fabricated on magnetic NPs; Mn–Fe NPs were supported on reduced graphene oxide (rGO) nano-sheets [12]. Graphene is a 2 dimensional (2D) crystal of high quality obtained from a single atom with unique electronic properties. It is a building block for other forms of carbon. Due to its flatness and semi-conductivity in addition to its high surface area, high mechanical rigidity, high thermal stability, superior thermal conductivity, and electrical conductivity, good biocompatibility, and easy functionalization [13].

Scope of Nanocatalysis

A large number of nanocatalysts have been applied for varieties of reaction catalysis. There has been tremendous improvement in the efficiency of nanocatalysts due to the advancement in the functionalization processes involved in the fabrication/decoration of NPs over advanced support materials. For example, the green synthesized Ag NPs were applied for the reduction of 4-nitrophenol to 4-aminophenol in the presence of $NaBH_4$. Similarly, spherical Ag NPs with an average particle size of 20 nm, exhibited superior activity against methylene blue dye [14].

Silver NPs have also been used for the epoxidation of ethylene to produce ethylene oxide. Interestingly, Ag NPs were found to be an effective catalyst at temperatures lower than 140° C for bringing about the catalysis of organic oxidation reactions. In addition, functionalized Ag NPs decorated on the diversified supports have been utilized for the epoxidation reactions [15].

Gold nanoparticles (Au NPs) have also attracted significant attention in the field of nanoscience and material science owing to their unique physicochemical properties for potential applications in catalysis [16]. Many researchers have reported the synthesis and catalytic applications of Au NPs in the recent past. Green Au NPs synthesized using green tea whereas applied towards the catalytic reduction of methylene blue in the presence of stannous chloride. In 2012, Ghosh et al. applied Au NPs for the reduction of 4-nitrophenol. On similar lines, many reports on the applications of the green Au NPs for the reduction of 4-nitrophenol or methylene blue have been reported. Green synthesized Au NPs for the degradation of 4-nitrophenol was described in 2016 by Borhamdin et al., Ghosh *et al.*and Lim *et al.* [14] The fabricated nano-gold fabricated heterogeneous catalyst was successfully used in many organic transformations reactions [15].

The reduction of para-nitro aniline by $NaBH_4$ usually takes a few months to complete, but the same reaction was found to occur within a few minutes in the presence of Au nanocatalyst. In the presence of small-sized Au NPs, the reaction is quicker due to the convenience of higher SAR [17].

Palladium nanoparticles (Pd NPs) synthesized using *Pulicaria glutinosa* plant extract, were successfully applied for the catalysis of Suzuki coupling reaction, which involves the effective conversion of bromobenzene to biphenyl compound [14]. Pd NPs decorated with polyoxometalate were also used as green electrocatalysts. In addition, It was reported that Pd NPs acted as efficient catalysts for the HER (hydrogen evolution reaction) process [18].

The degradation of bromothymol blue was achieved by the application of amorphous Fe NPs synthesized using the extract of *sorghum bran* extracts with the mediation of H_2O_2. Fe NPs synthesised using black tea, grape marc, and vine leaves aqueous extracts that were applied for the degradation of ibuprofen with the reported efficiency of 95% [14].

The green copper nanoparticles (CuNPs) obtained by using *Ginkgo biloba L.* extract, were applied for the cycloaddition of azides and alkynes at room temperature. These NPs were utilized as a catalyst for the cyanation of aldehydes, for the synthesis of aryl nitriles with the obtained yield of 85% [14].

Applications of Nanocatalysts

In the year 1989, nanosized Au NPs were proved to be catalytically active for the oxidation of CO for the first time. This observation on Au NPs proved its worth as a catalyst and thus opened the door to applications of many applications of nanosized materials in catalysis. The nanocatalysts have attracted great significance worldwide. The availability of the most advanced synthetic approaches accelerated the fabrication and functionalization of the nanocatalysts [19]. The various nanoforms of catalysts were also used in several chemical processes [20]. Metal and metal oxide NPs have also been applied in sensing the heavy metals, pollutants, and dye molecules in addition to their application in catalysis [21]. The other applications of noble metal NPs also include catalysis of coupling reactions, bio-optical sensing, and medical diagnosis [19].

The environmental pollution caused by organic pollutants and toxic chemicals exhibited great concern globally to eradicate toxic pollutants. The degradation of pollutants into nontoxic molecules by the application of pure single nanocatalysts or functionalized nanocomposites is considered as one of the alternative solutions to this global problem [22]. The applications of nanocatalysis include water purification, biodiesel production, fuel cell application, nanotoxicology, production of hydrogen for fuel cell *etc* [20]. In order to apply the nanocatalysts towards multifunctional applications (Fig. **5**), one should look for efficient combinations of nanomaterials and the support which can yield efficient and stable hybrid structures [23].

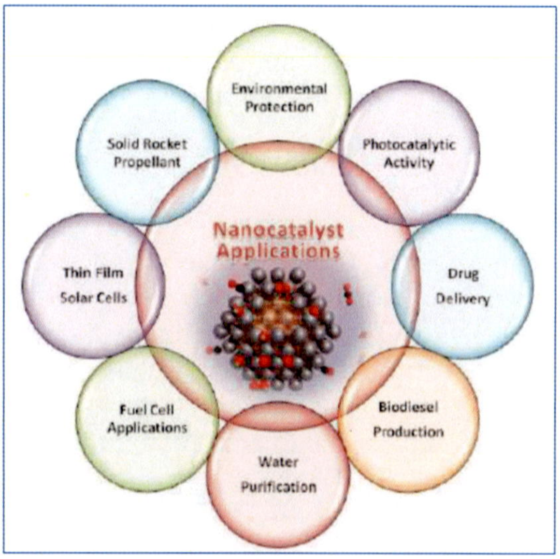

Fig. (5). The multifunctional applications of nanocatalyst [4].

Many reports in the recent past have revealed the application of Cu and Ag NPs towards the reduction of nitroaromatic compounds found in the industrial wastewater at room temperature. But the reduction of 4-nitrophenol is considered as a model reaction for testing the activity of nanoparticle catalysts [24].

Sensors and Application of Nanomaterials

A sensor, in general, is a device that provides information about the changes in its environment, whereas a chemical sensor converts chemical or physical properties into measurable signals. The sensors can be classified based on the property to be determined such as electrical, electrochemical, optical, mass-sensitive, magnetic, or thermal (thermometric). Owing to high sensitivity, long-term reliability, high accuracy, fast response, simplicity, low cost, and easy miniaturization electrochemical sensors have been received great attention for their applications in the analysis of species in biological, environmental, industrial, and pharmaceuticals. In the last few decades, various electrochemical sensors were developed from nanomaterials like metals, metal oxides, carbon-based nanomaterials, conductive polymers, metal-organic frameworks *etc.* [21].

Gas sensors based on semiconducting nanomaterials have received great interest in the recent past. Metal oxides have been considered a primitive component in gas sensors and commercial devices. SnO_2 sensors are the best-understood prototype of oxide-based gas sensors. Moreover, highly selective and sensitive SnO_2 sensors have not yet been reported. The major commercial applications of gas sensors include environment monitoring and food quality analysis. The fabrication and functionalization of nanomaterials and support are very significant towards the innovation of gas sensors and the evolution of gas sensor technology [21].

BIOSENSORS

The biosensors belong to one of the categories of chemical sensors, which finds application in clinical diagnostics, biotechnology, the food industry, and environmental protection [25]. Electrochemical sensors have shown great promise for sensitive glucose detection owing to their fast response, high selectivity, convenient operation, and low cost. The nanozyme based on different MnO_2 based nanomaterials were evaluated for the quantitative detection of biomolecules such as H_2O_2, glucose, glutathione (GSH), and ascorbic acid (AA) [27].

Metal NPs decorated on reduced graphene oxide was utilized as a better electrochemical sensor to detect nitrite ions. A few studies have reported the construction of electrochemical biosensors by decorating the noble metal NPs on the surface of reduced graphene oxide [28]. Fig. (**6**) illustrates the application of Ag NPs supported on PMMA - PVP (Polymethyl methacrylate-polyvinyl pyrrolidine) base for the catalytic activity of 4-NP. Carbon dioxide (CO_2) has been considered as a major greenhouse gas causing global warming as well as drastic climatic transformation. On the other hand, CO_2 is considered a renewable resource for the synthesis of useful chemicals (Fig. **7**) [29].

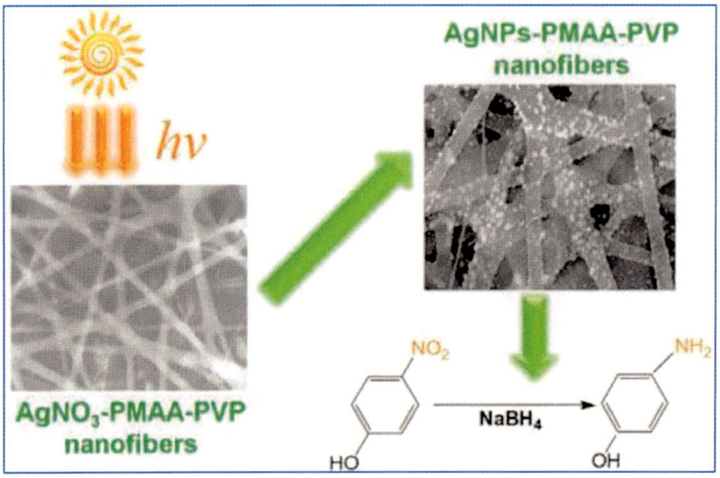

Fig. (6). The schematic of the *in situ* photosynthesis of electros nanofibers and their catalytic activity towards 4-NP [7].

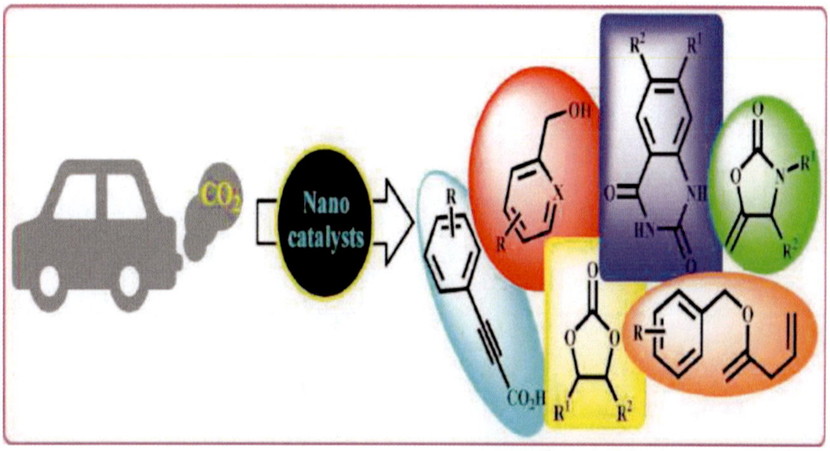

Fig. (7). The nanocatalysts for chemical transformation of carbon dioxide [29].

A large number of industrial products such as cosmetics, food products, dairy products, pharmaceuticals, detergents, plastics, petrochemicals *etc.*, were produced by the application of appropriate catalyst materials in the process. The role played by the catalyst in the process development is not only to accelerate the chemical reaction but also to enhance the productivity of the whole process, by reducing energy consumption and the formation of harmful products [14]. The summary of the scope and applications of metal /doped nanocatalysts in catalysis is displayed in Table **2**.

Table 2. Summary on Scope and application of nanocatalysis

Nanocatalysts	Reaction Catalyzed	Ref
Pd	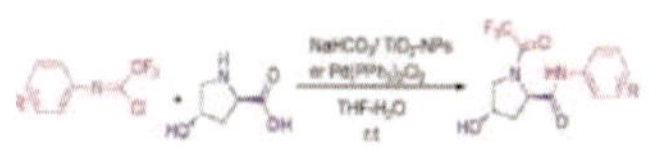	[31]
Fe	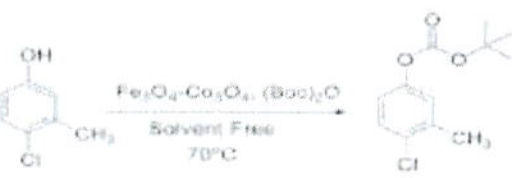	[32]
Cu	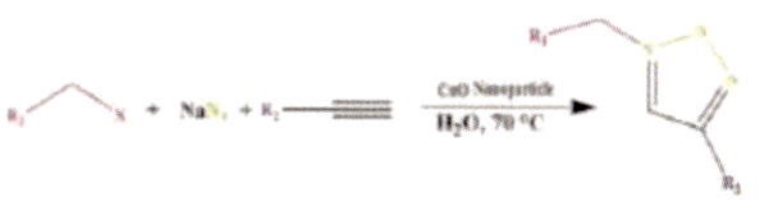	[33]
Cd/Zr	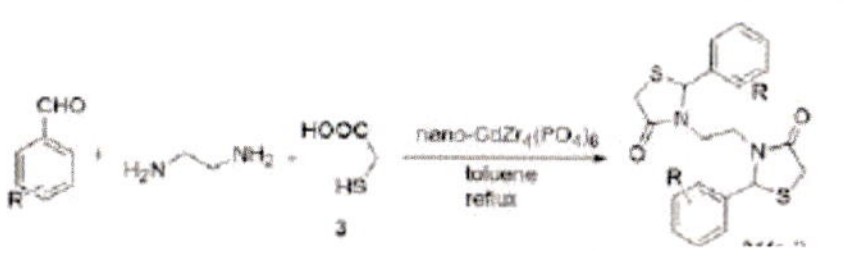	[28]

(Table 2) cont.....

Nanocatalysts	Reaction Catalyzed	Ref
Fe	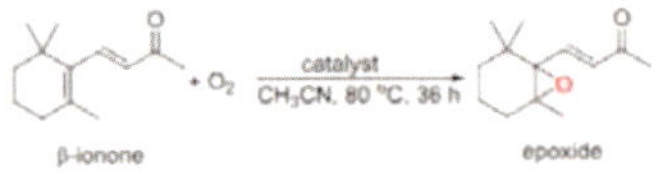	[34]
Au, Pd, Ag, or Pt, Ag		[35]
Pd	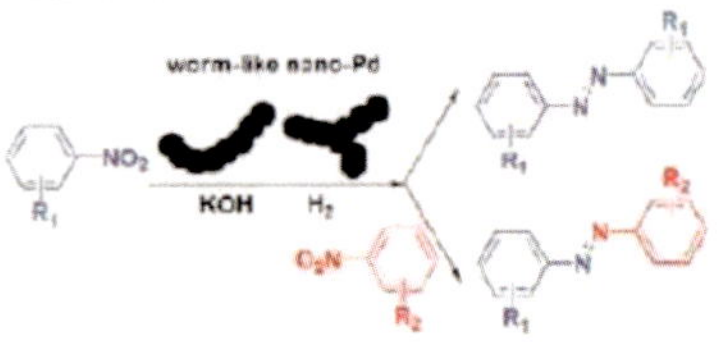	[3]
FePt@Cu	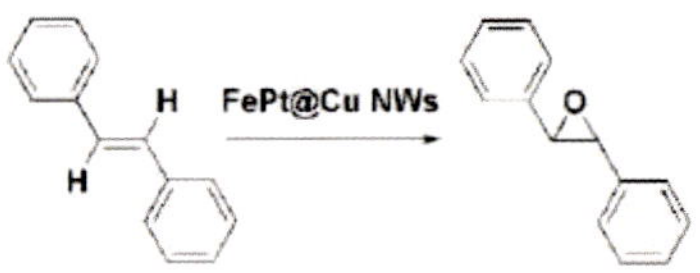	[3]
Co	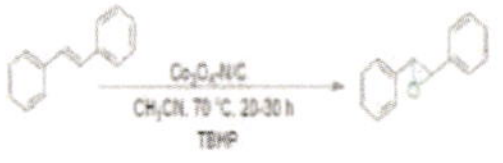	[18]
Pd and Au	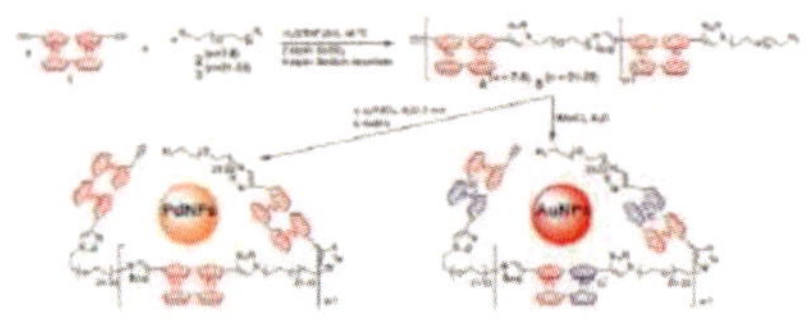	[36]
Ni and Mn	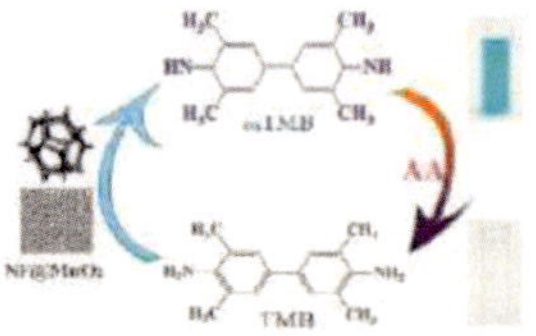	[37]
Ag	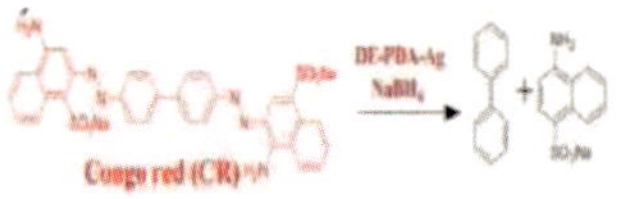	[38]

(Table 2) cont.....

Nanocatalysts	Reaction Catalyzed	Ref
Pt and Cu	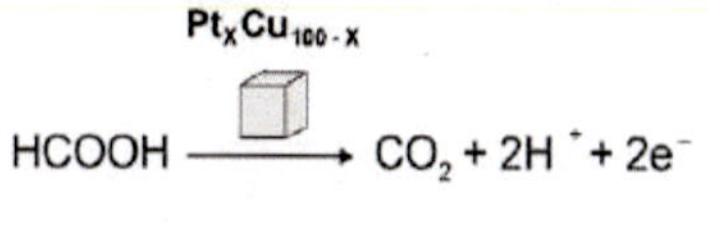	[34]
Rh	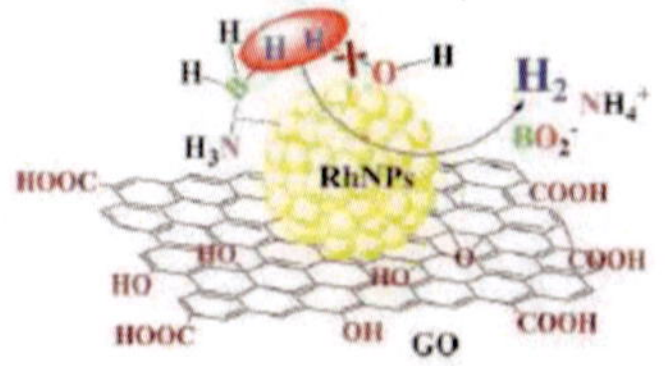	[8]
Au/Mn	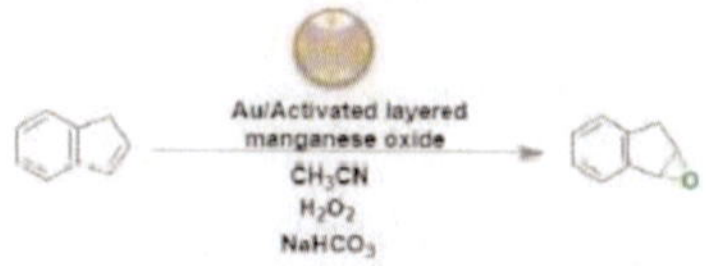	[15]
Rh	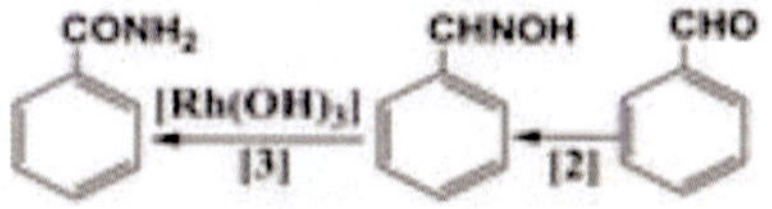	[10]
Pd	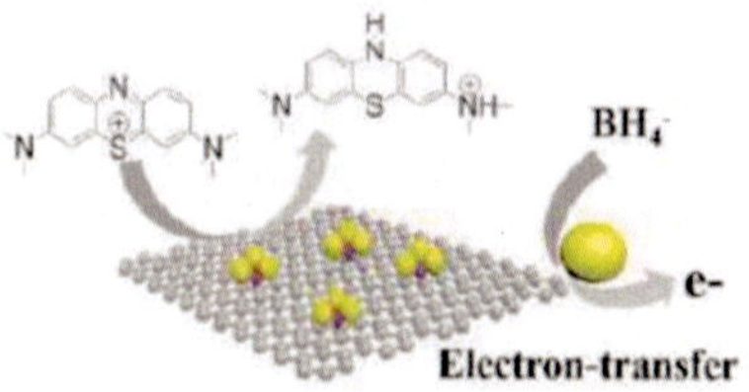	[39]

CONCLUSION

The nanocatalysts have valuable potentiality as homogenous and heterogeneous catalysts. The coupling, redox and electrochemical reactions are countable catalytic chemical reactions that occur by nanocatalysts. This book chapter provides an overview of the tremendous developments in the process of synthesis of varieties of nanocatalyst materials such as metal NPs, metal oxide NPs, doped metal oxide NPs, nanocomposites with the support of nanomaterials such as graphene, reduced graphene oxide, carbon nanotubes, *etc*. A detailed discussion on the need for nanocatalysts towards the advancement of economically favored industrial processes has been presented. Special importance has been given to catalytic and sensor applications of metal and metal oxide NPs.

CONSENT FOR PUBLICATION

Not Applicable.

CONFLICT OF INTEREST

The author declares no conflict of interest, financial or otherwise.

ACKNOWLEDGEMENTS

Authors are grateful to the support rendered by Adama Science and Technology University, Ethiopia.

REFERENCES

[1] Olveira, S.; Forster, S.P.; Seeger, S. Nanocatalysis: Academic Discipline and Industrial Realities. *J. Nanotechnol.*, **2014**, *2014*, 1-19. [http://dx.doi.org/10.1155/2014/324089]

[2] Philippot, K.; Serp, P. Concepts in nanocatalysis. In: *Nanomaterials in Catalysis*; Wiley-VCH Verlag GmbH & Co. KGaA: Weinheim, Germany, **2012**; pp. 1-54. [http://dx.doi.org/10.1002/9783527656875.ch1]

[3] Wang, J.; Gu, H. Novel metal nanomaterials and their catalytic applications. *Molecules,* **2015**, *20*(9), 17070-17092. [http://dx.doi.org/10.3390/molecules200917070] [PMID: 26393550]

[4] Somwanshi, S.B.; Somvanshi, S.B.; Kharat, P.B. Nanocatalyst: a brief review on synthesis to applications. *J. Phys. Conf. Ser.,* **2020**, *1644*(1), 012046. [http://dx.doi.org/10.1088/1742-6596/1644/1/012046]

[5] Polshettiwar, V.; Asefa, T. Introduction to nanocatalysis. In: *Nanocatalysis: Synthesis and Applications*; Asefa, T., Ed.; John Wiley & Sons, Inc, **2013**; pp. 1-9. [http://dx.doi.org/10.1002/9781118609811.ch1]

[6] Liu, H.; Guan, J.; Mu, X.; Xu, G.; Wang, X.; Chen, X. *Nanocatalysis, First Edit*; Wang, Z., Ed.; John Wiley & Sons, Inc, **2017**.

[7] Pmaa, A. A study of the catalytic ability of *in situ* prepared AgNPs–PMAA– PVP electrospun nanofibers. *New J. Chem.,* **2015**, *39*(12), 9518-9524. [http://dx.doi.org/10.1039/C5NJ01519G]

[8] Wang, C.; Ciganda, R.; Yate, L.; Tuninetti, J.; Shalabaeva, V.; Salmon, L.; Moya, S.; Ruiz, J.; Astruc, D. Redox synthesis and high catalytic efficiency of transition-metal nanoparticle–graphene oxide nanocomposites. *J. Mater. Chem. A Mater. Energy Sustain.,* **2017**, *5*(41), 21947-21954. [http://dx.doi.org/10.1039/C7TA06182J]

[9] Damasceno, V.; Maroneze, C.M.; Strauss, M.; Sigoli, F.A.; Mazali, I.O. Preparation of Supported AuPd Nanoalloys Mediated by Ionic Liquid-like Functionalized SBA-15 : Structural Correlations Concerning Its Catalytic Activity †. *New J. Chem.,* **2016**, *40*(8), 6636-6642. [http://dx.doi.org/10.1039/C6NJ00380J]

[10] Imtiaz, A.; Farrukh, M.A. Khaleeq-ur-rahman, M.; Adnan, R. Micelle-Assisted Synthesis of Al_2O_3. CaO Nanocatalyst : Optical Properties and Their Applications in Photodegradation of 2, 4, 6-Trinitrophenol. *ScientificWorldJournal,* **2013**, 11.

[11] Simon, P.; Olveira, S.; Forster, S.P.; Seeger, S. Nanocatalysis : Academic Discipline and Industrial Realities Nanocatalysis : Academic Discipline and Industrial Realities. *J. Nanotechnol.,* **2014**, 19.

[12] Pereira Lopes, R.; Astruc, D. Biochar as a Support for Nanocatalysts and Other Reagents: Recent Advances and Applications. *Coord. Chem. Rev.,* **2021**, *426*, 213585. [http://dx.doi.org/10.1016/j.ccr.2020.213585]

[13] Wang, Y.; Zhong, J.; Ding, F.; Zhao, Q.; Zhang, Z.; Liu, X.; Liu, Y.; Rao, H.; Zou, P.; Wang, X. A Bifunctional $NiCo_2$ S_4/Reduced Graphene Oxide@polyaniline Nanocomposite as a Highly-Efficient Electrode for Glucose and Rutin Detection. *New J. Chem.,* **2018**, *42*(12), 9398-9409. [http://dx.doi.org/10.1039/C8NJ00663F]

[14] Fierascu, ; Ortan, ; Avramescu, ; Fierascu, Fierascu; ortan; avramescu; fierascu. phyto-nanocatalysts: green synthesis, characterization, and applications. *Molecules,* **2019**, *24*(19), 3418. [http://dx.doi.org/10.3390/molecules24193418]

[15] Anuj, S Sharmaa; Vinay , S Sharmab; Harjinder Kaur, R. S. V. Supported Heterogeneous Nanocatalysts in Sustainable, Selective and Eco- Friendly Epoxidation of Olefins. *R. Soc. Chem.,* **2020**, 1-76. [http://dx.doi.org/10.1039/D0GC01927E]

[16] Li, R.; Li, Z.; Wu, Q.; Li, D.; Shi, J.; Chen, Y.; Yu, S.; Ding, T.; Qiao, C. One-step synthesis of monodisperse aunps@pani composite nanospheres as recyclable catalysts for 4-nitrophenol reduction. *J. Nanopart. Res.,* **2016**, *18*(6), 142. [http://dx.doi.org/10.1007/s11051-016-3452-8]

[17] Liu, X.; Rapakousiou, A.; Deraedt, C.; Ciganda, R.; Wang, Y.; Ruiz, J.; Gu, H.; Astruc, D. Multiple applications of polymers containing electron-reservoir metal-sandwich complexes. *Chem. Commun. (Camb.),* **2020**, *56*(77), 11374-11385. [http://dx.doi.org/10.1039/D0CC04586A] [PMID: 32990300]

[18] Narayanan, R. Synthesis of green nanocatalysts and industrially important green reactions. *Green Chem. Lett. Rev.,* **2012**, *5*(4), 707-725. [http://dx.doi.org/10.1080/17518253.2012.700955]

[19] Wang, D.; Xie, T.; Li, Y. Nanocrystals: Solution-Based Synthesis and Applications as Nanocatalysts. *Nano Res.,* **2009**, *2*(1), 30-46. [http://dx.doi.org/10.1007/s12274-009-9007-x]

[20] Chaturvedi, S.; Dave, P.N.; Shah, N.K. Applications of nanocatalyst in new era applications of nano-catalyst in new era. *J. Saudi Chem. Soc.,* **2012**, *16*(3), 307-325. [http://dx.doi.org/10.1016/j.jscs.2011.01.015]

[21] Prinsen, P.; Luque, R. *Introduction to Nanocatalysts*; Chapter 1. , **2019**, pp. 1-36. [http://dx.doi.org/10.1039/9781788016292-00001]

[22] Fang, H.; Guo, Y.; Wu, T.; Liu, Y. Biomimetic synthesis of urchin-like cuo/zno nanocomposites with excellent photocatalytic activity. *New J. Chem.,* **2018**, *42*(15), 12779-12786. [http://dx.doi.org/10.1039/C8NJ02052C]

[23] Song, X.; Su, Q.; Li, S.; Liu, S.; Zhang, N.; Meng, Y.; Chen, X.; Tan, Z. Triple-Shelled CuO/CeO2 Hollow Nanospheres Derived from Metal–Organic Frameworks as Highly Efficient Catalysts for CO Oxidation. *New J. Chem.,* **2019**, *43*(40), 16096-16102. [http://dx.doi.org/10.1039/C9NJ04244J]

[24] Naganath, G. Patil, N. B. B. and A. V. A. Copper and Silver Nanoparticles Stabilized by Bistriazole-Based Dendritic Amphiphile Micelles for 4-Nitrophenol Reduction. **2017**, 4546-4554. [http://dx.doi.org/10.1039/C7NJ00605E]

[25] Abdel-Karim, R.; Reda, Y.; Nanostructured Materials-Based Nanosensors, A.A-F. J. Electrochem. Soc. **2020**, *11*, 12.

[26] Li, M.; Guo, Q.; Xie, J.; Li, Y.; Feng, Y. CuO Nanoparticles supported on nitrogen and sulfur co-doped graphene nanocomposites for non-enzymatic glucose sensing. *J. Nanopart. Res.,* **2017**, *11*(1), 12.

[http://dx.doi.org/10.1007/s11051-016-3712-7]

[27] Rukiye, Babacan; Tosun, Ç.K. A. T. Polymeric template assisted synthesis of monodisperse-porous manganese oxide Microspheres: A new nanozyme with oxidase-like activity allowing biomolecule determination *via* bimodal sensing†. *New J. Chem.,* **2019**, *43*, 18505-18516. [http://dx.doi.org/10.1039/C9NJ03527C]

[28] Zou, H.; Zhang, F.; Wang, H.; Xia, J.; Gao, L.; Wang, Z. a Haiyan Wang, a Jianfei Xia, a L. G. and Z. W. Au nanoparticles supported on functionalized two-dimensional titanium carbide for the sensitive detection of nitrite. *New J. Chem.,* **2019**, *43*(6), 2464-2470. [http://dx.doi.org/10.1039/C8NJ05999C]

[29] Vessally, E.; Babazadeh, M.; Hosseinian, A.; Arshadi, S.; Edjlali, L. Nanocatalysts for chemical transformation of carbon dioxide. *J. CO_2 Util,* **2017**, *21*, 491-502. [http://dx.doi.org/10.1016/j.jcou.2017.08.014]

[30] Chen, M.; Mo, L.; Cui, Z.; Zhang, Z. Magnetic nanocatalysts: synthesis and application in multicomponent reactions. *Curr. Opin. Green Sustain. Chem.,* **2019**, *15*, 27-37. [http://dx.doi.org/10.1016/j.cogsc.2018.08.009]

[31] Darehkordi, A.; Ramezani, M. One-pot synthesis of novel catalysts and investigation of their biological activities. *Mol. Divers.,* **2017**. [http://dx.doi.org/10.1007/s11030-017-9726-y] [PMID: 28190224]

[32] Gade, V.B.; Goswami, A.; Varma, R.S.; Shelke, S.N.; Gawande, M.B. Iron oxide-cobalt nanocatalyst for *O-tert*-Boc protection and *O*-arylation of phenols. *Nanomaterials (Basel),* **2018**, *8*(4), 246. [http://dx.doi.org/10.3390/nano8040246] [PMID: 29673159]

[33] Rabiee, N.; Bagherzadeh, M.; Kiani, M.; Ghadiri, A.M.; Etessamifar, F.; Jaberizadeh, A.H.; Shakeri, A. Biosynthesis of copper oxide nanoparticles with potential biomedical applications. *Int. J. Nanomedicine,* **2020**, *15*, 3983-3999. [http://dx.doi.org/10.2147/IJN.S255398] [PMID: 32606660]

[34] Rodrigues, T.S.; Silva, A.G.M. Nanocatalysis by noble metal nanoparticles: controlled synthesis for the optimization and understanding of activities. *J. Mater. Chem. A Mater. Energy Sustain.,* **2019**, *7*(11), 5857-5874. [http://dx.doi.org/10.1039/C9TA00074G]

[35] Walker, J. M.; Zaleski, J. M. A simple route to diverse noble metal-decorated iron oxide nanoparticles for catalysis. *R. Soc. Chem.,* **2015**, 1-28. [http://dx.doi.org/10.1039/C5NR06700F]

[36] Liu, X.; Rapakousiou, A.; Deraedt, C.; Ciganda, R.; Wang, Y.; Ruiz, J.; Gua, H. D. A. Multiple applications of polymers containing electron-reservoir metal-sandwich complexes. *R. Soc. Chem.,* **2020**. [http://dx.doi.org/10.1039/D0CC04586A]

[37] Niua, Q.; Jing, Z.; Liub, L.; Xua, J.; Alsulamic, H.; Kutbic, M. A. Nanostructured MnO_2 nanosheets grown on nickel foam: an efficient and readily recyclable 3D artificial oxidase for colorimetric detection of ascorbic acid. *R. Soc. Chem,* **2020**. [http://dx.doi.org/10.1039/D0NJ01909G]

[38] Uthappa, U.T.; Kigga, M.; Sriram, G.; Ajeya, K.V.; Jung, H.; Neelgund, G.M.; Kurkuri, M.D. Facile green synthetic approach of bio inspired polydopamine coated diatoms as a drug vehicle for controlled drug release and active catalyst for dye degradation. *Microporous Mesoporous Mater.,* **2019**, *109572*, 109572. [http://dx.doi.org/10.1016/j.micromeso.2019.109572]

[39] Xia, S.; Yang, Y.; Lu, C. Quaternized POSS modified rgo-supported pd nanoparticles as a highly efficient catalyst for reduction and suzuki coupling reactions †. *New J. Chem.,* **2019**, *43*(47), 18601-18610. [http://dx.doi.org/10.1039/C9NJ04491D]

SUBJECT INDEX

A

B

C

Vijai K. Rai, Manorama Singh and Ankita Rai

D

E

N

O

P

Q

R

S

T

U

V

W

X

Z

Made in the USA
Las Vegas, NV
06 April 2026

ce6310c2-62cf-4a3d-915a-1e66b75dea4eR02